Bio-Inspired Approaches—a Leverage for Robotics

Bio-Inspired Approaches—a Leverage for Robotics

Guest Editors

Swaminath Venkateswaran
Damien Chablat
Poramate Manoonpong
Julien R Serres

MDPI

Basel • Beijing • Wuhan • Barcelona • Belgrade • Novi Sad • Cluj • Manchester

Guest Editors
Swaminath Venkateswaran
De Vinci Higher Education
De Vinci Research Center
Paris
France

Damien Chablat
Ecole Centrale de Nantes
Université de Nantes
LS2N UMR CNRS 6004
Nantes
France

Poramate Manoonpong
SDU Biorobotics
University of Southern Denmark
Odense
Denmark

Julien R Serres
Institut des Sciences du Mouvement
Aix-Marseille Université
Marseille
France

Editorial Office
MDPI AG
Grosspeteranlage 5
4052 Basel, Switzerland

This is a reprint of the Special Issue, published open access by the journal *Biomimetics* (ISSN 2313-7673), freely accessible at: www.mdpi.com/journal/biomimetics/special_issues/4679UQR90R.

For citation purposes, cite each article independently as indicated on the article page online and as indicated below:

Lastname, A.A.; Lastname, B.B. Article Title. *Journal Name* **Year**, *Volume Number*, Page Range.

ISBN 978-3-7258-4546-0 (Hbk)
ISBN 978-3-7258-4545-3 (PDF)
https://doi.org/10.3390/books978-3-7258-4545-3

Contents

About the Editors

Swaminath Venkateswaran

Dr. Swaminath Venkateswaran obtained his Doctorate of Philosophy with honours in Mechanical Engineering from École Centrale de Nantes, France, in 2020. After his PhD, Dr. Venkateswaran worked as a Research Associate at the G-SCOP laboratory of Grenoble Institute of Technology (Grenoble-INP) from 2020 to 2021. Currently, Dr. Venkateswaran works as an Assistant Professor in Mechatronics and serves as the head of the Industry & Robotics program at the Leonardo da Vinci engineering school (ESILV), Paris. His research areas/expertise include product design and analysis, kinematics of mechanisms, control of robots, cobots for the circular economy, and Industry 4.0. His teaching activities are centred around the domain of mechatronics and industrial engineering for the Bachelor's and Master's levels. His research activities are affiliated with the group "New Materials, Intelligent Systems, and Innovative Companies" of the Da Vinci Research Center (DVRC). Dr. Venkateswaran has published in and contributed to around 30 international journals (*Mechanism and Machine Theory*, *Journal of Mechanisms and Robotics*) and conferences (ASME-IDETC, ASME-IMECE, and IFToMM). He has over 200 citations with an h-index of 8 (Google Scholar and Scopus).

Damien Chablat

Dr. Damien Chablat obtained his PhD in Mechanical Engineering at the University of Nantes/Ecole Centrale de Nantes, France, in 1998. After finishing his PhD, Dr. Chablat worked for one year at the famous Centre for Intelligent Machines at McGill University, Canada, with Prof. Jorge Angeles. He came back to France and started to work at the Centre National de la Recherche Scientifique (CNRS) in 1999. He became a CNRS senior researcher in 2011 and now works in the ROMAS team (Robots and Machines for Manufacturing, Society and Services) at Laboratoire des Sciences du Numérique de Nantes. His research interests include robotics, the design of parallel manipulators, human fatigue evaluation for industrial tasks, and the design of devices for medical surgeries. Under his supervision, 22 PhD degrees have been awarded, and he is currently leading 6 PhD students within four different research projects. As evidence of his exceptional research career, Dr. Chablat is in the Stanford University Names World's Top 2%. Dr. Chablat is highly visible at the national level and actively contributes to the international scientific community via his presence on the editorial boards of journals such as the IFToMM's *Mechanism and Machine Theory*, ASME's *Journal of Mechanisms and Robotics*, and ASME's *Letters in Translational Robotics*. Dr. Chablat has published over 300 papers, with more than 70 appearing in high-ranking journals (*Machine and Mechanism Theory*, *Journal of Mechanisms and Robotics*, *Journal of Mechanical Design*, etc.). He also has over 8200 citations and an H-index of 44 (Google Scholar) and 32 (Scopus).

Poramate Manoonpong

Dr. Poramate Manoonpong is a Professor at the University of Southern Denmark (SDU), Denmark. He has published over 120 publications in journals (e.g., *Nature Physics*, *Nature Machine Intelligence*, and *Advanced Science*) and conferences. His articles have been cited more than 4,000 times in total. His H-index is 34 (from Google Scholar). Dr. Manoonpong has been a Principal Investigator (PI) or co-PI on more than 10 projects, including projects funded by EU Horizon 2020, HFSP, and Doctoral Networks - Marie Skłodowska-Curie Actions. The central goal of his research is to understand how brain-like mechanisms and biomechanics can be realized in robots so that they can become more intelligent, like living creatures. Accordingly, his team has developed various bio-inspired robots with general bio-inspired neural control, learning, and memory models and has shown that these robots can perform complex behaviors with continuous adaptation, like animals. Additionally, his team also focuses on transferring biomechanical and neural developments of robots to other real-world applications, such as exoskeleton-based mobility, inspection, and service.

Julien R Serres

Dr. Julien R. Serres has been a Senior Lecturer at Aix Marseille University since 2014 and a Full Professor since 2024. Dr. Serres gained his master's degree in Medical Imaging and his master's degree in Electronics, Electrotechnics, and Automatic Control Engineering from Paris-Saclay University and the École Normale Supérieure Paris-Saclay, France, in 2003. Later that year, Dr. Serres joined the Biorobotics Lab at the Movement and Perception Institute, CNRS and Aix Marseille University, Marseille, France, as a Ph.D. student under the supervision of Dr. Nicolas Franceschini. Dr. Serres obtained his Ph.D. at the University of Montpellier, France, in 2008. Dr. Serres is a member of the Bio-Inspired Systems team at the Institute of Movement Sciences, Étienne-Jules Marey, France. He is the author of more than 100 publications, including 4 patents and articles published in 34 indexed and peer-reviewed journals, as well as the recipient of several honors (The National Trophy IoT Industry & Services in 2019, The 7th International Bionic Award 2020, The 5th National Prize for Research in Provence in 2021, and the Medal of Institut Universitaire de France in 2022). He has been appointed as a Senior Member of Institut Universitaire de France from 2022 to 2027, where he holds an Innovation Chair on Bio-Inspired Robotics and Sensorimotor Control.

biomimetics

MDPI

Editorial

Bio-Inspired Approaches—A Leverage for Robotics

Swaminath Venkateswaran [1,*], Damien Chablat [2,3,*], Poramate Manoonpong [4,5,*] and Julien R. Serres [6,7,*]

1 De Vinci Research Center, De Vinci Higher Education, 92616 Paris, France
2 Research Center for Industrial Robots Simulation and Testing, Technical University of Cluj-Napoca, Memorandumului 28, 400114 Cluj-Napoca, Romania
3 Nantes Université, CNRS, LS2N, UMR 6004, F-44000 Nantes, France
4 Embodied AI and Neurorobotics Laboratory, SDU Biorobotics, The Mærsk Mc-Kinney Møller Institute, The University of Southern Denmark, 5230 Odense, Denmark
5 Bio-Inspired Robotics and Neural Engineering Laboratory, School of Information Science and Technology, Vidyasirimedhi Institute of Science and Technology, Rayong 21210, Thailand
6 Aix Marseille University, CNRS, ISM, 13284 Marseille, France
7 Institut Universitaire de France, IUF, 75231 Paris, France
* Correspondence: swaminath.venkateswaran@devinci.fr (S.V.); damien.chablat@cnrs.fr (D.C.); poma@mmmi.sdu.dk (P.M.); julien.serres@univ-amu.fr (J.R.S.)

The field of bio-inspired approaches (also known as biomimetics or biomimicry) is a design approach whereby a product or process is inspired by elements of nature, such as plants or animals. Bio-inspired approaches serve as inspiration and motivation for many engineers and designers in their efforts to identify unexpected solutions to problems. These approaches have resulted in significant innovations in the aerospace, marine, and automotive industries. The domains of bio-inspiration and bio-mimetics have also been the focus of several studies in the domain of robotics. There are several examples of their use in the literature, including their implementation in snake-type robots for underwater inspection or in worm-type systems for industrial pipeline inspections. This Special Issue presents the recent advancements in the domain of bio-inspired robotics and their potential applications in industry. This will help researchers from all communities to understand the relevance of bio-inspiration in robotics and will serve as a platform for the application of these cutting-edge approaches in other fields. This Special Issue comprises 11 articles in the domain of biomimetics, examining the study of this concept and its application in robotic systems.

In paper [1], the authors exploit the locomotion of bio-primates, which include continuous brachiation and ricochetal brachiation. The design mimics the transverse movements of sports climbers holding onto horizontal wall ledges. Based on a two-hand release design, a novel transverse ricochetal brachiation mechanism exploits inertial energy storage to enhance moving distance. Experimental prototypes helped to predict the success of subsequent locomotion cycles.

In paper [2], the authors present an angle sensor based on step-index profile plastic optical fiber (SI-POF), which was cost-effective and durable. The performance of the POF sensor was evaluated by measuring sensitivity and resolution in order to verify its reliability under extreme conditions, especially underwater. This study is vital to the development of biomimetic robot industry where existing sensors are difficult to deploy.

Paper [3] presents a novel whisker-sensing disk designed for 3D mapping in unstructured environments. The modeling is performed based on analytical and data-driven approaches to predict rotation angles based on magnetic field measurements. Experiments were conducted to validate the modeling, and the results highlight the effectiveness of 3D mapping in complex environments for robotic platforms in the future.

 https://doi.org/10.3390/biomimetics10070417

Paper [4] presents a review of robot control in spaces with low material and structural stiffness, which is usually challenging. Based on an in-depth review of scientific articles, the conclusions of this study indicate three top performing methods. The methods are based on minimizing control effort usage, tracking error mean, and tracking error deviation with some improvements in performance measures.

In paper [5], the authors propose the use of impedance control for the regulation of dynamic response of pneumatic soft robots, an approach already in existence for rigid robots. A non-linear discrete sliding mode impedance controller is formulated to control soft pneumatic robots. The controller does not require the manual tuning of parameters and can automatically calculate them based on impedance value. Experiments showed that the proposed controller can effectively limit the amplitude of undesirable vibrations.

Paper [6] presents a distributed model predictive controller based on leader–follower approach for addressing the collaborative transportation control issue of dual humanoid robots. Network latency issue is a prominent problem due to unstable network conditions that affects the consistency of collaboration. A socket communication was constructed to resolve the latency issue. A distributed model predictive control helped consider cumulative errors that lead to the enhanced position tracking accuracy of dual-robot collaborative control.

In paper [7], the authors measured hand movements in stroke patients using MediaPipe and Fahrenheit to assess their criterion-related validity. Consistent results were observed in peak angle and velocity comparisons across severity stages. The study highlighted the importance of MediaPipe in paralysis estimation.

Paper [8] presents important research on malfunctioning joints that lead to a high degree of compactness, eventually resulting in benefits such as low mass, low moment of inertia, and low drag. It was found that this multifunctionality was achieved through various means, such as multiple degrees of freedom, multifunctioning parts, over-actuation, and reconfiguration. This research suggests that the agility of robots could be improved using multifunctioning to reduce the size and mass of robotic joints.

In paper [9], the authors present a motion prediction model developed using convolutional neural networks (CNNs) for the efficient identification of motion types at the initial states. A multi-axial robotic arm integrated with a motion identification platform was developed to interact with humans by emulating their movements. A control strategy for addressing non-linearities and cross-coupled dynamics of the robotic system was applied. The results show that the robotic arm was able to achieve adequate controlled outcomes, thereby validating the feasibility of such an interactive robotic system in effective bio-inspired motion emulation.

Paper [10] presents a novel bionic amphibious robot, AmphiFinbot-II. The robot's swimming and walking components involved a compound drive mechanism that allows for the simultaneous control of the rotation of the track and the wave-like motion of the undulating fin. The performance was tested via different motion patterns through computational fluid dynamics simulations. Experiments were also conducted on land and underwater, and the results were consistent with the simulation. The findings showed that the proposed robot possesses excellent amphibious motion capabilities through a unified control approach.

Finally, paper [11] presents the empirical kinematic control and data-driven modeling of a soft swimming robot. The robot consists of six serial connected segments that can bend individually using segmental pneumatic artificial muscles. Experiments were conducted to gather position and velocity of spatially digitized points using Qualisys Tracking Manager 1.6.0.1. Through offline analysis, a new complex variable algorithm was proposed to extract a linear approximative model. The proposed algorithm helped in extracting linear and chaotic modes and aided in linearizing the overall system dynamics.

The guest editors would like to sincerely thank all the contributors who took a keen interest in our Special Issue. All the articles went through rigorous peer review to ensure a high-quality publication. We also thank all the reviewers for their valuable comments and revisions. In addition, we would like to thank the editors from MDPI for their support in the organization and publication of this Special Issue.

We believe that this Special Issue will help fellow researchers and industrial experts to gain an in-depth understanding of the importance of biomimetics and how it could be implemented for various applications.

Author Contributions: Conceptualization, S.V., D.C., P.M. and J.R.S.; writing—original draft preparation, S.V.; writing—review and editing, D.C., P.M. and J.R.S. All authors have read and agreed to the published version of the manuscript.

Conflicts of Interest: The authors declare no conflict of interest.

References

1. Lin, C.Y.; Lee, J.M. Multi-locomotion design and implementation of transverse ledge brachiation robot inspired by sport climbing. *Biomimetics* **2023**, *8*, 204. [CrossRef] [PubMed]
2. Lee, H.W.; Kim, D.H.; Shin, S. Low-Cost Angle Sensor for Robotics Applications Using Plastic Optical Fiber Based on Optical Loss Mechanism. *Biomimetics* **2023**, *8*, 567. [CrossRef] [PubMed]
3. Gomez, V.; Remmas, W.; Hernando, M.; Ristolainen, A.; Rossi, C. Bioinspired Whisker sensor for 3D mapping of underground mining environments. *Biomimetics* **2024**, *9*, 83. [CrossRef] [PubMed]
4. Sands, T. Bio-inspired space robotic control compared to alternatives. *Biomimetics* **2024**, *9*, 108. [CrossRef]
5. Khan, A.H.; Li, S. Discrete-Time Impedance Control for Dynamic Response Regulation of Parallel Soft Robots. *Biomimetics* **2024**, *9*, 323. [CrossRef]
6. Wen, S.; Shi, Z.; Li, H. Coordinated Transport by Dual Humanoid Robots Using Distributed Model Predictive Control. *Biomimetics* **2024**, *9*, 332. [CrossRef] [PubMed]
7. Suwabe, R.; Saito, T.; Hamaguchi, T. Verification of Criterion-Related Validity for Developing a Markerless Hand Tracking Device. *Biomimetics* **2024**, *9*, 400. [CrossRef] [PubMed]
8. Burgess, S.C. How Multifunctioning Joints Produce Highly Agile Limbs in Animals with Lessons for Robotics. *Biomimetics* **2024**, *9*, 529. [CrossRef]
9. Cheng, M.H.; Huang, P.L.; Chu, H.C. Bio-Inspired Motion Emulation for Social Robots: A Real-Time Trajectory Generation and Control Approach. *Biomimetics* **2024**, *9*, 557. [CrossRef] [PubMed]
10. Xia, M.; Zhu, Q.; Yin, Q.; Lu, Z.; Zhu, Y.; Luo, Z. Hydrodynamic Simulation and Experiment of a Self-Adaptive Amphibious Robot Driven by Tracks and Bionic Fins. *Biomimetics* **2024**, *9*, 580. [CrossRef]
11. Sayahkarajy, M.; Witte, H. Empirical Data-Driven Linear Model of a Swimming Robot Using the Complex Delay-Embedding DMD Technique. *Biomimetics* **2024**, *10*, 60. [CrossRef]

biomimetics

MDPI

Article

Multi-Locomotion Design and Implementation of Transverse Ledge Brachiation Robot Inspired by Sport Climbing

Chi-Ying Lin * and Jhe-Ming Lee

Department of Mechanical Engineering, National Taiwan University of Science and Technology, No. 43, Keelung Rd., Sec. 4, Taipei 106, Taiwan; a25896365@gmail.com
* Correspondence: chiying@mail.ntust.edu.tw; Tel.: +886-2-2737-6484

Abstract: Brachiation robots mimic the locomotion of bio-primates, including continuous brachiation and ricochetal brachiation. The hand-eye coordination involved in ricochetal brachiation is complex. Few studies have integrated both continuous and ricochetal brachiation within the same robot. This study seeks to fill this gap. The proposed design mimics the transverse movements of sports climbers holding onto horizontal wall ledges. We analyzed the cause-and-effect relationship among the phases of a single locomotion cycle. This led us to apply a parallel four-link posture constraint in model-based simulation. To facilitate smooth coordination and efficient energy accumulation, we derived the required phase switching conditions as well as joint motion trajectories. Based on a two-hand-release design, we propose a new style of transverse ricochetal brachiation. This design better exploits inertial energy storage for enhanced moving distance. Experiments demonstrate the effectiveness of the proposed design. A simple evaluation method based on the final robot posture from the previous locomotion cycle is applied to predict the success of subsequent locomotion cycles. This evaluation method serves as a valuable reference for future research.

Keywords: sport climbing; transverse ledge brachiation; ricochetal brachiation; multi-locomotion design; parallel four-link posture

1. Introduction

Transverse ledge brachiation robots are bio-inspired climbing robots that mimic climbing athletes [1–3]. They move sideways across walls by grabbing wall protrusions [4], swinging their lower limb(s), and using alternating handholds. Such robots can take the place of humans in hazardous work such as monitoring, maintenance, and rescue [5–9] in situations involving wall ledges such as rain deflectors, exterior window sills or trim, flat/pitched roofs [10], or friezes. For instance, densely-populated urban building areas are excellent implementation venues for transverse ledge brachiation robots.

Transverse ledge brachiation robots first swing their lower limb(s) and body to accumulate adequate kinetic energy. They then release one hand to grab the target ledge while simultaneously adjusting their posture so that the other hand can return to the original double handhold position under the guidance of inertia. This facilitates the beginning of swing excitation in the next locomotion cycle. As the processes of energy storage/release repeat and alternate, the robot traverses along the wall [11]. This unique locomotion gait is called transverse ledge brachiation (TLB). In events where swift locomotion is needed, or the distance between climbing ledges is greater than the robot's arm span [12,13], climbers can let go of the ledge with one hand "in the instant before" the other hand grabs the target ledge. This results in a leap during which neither hand is in contact with a ledge. This completely releases the energy that the robot has accumulated during the swinging process to obtain better locomotion efficiency. This type of locomotion is called transverse ledge ricochetal brachiation (TLRB) [1].

Humans and apes are both primates, so brachiation robots modeled on humans and apes are similar [14]. However, the target handholds of ape-inspired robots are round ladder rungs [15,16] or cables [17,18], whereas TLB robots move transversely at 90° to ledges [2]. Conventional brachiation robots mostly focus on efficiently accumulating energy during the swinging process and planning joint trajectories to facilitate continuous brachiation between spaced rungs [19–21]. However, in addition to the swing actions, special consideration must also be given to the grasping action during the gait process of ricochetal brachiation [22,23]. The complexity of the required hand-eye coordination [1,2] means research on such robots is rare. Even fewer studies include comprehensive practical assessments.

Transverse brachiation requires good multi-limb coordination [24]. A simple two-link brachiation robot [12,13,15,17–19] is only capable of performing status switches between the swing phase and the grabbing phase [25]. Furthermore, fulfilling the grabbing conditions of rectangular ledges is not as easy as that of round bars. Impact dynamics can make it difficult to maintain a good grip posture when grabbing a ledge [1]. It is also difficult to perform three-dimensional posture compensation in robots with a low degree of freedom (DOF) in their actuators.

Most existing studies of transverse brachiation robots used multi-limbed robot designs based on human climbing motions [20,26–28]. Lin and Yang [1] employed a four-link arm-body-tail robot configuration, dividing transverse ricochetal brachiation into a swing phase, a flight phase, and a landing phase. During the leap process, or "flight phase," they used a posture compensation control algorithm to maintain a vertical body posture. This means both arms can grab horizontal ledges smoothly while obtaining longer flight time and longer grasping distance. In addition, using a four-link arm-body-tail robot configuration, Lin and Tien [2] designed a transverse brachiation locomotion gait for situations involving horizontal ledges with large spans or different elevations. Their design includes the following four phases: release, body reversal, swinging up, and grasping. However, due to mechanical hardware limitations, their robot could only complete one locomotion cycle with this gait. Drawing inspiration from the climbing motions of athletes with greater muscle strength, Lin and Hu [10] proposed a five-DOF design under the premise of relying only on upper body strength (i.e., no lower limbs or tail); the design included a body, grippers, and shoulders and had a locomotion gait that could climb back and forth along a ledge. We refer to this gait as transverse ledge climbing. Their locomotion control strategy is based on the central pattern generator (CPG) algorithm and infrared sensors. It successfully achieves continuous climbing on horizontal and sloped ledges. However, this transverse climbing motion is not ergonomic, nor is it a type of brachiation; thus, it is more suited for situations where the lower limbs cannot swing.

Table 1 compares studies on transverse ledge climbing/brachiation robots. As can be seen, the primary limitations of existing studies include (1) a single locomotion gait and (2) the inability to perform multiple cycles. The final posture of a robot's grippers as it transitions from the swing phase to the grabbing phase is inevitably affected by the motions of the previous phase. Thus, a non-negligible offset accumulates, which means the pendulum-like boundary conditions of the next locomotion cycle cannot be met. An inadequate grip affects system excitation and the subsequent movement phases, thereby preventing a continuous brachiation gait (via swing). Another crucial factor neglected by existing studies is the influence of the motions in each phase on those in the next phase, which affects the smoothness of transitions. Athletes are characterized by their ability to make graceful movements that exploit inertial energy [29]. In the event of external disturbances, motions are likely to become uncoordinated to the point of gait failure.

In this study, we refer to the unique movements of athletes climbing transversely along wall ledges to propose a design process that takes the cause-and-effect relationships in each phase of movement into account. This process enabled us to realize multiple locomotion gaits for coordinated bio-inspired robots. We used a parallel four-link posture as the constraint. Using model-based dynamic analysis, we derived the design parameters and joint trajectories for smooth transitions. Experiments demonstrate the efficacy of the proposed gait design method. Based on these results, we propose an evaluation index to analyze whether the grip posture of the grippers meets the boundary conditions of the swing phase to predict the success rate of achieving the next locomotion cycle. This evaluation tool serves as a valuable reference for future research.

Table 1. Comparison of bio-inspired transverse ledge brachiation/climbing robots.

	Current Study	Transverse Ledge Ricochetal Brachiation Robot [1] (Lin and Yang, 2020)	Transverse Ledge Climbing Robot [10] (Lin and Hu, 2022)	Transverse Ledge Brachiation Robot [2] (Lin and Tien, 2022)
Number of robot link	4	4	5	4
Robot configuration	Arm-body-tail	Arm-body-tail	Arm-body	Arm-body-tail
System classification	Arm-body-tail	Arm-body-tail	Arm-body	Arm-body-tail
Multi-locomotion	Yes (TLB/TLRB)	No (TLRB)	No (TLC)	No (TLB)
Application scenario	Horizontal brachiation (continuous/gapped ledges)	Horizontal brachiation (continuous/gapped ledges)	Horizontal climbing (continuous/sloped ledges)	Horizontal and non-level horizontal brachiation (ledges at different elevations)
Continuous brachiation (via swing)	Yes	No	N/A	No
Lower limbs	Yes	Yes	No	Yes

The remainder of this paper is organized as follows. Section 2 presents locomotion analysis and classified phases for the two types of brachiation. Section 3 summarizes the dynamic modeling results used for multi-locomotion design. Section 4 offers an analysis of phase switching parameters and required joint trajectories for a smooth transition. Section 5 presents the transverse brachiation experiments based on a four-link arm-body-tail configuration robot. Finally, concluding remarks are made.

2. Locomotion for Transverse Ledge Brachiation

In this section, we discuss the locomotion gaits associated with continuous transverse brachiation and hybrid transverse brachiation. Specifically, we explore design requirements based on the movement phases. The complete gait design process is presented in subsequent sections.

2.1. Transverse Ledge Brachiation Based on Athletes' Locomotion

Sport climbers transverse brachiate using a periodic transverse movement behavior [11]. Figure 1a dissects this locomotion mode. First, the athlete swings his lower limbs to increase system energy and the overall swing range. Once enough energy has been accumulated; the athlete is able to grasp the ledge with his right hand. In this phase, the system energy is converted from kinetic to potential energy. We found that the athlete does not grasp the ledge as far as possible but maintains a suitable distance between his hands. This strategy is critical to whether subsequent movement phases can be completed successfully. After his right hand has grasped the wall ledge, the swinging of the lower limbs rotates the upper limbs and torso, which adjusts the athlete's posture. Finally, his left-hand grasps the ledge to the right, completing a single transverse brachiation locomotion cycle. Based on the gait analysis above, we can divide the locomotion gait into the four phases shown in Figure 1b: a swing phase, a target-approaching phase, a transition phase, and a hands-on phase. Each phase must be completed, as shown in Figure 1a, in order to enter the next phase. If any movement in a phase is incorrect or if the timing of phase switching is off, then the motion will immediately become uncoordinated and cause the locomotion gait to fail. The most common consequence is that the athlete will fall to the ground. This outcome is unacceptable for a ledge-climbing robot. Thus, in this study, we focus on the means of planning the motion processes of each phase and the postures needed for the robot to transition from one movement phase to the next smoothly.

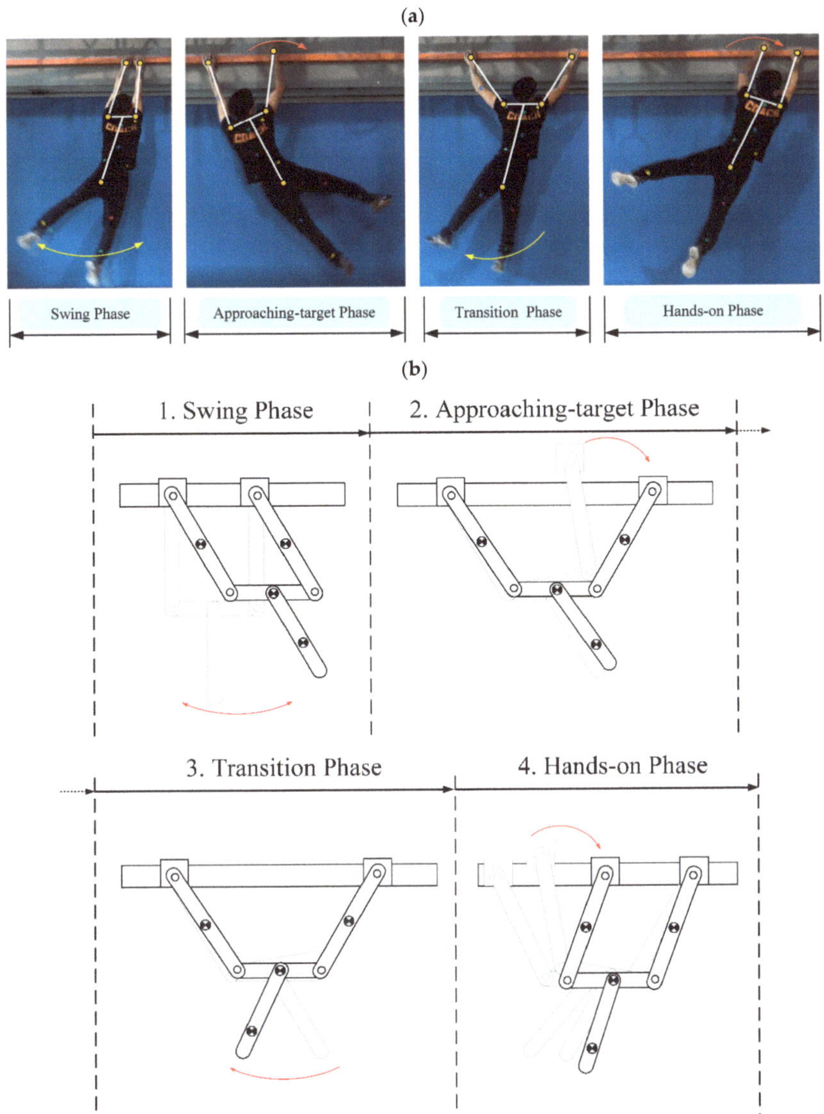

Figure 1. Phase classification for transverse ledge brachiation: (**a**) locomotion analysis of a sport climber [11]; (**b**) robot locomotion design. Yellow and red arrows indicate moving direction.

2.2. Transverse Ledge Ricochetal Brachiation Based on Athletes' Locomotion

Transverse ricochetal brachiation is a movement strategy applied by climbing athletes when the target ledge is out of reach. In a single locomotion cycle, this gait can cover a longer distance and enables movement when there are no handholds. Figure 2a dissects the process of this gait. The athlete begins by swinging his lower limbs to accumulate energy. Once there is enough inertia, the athlete is able to reach out with his right hand. When this hand is about to touch the target ledge, he lets go with his left hand to release the accumulated energy completely. The duration of the flight phase in which neither hand is touching a ledge is extremely short (<0.1 s), so it is difficult to see this motion in the images captured in Figure 2a. Finally, the athlete grasps the target ledge with his right hand, and his left hand also swiftly resumes its position above the athlete's head. This completes a single cycle of the transverse ricochetal brachiation gait. Based on the gait analysis results above, we divided this gait into the four phases shown in Figure 2b: swinging, approaching target, two-hand release, and hands-on.

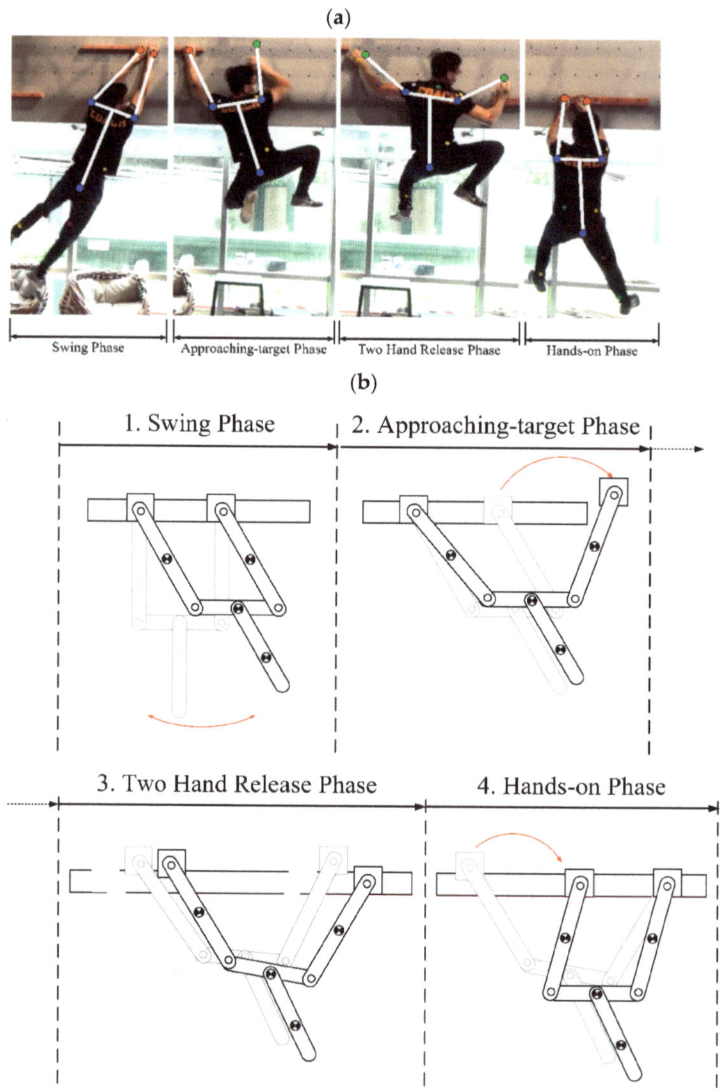

Figure 2. Phase classification for transverse ledge ricochetal brachiation: (**a**) locomotion analysis of climbing athlete where red indicates holding, green indicates releasing and blue indicates the actuating joints [11]; (**b**) robot locomotion design.

Two common modes of locomotion can serve as a simple analogy to describe the two brachiation gaits above: walking and running. Walking is slower, but the entire process is steadier and saves energy. In contrast, running obviously has higher movement efficiency, but it consumes more energy during the process, and the posture of the body when both feet leave, and land on the ground influences the effectiveness and stability of this mode of locomotion. From the perspective of bionics, ricochetal brachiation can, to a certain extent, be compared to the trot gait of multiped robots. However, the gaits of brachiation robots must be paired with robust and coordinated ledge grip motions, as dissimilar to multiped robots, which can stand up and continue even after falling down [30], it is impossible for brachiation robots to return to a ledge once they fall. This is the greatest challenge in realizing continuous cycles in transverse brachiation.

2.3. Problem Statement

Past studies [1,5] have highlighted two main challenges in the development of transverse brachiation robots. The first is that it is difficult to achieve two or more continuous cycles because brachiation robots generally rely on inertia to turn their bodies to facilitate ledge grasping [2]. However, this motion tends to create interference, which is a problem that must be considered carefully in mechatronic design. Furthermore, although the design in which parallel linkages are used in both arms [1] (as shown in Figure 3a) ingenuously keeps the robot's body vertical to the ledge, it also reduces the freedom of body motions. If the body cannot smoothly rotate due to mechanism restrictions after the robot has grasped the ledge, the entire motion will be forced to stop. If the robot enters the hands-on phase in an awkward position, canceling out the energy accumulated during the swinging process. To delve deeper into this issue, we included the system dynamics of all of the phases in our robot gait analysis and redesigned suitable locomotion trajectories and the switching conditions to smoothly connect the various phases in accordance with the parallel four-link posture constraints (as shown in Figure 3b). The gait design objectives include enabling smooth transitions between movement phases, reducing the energy consumed to switch from one phase to the next, and maintaining the parallel four-link posture in the hands-on phase.

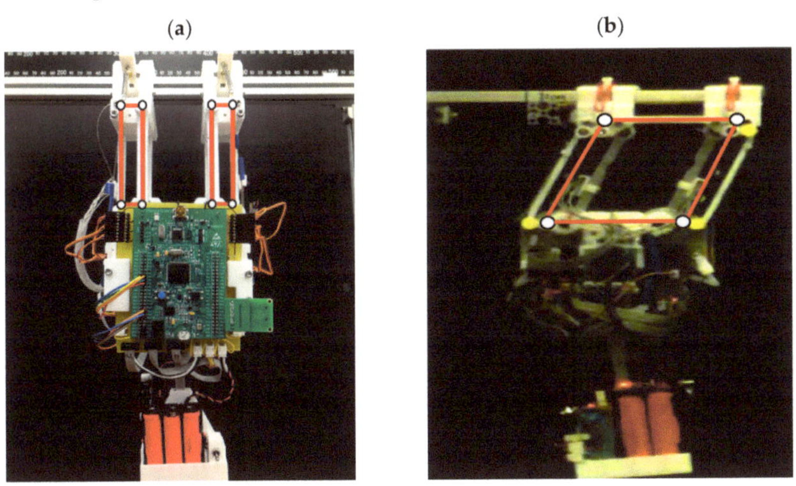

Figure 3. Mechanical design issues in existing transverse brachiation robots: (**a**) parallel links in dual arms; (**b**) parallel links constructed by two arms, horizontal ledge, and robot body.

Another focus of this study was to enable the same robot configuration to realize a hybrid locomotion gait, including transverse brachiation and transverse ricochetal brachiation. This mechatronic realization problem has yet to be solved for ape-like brachiation robots. From Figures 1 and 2, we can see that the main difference between the two types of brachiation lies in the third phase. With the transverse brachiation gait, both arms return to their original position in this phase for posture adjustment. Therefore, this phase is called the transition phase (Figure 1b). As for the transverse ricochetal brachiation gait, both grippers release at the same time during this stage for complete energy release, and inertia enables the fore gripper to grab the ledge. We defined this as the two-hand-release phase

(Figure 2b). The other phases are similar; only the timing of switching between phases and the joint motion planning strategies differ slightly. This greatly increases the chance of successfully achieving these two types of gaits using the same robot configuration.

3. System Modeling

This study presents locomotion gait designs for transverse brachiation and transverse ricochetal brachiation with a locomotion cycle comprising four phases. The first two phases are the swinging phase and the approaching-target phase. The method of energy release in the third phase differentiates the two locomotion gaits. After both grippers have grasped the ledge, the robot adjusts its posture to help it enter the initial swing phase of the next locomotion cycle. The gait design contains a total of five different movement phases. We next give a brief introduction to the processes of establishing the dynamic model for each phase and the differences in the mathematical models of the phases. This serves as the foundation for gait parameter analysis and design.

We consider the example of a robot performing transverse brachiation to the right. Using the Euler–Lagrange method, we first define suitable origins for the coordinate system based on the various movement phases. Next, we select a state vector $q \in \mathbb{R}^{n\times 1}$ that can present the system dynamics completely. The external force input vector corresponding to the various states is $u \in \mathbb{R}^{n\times 1}$. With these settings, we can calculate the Lagrangian $\mathcal{L}$, including the kinetic energy T and potential energy V of the system, as follows:

$$\begin{array}{l} \frac{d}{dt}\left(\frac{\partial \mathcal{L}}{\partial \dot{q}}\right) - \frac{\partial \mathcal{L}}{\partial q} = u \\ \mathcal{L} = T - V \end{array} \quad (1)$$

Based on the definition above, we can derive the dynamic equation of general standard robots as shown:

$$M(q)\ddot{q} + C(q,\dot{q}) + G(q) = u \quad (2)$$

where $M(q) \in \mathbb{R}^{n\times n}$ is the inertia matrix, $C(q,\dot{q}) \in \mathbb{R}^{n\times 1}$ denotes the dynamic torques associated with the centripetal and Coriolis forces, and $G(q) \in \mathbb{R}^{n\times 1}$ represents the gravity torques. Figure 4 and Table 2, respectively, present the schematics of the proposed robot in each movement phase and the parameters associated with the system dynamic model. Next, we define the coordinate origin, system states, and input vectors of each phase to derive Equation (2) and explain the physical meaning and objective of the motions in each phase.

Table 2. Robot model parameters.

Symbol	Physical Meaning	Symbol	Physical Meaning
m_L	Center of mass of left arm	l_{RC}	Distance between right wrist joint and center of mass of right arm
m_R	Center of mass of right arm	l_{BC}	Distance between left shoulder joint and center of mass of body
m_B	Center of mass of body	l_{LC}	Distance between left wrist joint and center of mass of left arm
m_T	Center of mass of tail	l_T	Tail length
I_L	Moment of inertia of left arm	l_{BW}	Body width
I_R	Moment of inertia of right arm	L	Distance between two grippers
I_B	Moment of inertia of body	θ_L	Joint angle of left arm
I_T	Moment of inertia of tail	θ_R	Joint angle of right arm
l_L	Length of left arm	θ_B	Body joint angle
l_R	Length of right arm	θ_T	Tail joint angle

1. Swinging phase

In this phase, both grippers are gripping the ledge, and the arms form a parallel four-link geometric configuration with the environment. By swinging the tail, adequate energy is accumulated to facilitate the subsequent transverse movement. Based on the assumption of transverse locomotion to the right, we set the coordinate origin of this phase at the joint of the left wrist, O_I, as shown in Figure 4a. Suppose the distance between the grippers in this phase is L, and the grippers are gripping the ledge firmly with no slippage occurring. Through the geometric relationships in the four-bar linkage mechanism [31], we can derive the relations of joint angles θ_R, θ_L, θ_B as follows:

$$\begin{aligned} \theta_R(\theta_L) &= 2\tan^{-1}\left(\frac{-k_1+\sqrt{k_1{}^2+k_2{}^2+k_3{}^2}}{k_3-k_2}\right) \\ \theta_B(\theta_R,\theta_L) &= \tan^{-1}\left(\frac{L-l_L\sin\theta_L+l_R\sin\theta_R}{l_R\cos\theta_R-l_L\cos\theta_L}\right) \end{aligned} \tag{3}$$

where $k_1 = 2l_RL - 2l_Ll_R\sin\theta_L$, $k_2 = -2l_Ll_R\cos\theta_L$, $k_3 = L^2 + l_L^2 + l_R^2 - l_{BW}^2 - 2l_LL\sin\theta_L$.

Equation (3) shows that right arm joint angle θ_R and body joint angle θ_B can be expressed as functions of left arm joint angle θ_L. Thus, we can select the generalized coordinate vector $q_{\text{swing}} = \begin{bmatrix}\theta_L & \theta_T\end{bmatrix}^T$ and use two degrees of freedom to express the dynamics of the robot in this phase completely. This phase only uses the tail joint to perform system excitation, making it the equivalent of a two-link under-actuated system. Thus, we can let the input vector be $u_{\text{swing}} = \begin{bmatrix}0 & \tau_T\end{bmatrix}^T$.

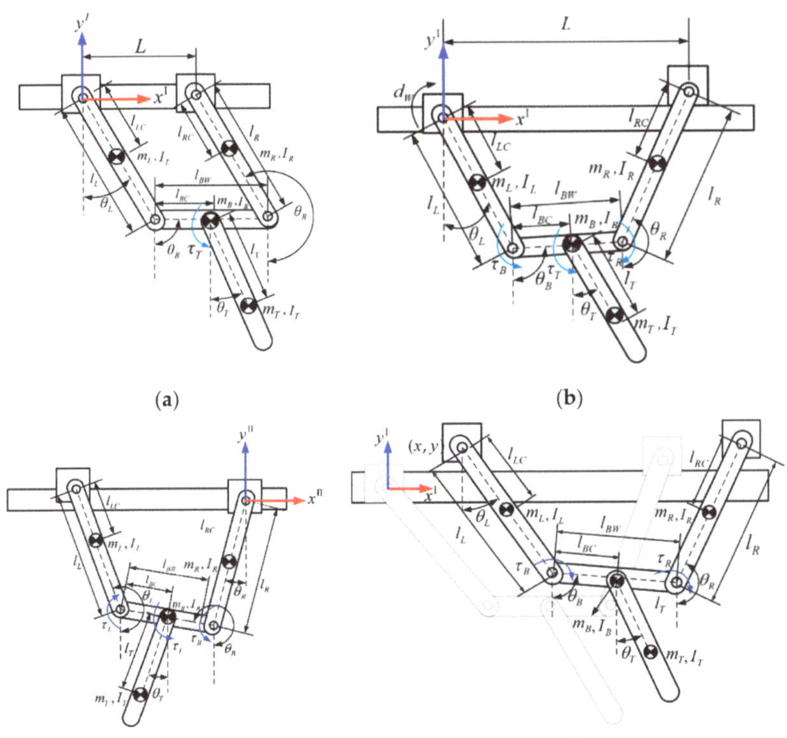

Figure 4. Schematic diagram of proposed transverse brachiation robot for classified phases Table 2. Robot model parameters. (**a**) Swing/Transition Phase. (**b**) Approaching-target Phase. (**c**) Hands-on Phase. (**d**) Two-Hand Release Phase. The red and blue arrows indicate the used coordinate system.

2. Approaching-target phase

After the tail has accumulated enough energy by swinging back and forth, the robot can switch to this phase at an appropriate time and reach for the target ledge; that is, the right hand can let go of the ledge and grasp the ledge to the right as shown in Figure 4b. In this phase, the robot's left gripper still has a firm grip on the ledge, and assuming there is no slipping between the gripper and the ledge, the origin of the coordinate system is set at the left wrist joint, O_I, as exhibited in Figure 4b. The robot's right gripper has let go in this phase, so the right arm is not subject to the geometric constraints as it was in the previous phase. Thus, we select a generalized coordinate vector $q_{\text{target}} = \begin{bmatrix} \theta_L & \theta_B & \theta_R & \theta_T \end{bmatrix}^T$. In this phase, the left wrist has no actuation torque. During the movement, the shoulder joints and the tail joint provide the actuation torque (τ_B, τ_R, τ_T), which, along with the energy accumulated in the swing phase, is used to reach for and grasp the target ledge. Thus, the input vector is defined as $u_{\text{target}} = \begin{bmatrix} 0 & \tau_B & \tau_R & \tau_T \end{bmatrix}^T$.

3. Transition phase

After the robot has successfully grasped the ledge in the previous approaching-target phase, it returns to having both grippers gripping the ledge at the same time. Here, we designed a posture transition phase to help the robot return smoothly to its original posture at the beginning of a gait cycle. During this stage, the upper limbs similarly form a four-link configuration with the ledge as they perform in the swing phase, and the robot relies on swinging its tail to correct its overall posture and help the other gripper move forward to its original position. As shown in Figure 4a, the coordinate system, generalized coordinates, and input vectors are identical to those in the swing phase, which means that the mathematical dynamic model is the same. However, essential differences exist between the joint motion trajectories and planning objectives of these two phases.

4. Hands-on phase

Once the robot adjusts to a certain posture in the previous phase, the left gripper releases the ledge and moves to the right back to its original position so that the distance between the two grippers returns to the set distance in the initial swing phase, the posture of the body and the tail is coordinated, and the robot can again begin the excitation and energy storage phase needed for the next locomotion cycle. This can be regarded as a successful locomotion gait cycle. In this phase, we assume that the left gripper releases the ledge as the right gripper has a firm grip on the ledge with no slippage. Thus, the coordinate origin is set at the right wrist joint O_{II}, as shown in Figure 4c. The generalized coordinate variable of gripping with one gripper is identical to that in the target-approaching phase, that is, $q_{\text{hbo}} = \begin{bmatrix} \theta_L & \theta_B & \theta_R & \theta_T \end{bmatrix}^T$ and the external force input vector is set $u_{\text{hbo}} = \begin{bmatrix} \tau_L & \tau_B & 0 & \tau_T \end{bmatrix}$ based on the actuation joint torque.

Reminder: The system state variable q_{hbo} in this phase is defined based on the right wrist joint coordinate system (O_{II}) (Figure 4c) and is different from the status variable q_{target} in the second phase, which was defined based on the left wrist joint coordinate system (O_I) (Figure 4b). The conversion formulas of the joint angles with regard to these two coordinate systems are as follows:

$$\begin{cases} \theta_L^{\mathrm{II}} = \theta_L^{\mathrm{I}} + \pi \\ \theta_B^{\mathrm{II}} = \theta_B^{\mathrm{I}} + \pi \\ \theta_R^{\mathrm{II}} = \theta_R^{\mathrm{I}} - \pi \\ \theta_T^{\mathrm{II}} = \theta_T^{\mathrm{I}} \end{cases} \tag{4}$$

5. Two-hand-release phase

In this study, the four movement phases above can be used to express the entire dynamics of the transverse brachiation locomotion gait. However, ricochetal brachiation requires a description of the dynamics when both grippers have released the ledge, and the robot is airborne. We use the two-hand-release phase in Figure 4d to represent this. In ricochetal brachiation to the right, we chose to set the coordinate origin at the left wrist joint O_I. While the robot is airborne, there are no geometric constraints for its joints, and there is additional two-dimensional translation along the xy plane. Thus, the state variable that we chose for this phase was $q_{\text{thr}} = \begin{bmatrix} \theta_L & \theta_B & \theta_R & \theta_T & x & y & \end{bmatrix}^T$. The goal of this phase is to use the actuators at the shoulder joints (τ_B, τ_R) to enable the two shoulders of the robot to move toward the target ledge during the extremely short leap. During this process, the tail actuator (τ_T) maintains a certain posture (i.e., a fixed tail joint angle) to reduce its impact on the swinging motions of both shoulders during flight and thereby increase the landing success rates of both grippers. The external force input vector is defined as $u_{\text{thr}} = \begin{bmatrix} 0 & \tau_B & \tau_R & \tau_T & 0 & 0 \end{bmatrix}^T$.

Based on the coordinate system, state variable, and input vector defined above, we can derive the system dynamic equation of each phase (Equation (2)), where Phases 1-2-3-4 represent a transverse brachiation locomotion cycle and Phases 1-2-5-4 represent a transverse ricochetal brachiation locomotion cycle. In the next section, we analyze the design parameters that meet the dual-arm robot configuration conditions to enable smooth transitions.

4. Robot Locomotion Design

Based on the derived dynamic model, this section analyzes the design goals needed to complete an entire locomotion cycle. On the premise of the causal relationship, we explain the means of selecting switch parameters to achieve smooth transitions between phases. We then propose corresponding motion control strategies. We referred to the modified version of a transverse brachiation robot prototype already developed by our lab [1,2] for the required design parameters.

4.1. Locomotion for Transverse Brachiation

To achieve the goal of continuous transverse locomotion, we set two design goals for the robot locomotion gait:

(1) each movement phase must be executed successfully, and two continuous locomotion gait cycles must be completed; in other words, eight movement phases must be completed continuously;

(2) the energy connecting the motion processes of each phase must be maintained and not return to zero at the end of which phase, which would cause uncoordinated motions.

The proposed process for the design of transverse brachiation locomotion gaits is presented in the flow chart in Figure 5. For the sake of convenience, we use the following abbreviations for each phase in the parameters and explanation: swinging phase (S), approaching-target phase (A), transition phase (T), and hands-on phase (H). As seen in the figure, this design method does not analyze the switching conditions in the S→A→T→H phase order but first defines a flexibility index first to obtain the maximum relative distance between the two grippers (the reason for which is explained later). Once the maximum gripper distance and upper limb assumptions have been obtained, mechanical geometric relationships are used to derive the arm joint motion trajectories for the approaching-target phase and the switching conditions for posture adjustment. The next step is to analyze the swinging phase parameters while meeting the aforementioned assumptions to obtain the appropriate timing to initiate reaching and gripping. The dynamic simulation results obtained from the switching conditions of the first two phases are substituted into the dynamic model of the posture transition phase. This facilitates the selection of the operating parameters needed to switch to the next phase. Finally, the posture condition parameters of the motions completed in the previous phase are substituted into the dynamic model of the hands-on phase to derive the position where the left arm needs to go so that it is in the same position as it was in the initial cycle, as well as the switching conditions for the swinging phase of the next cycle. In the next section, we explain the logic of this design process in detail and the reasons underlying the selection of the switching conditions.

4.1.1. Design Requirements and Analysis Procedure for Transverse Brachiation

In the gait analysis in Section 2, we observed that the athlete did not grasp the furthest position in the approaching-target phase but maintained a suitable distance between his hands. This creates the necessary posture for the next phase. Thus, we first analyzed the influence of the distance between the two hands in the transition phase on the entire locomotion gait. As shown in Figure 6, we first defined the distance between the two grippers after the robot switches from the approaching-target phase to the posture transition phase as the design parameter L^T. Note that a greater distance is not superior; an overly-far grasping distance increases the rigidity of the entire robot and reduces the range at which the robot's body can rotate as the tail swings for posture adjustment. In the worst-case scenario, the body will be completely unable to adjust its posture, thereby forcing the entire locomotion gait to stop.

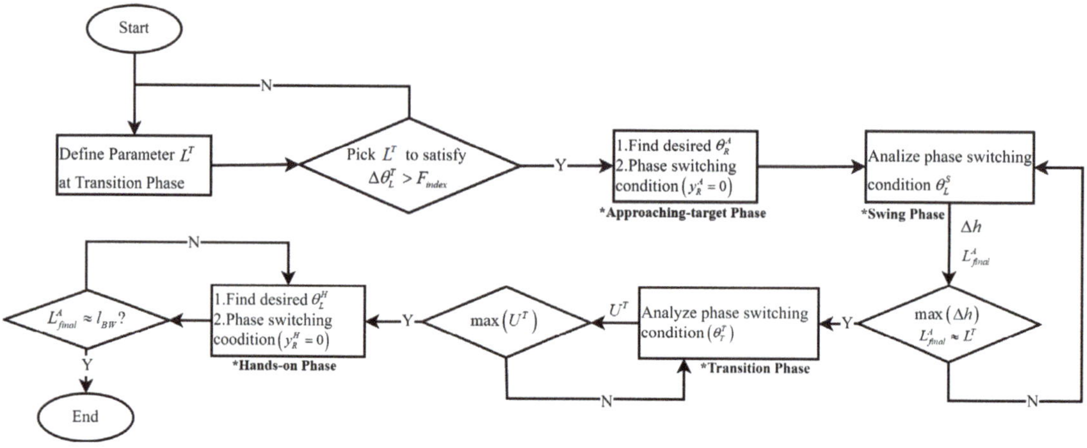

Figure 5. Flow chart of locomotion control design strategy for transverse brachiation. * use to highlight each different phase.

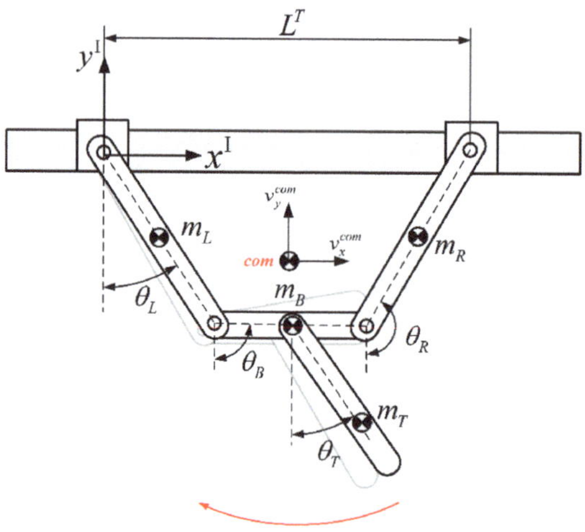

Figure 6. Schematic diagram of transition phase where design parameter L^T is related to the range of motion of upper limb ($\Delta\theta_L$). Red arrow indicates moving direction.

We assumed that the upper limbs for an isosceles trapezoid with the ledge at the beginning of the transition phase; that is $\theta_B = 90°$ in Figure 6. If we issue a swinging command to the lower limb, we can see a clear connection between the L^T distance between the two grippers and $\Delta\theta_L$ the yaw angle of the arms, which is defined as

$$\Delta\theta_L \triangleq \left|\theta_{L\_max}^T - \theta_{L\_min}^T\right| \geq F_{index} \tag{5}$$

where $\theta^T_{L\_max}$ and $\theta^T_{L\_min}$ are the maximum and minimum values of θ_L the angle of the left arm as the tail is swinging in the transition phase. Thus, the difference $\Delta\theta_L$ is the permissible swing range of the left arm. The larger this range is, the more flexible the motions of the upper limbs are and the less likely it is for dead-point positions or locked chains in this phase. In Equation (5), we define the flexibility index F_{index} to represent the range in which the upper limbs of the robot can move with different distances between the two grippers. When the upper limbs and the ledge form a parallel four-link posture (i.e., the initial state of the swing phase), the robot is the most flexible; however, if the relative distance between the grippers increases after the robot enters the transition phase, the upper limbs will become less flexible. To increase the grasping distance as much as possible while maintaining a certain range in which the arms can move to facilitate smooth transitions to subsequent movement phases, we analyzed several different L^T values, and after performing mechatronic system tests, we set F_{index} at 20°. As shown in Figure 7, the dark blue bars represent results that meet the flexibility index, whereas the light purple bars represent results that do not meet the flexibility index. As can be seen, when L^T is equal to 170 mm, the range in which the arms can move is 22°, which is not too different from the 26° that the arms can move when L^T is equal to 160 mm. However, if the distance between the two grippers increases to 180 mm, the range in which the arms can move drops significantly to 13°. Thus, the design parameters obtained using F_{index} = 20° represent a good compromise. Following the selection of this parameter value, we introduce the switching conditions and motion planning strategies of the approaching-target phase, the swinging phase, the transition phase, and the hands-on phase according to the steps shown in Figure 5.

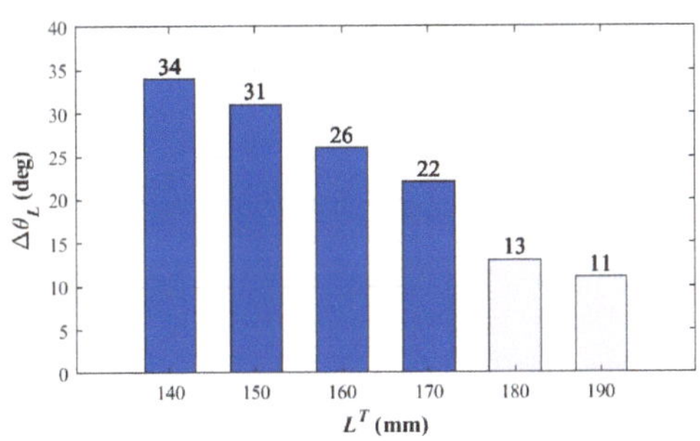

Figure 7. Relationship between yaw angle ($\Delta\theta_L$) and L^T. Deep blue bars indicate the results substantially distinguished from others.

4.1.2. Motion Control Strategy Associated with Approaching-Target Phase for Transverse Brachiation

The analysis in the previous section shows that at the beginning of the transition phase, the desired distance between the two grippers, L^T, is 170 mm, and the upper limbs form an isosceles trapezoid. To meet these conditions, we planned the joint motion trajectories of the robot in the previous movement phase as follows:

$$\begin{cases} \theta^A_R(t) = 180^\circ - \tan^{-1}\frac{\sin\theta^A_L(t_f)}{\sqrt{1-\sin^2\theta^A_L(t_f)}} \\ \tau^A_L(t) = 0 \\ \theta^A_T(t) = \theta^S_T(t_f) \\ \theta^A_B(t) = 90^\circ \\ \sin\theta^A_L(t_f) = \frac{L^T - L_{BW}}{2l_L} \end{cases} \tag{6}$$

In the approaching-target phase, the transmission of motor torque can be stopped by switching the clutch on the left wrist. Thus, $\tau^A_L(t) = 0$, which allows the four-link robot to freely rotate its left wrist and coordinate with the other three actuated joints to achieve the expected grip posture using proportional-derivative (PD) control. During the grasping process, we had the tail joint angle remain the same as it was at the end of the swing phase (t_f); in other words, $\theta_T^A(t) = \theta_T^S(t_f)$. With the isosceles trapezoid restriction, the conditions of the body joint angle θ_B and right arm joint angle θ_R are $\theta_B^A(t_f) = 90^\circ$ and $\theta_R^A(t_f) + \theta_L^A(t_f) = 180^\circ$, where θ_L is the feedback value of the left arm joint angle obtained in the event of no torque input. Based on this geometric restriction, we can obtain $\theta_R^A(t)$, the reference command of θ_R given during the grasping process, as shown in Equation (6). After the right gripper has grasped the ledge (i.e., the coordinate value $y_R^A = 0$ in the vertical direction of the right gripper in Figure 4b), the robot can switch to the next phase (i.e., transition). The switching condition can be expressed as

$$y_R^A = -l_L \cos\theta_L - l_{BW} \cos\theta_B - l_R \cos\theta_R = 0 \tag{7}$$

4.1.3. Motion Control Strategy Associated with Swinging Phase for Transverse Brachiation

Based on the aforementioned assumption and switching condition in Equation (7), we analyzed the parameters of the swinging phase and the conditions needed for motion planning. We sought to determine the appropriate timing to switch from the swinging phase to the approaching-target phase. Figure 8 presents a schematic diagram of the transition between the two phases. As mentioned previously, the goal of the swing phase is to swing the lower limb back and forth while the two arms are in a parallel ledge-gripping posture to store enough system kinetic energy to reach the target grasping distance. At this point, the distance between the two grippers is $L^S = L_{BW}$. By switching the clutches in both wrists, there is no torque input in both arm joints so they can rotate freely. The reference trajectory given to the tail joint motor is a sine function, and closed-loop system control is realized using the PD control law. Due to mechanical restrictions, we set the maximum amplitude of the sine function to 45°. Using a swept sine, we can derive that the resonant frequency of the robot system in the swing phase is approximately 1.5 Hz. The simulation results show that in the fourth swing, the system will enter a resonant state. Thus, three swings were performed before the conditions derived later were used to switch to the target-approaching phase during the fourth swing.

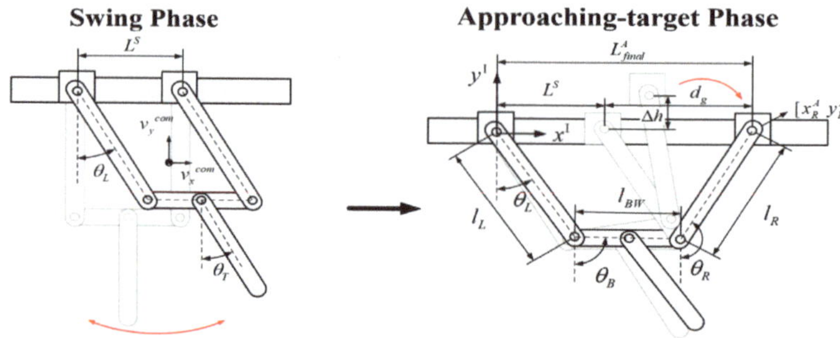

Figure 8. Schematic diagram of transition from swinging to approaching-target phases: robot releases its right hand to approach the target position as the swung arm satisfies phase switching condition $\theta_L = 28.8^\circ$.

As shown in Figure 8, a greater height difference (Δh) between the highest point of the right gripper after releasing the ledge and the height of the target ledge means a longer duration of flight as well as more flexibility in the designs of joint motion planning. As the direction of the speed is the same as the direction in which the system is moving, the y-direction speed component of the centroid at the instance of transition from the swing phase to the target-approaching phase should be greater than 0 ($v_y^{com} > 0$), and the direction of the angular velocity of the left arm joint should be counterclockwise ($\dot{\theta}_L > 0$). This enables the right arm to leave the ledge the right gripper was originally gripping and reach the right to grasp the target ledge successfully. With the target distance between the two grippers (L^A_{Target}) in the target-approaching phase meeting the previously set parameter L^T= 170 mm minus the initial distance between the two grippers (L^S= 100 mm), the target moving distance of the arms (d_g) is approximately 70 mm. We, therefore, chose the arm joint angle θ_L during the swing phase as the basis for the choice of when to switch to the target-approaching phase. Based on the aforementioned constraints, the height of the right gripper in the flight phase after it releases the ledge, must be as high as possible. We can therefore categorize the following optimization problems:

$$\begin{aligned} &\underset{\theta_L}{\arg\max}\{\Delta h\ (\theta_L, v_y^{com},\ \dot{\theta}_L)\} \\ &\text{subject to } v_y^{com} > \ 0\,, \dot{\theta}_L > 0,\ L^A_{\text{Target}} \approx 170\ (\text{mm}) \end{aligned} \tag{8}$$

Using dynamic simulation analysis, we derived that the phase switching condition (S→A) satisfying Equation (8) is for the right gripper to release the ledge to enter the approaching-target phase when the arms swing to $\theta_L = 28.8^\circ$ in the swinging phase. Figure 9 presents the relationship between the switching condition (θ_L) and the grasping distance (d_g) that can be completed in the target-approaching phase. The results where phase switching cannot be achieved are in the gray area. As can be seen, θ_L must be greater than about 6° in order for the robot's right gripper to leave the swinging phase successfully. The red horizontal dashed line represents the target movement distance for the right gripper (70 mm). We relaxed this constraint slightly to 69 mm (i.e., $L^A_{\text{final}} = 169$ mm) so that more energy can be accumulated during the swinging phase to compensate for any energy consumption that the dynamic model does not take into account. Thus, the corresponding phase switching condition is $\theta_L = 28.8^\circ$.

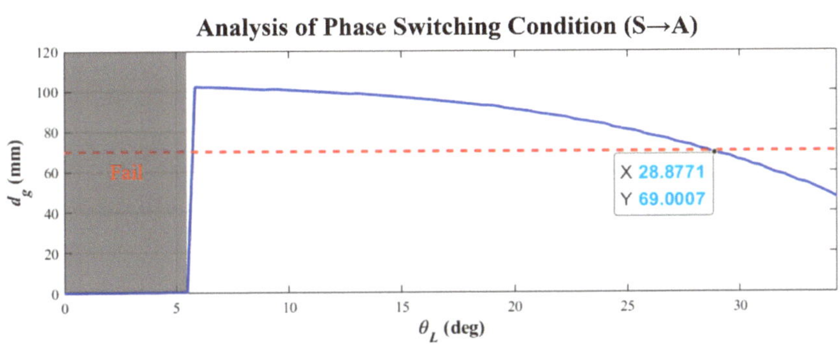

Figure 9. Relationship between phase switching (S→A) condition θ_L and gripper grasping distance d_g.

4.1.4. Motion Control Strategy Associated with Transition Phase for Transverse Brachiation

As shown in Figure 6, once the right gripper grasps the ledge, the goal of the posture transition phases is added, in which the posture of the upper limbs is adjusted by swinging the tail, and the potential energy of the entire system is also increased to help the left arm return to its original position. To this end, we used the same sine function as that used in the swinging phase as reference input of the tail joint angle ($\theta^T_T(t)$). Increasing the potential energy U of the entire system was the primary goal, and the tail swing angle θ_T was chosen as the switching parameter determining when to transition to the next phase (T→H). U can be derived using the following equation:

$$U = m_L g \vec{p}^{\,y}_L + m_B g \vec{p}^{\,y}_B + m_R g \vec{p}^{\,y}_R + m_T g \vec{p}^{\,y}_T \tag{9}$$

where g is the gravitational acceleration constant and $\overrightarrow{p}_i^y (i = L, B, R, T)$ is the vertical component of the position vectors from the origin of the coordinate system in this phase to the centroids of each link. Similarly, the y component of the system centroid speed at the instance of phase switching should also be greater than 0 ($v_y^{com} > 0$) to prevent the robot from falling when it releases the ledge. Furthermore, the direction of the angular velocity of the left arm joint should be clockwise ($\dot{\theta}_L < 0$) to coordinate with the movement of the left arm to the right to return to its original position.

Combining the above conditions, we can establish an optimization problem as follows:

$$\begin{aligned} &\underset{\theta_T}{\arg\max}\{U^T(\theta_T, v_y^{com}, \dot{\theta}_L)\} \\ &\text{subject to } v_y^{com} > 0, \dot{\theta}_L < 0 \end{aligned} \tag{10}$$

where U^T denotes the potential energy of the system in the transition phase. Figure 10 shows the parameter analysis results derived from the computer simulation; the X, Y, and Z axes, respectively, represent the phase-switching condition parameter (θ_T), the y component of the system centroid speed (v_y^{com}), and the potential energy of the system (U). From the figure, we can see that the critical value of potential energy at which the robot can maintain an upward overall centroid speed is approximately −2.455 J. At this point, the lower limb swinging to $\theta_T = -21.08^\circ$ (a negative value indicates the clockwise direction) is the appropriate time to switch to the next movement phase.

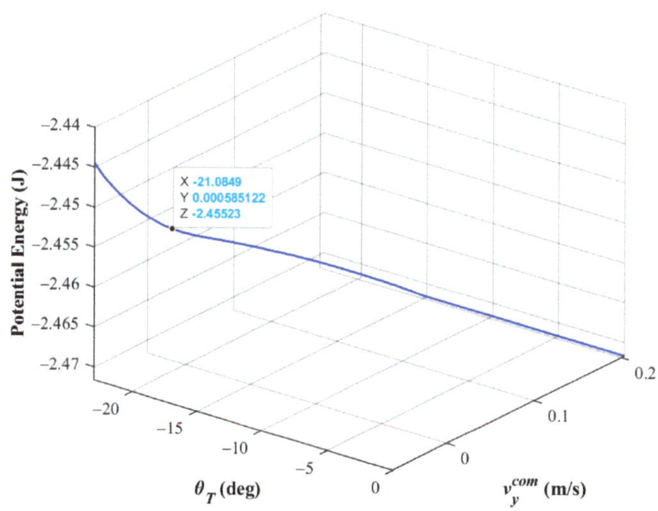

Figure 10. Parameter analysis of phase switching condition from the transition phase to the hands-on phase where the robot shifts to the next phase when the joint angle of the tail link swinging from right to left is $\theta_T = -21.08^\circ$.

4.1.5. Motion Control Strategy Associated with Hands-on Phase for Transverse Brachiation

Following the analysis of the first three phases, we substitute the known operating parameters into the dynamic model of the hands-on phase for analysis to complete the gait design. The motions of this phase are shown in Figure 11. The primary design goals of this phase are (1) enabling the robot to complete this transverse locomotion cycle successfully and (2) making the geometric structure of the upper limbs after the left gripper grasps the ledge as close to a parallel four-link posture (i.e., the distance between the two grippers $L_{final}^H \approx l_{BW} = 100$ mm) as possible so that the robot can again perform the next locomotion cycle and achieve continuous brachiation. Thus, the various joints must meet the following conditions:

$$\begin{cases} \tau_R^H(t) = 0 \\ \theta_T^H(t) = \theta_T^T(t_f) \\ \theta_L^H(t) = \theta_L^H(t_f) \\ \theta_B^H(t) = -90^\circ \\ \theta_L^H(t_f) - \theta_R^H(t_f) = 180^\circ \end{cases} \quad (11)$$

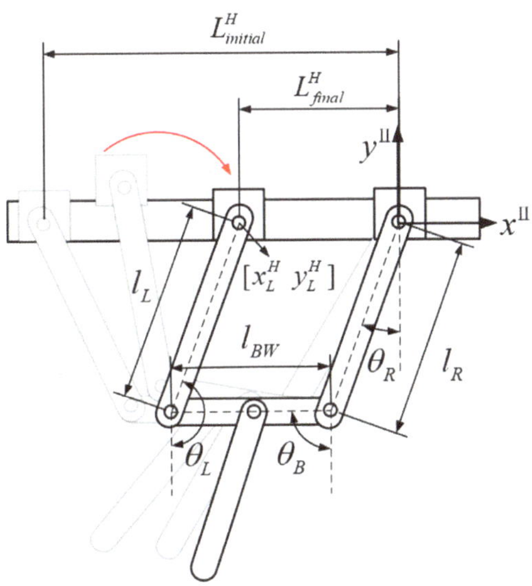

Figure 11. Schematic diagram of the hands-on phase where grey indicates the motion process and black indicates the robot has returned to the initial posture (i.e., swinging phase).

In this phase, the right wrist joint can rotate freely (by switching the clutch); therefore, $\tau_R^H(t) = 0$. The tail joint angle command ($\theta_T^H(t)$) is set at the angle it was at the end of the previous phase. $\theta_B^H(t_f) = -90^\circ$ and $\theta_L^H(t_f) - \theta_R^H(t_f) = 180^\circ$ are the joint position command conditions needed for the body joint angle θ_B and left arm joint angle θ_L to form the parallel four-link posture at the end of this phase (t_f), whereas the right arm joint angle θ_R equals the feedback value transmitted back via the encoder where there is no torque input. Thus, the angle command given to the left arm joint θ_L is $\theta_L^H(t_f)$. Note that the coordinate origin in this phase is located at the wrist joint of the right gripper. The positive and negative joint angles in this phase differ from those in the other phases (counterclockwise direction defined as positive). Furthermore, the initial distance between the two grippers in this phase is $L^H_{initial} = L^T$.

To achieve the posture in Equation (11), performing simulation analysis using the dynamic model of this phase produces the relationship between the final value of the left arm joint angle $\theta_L(t_f)$ and the distance between the two grippers after gripping, as shown in Figure 12. As can be seen from this figure, setting $\theta_L(t_f) = 181^\circ$ results in $L^H_{final} = L_{BW} = 100$ mm. This means that after the left gripper has grasped the ledge again, the geometric posture of the upper limbs is closest to a parallel four-link posture, as shown in Figure 11. When the y coordinate of the wrist of the left gripper is 0 ($y_L^H = 0$), the robot can return to the swinging phase to store energy for the next locomotion cycle; the phase switching (H→S) condition is

$$y_L^H = -l_R \cos\theta_R - l_{BW} \cos\theta_B - l_L \cos\theta_L = 0 \quad (12)$$

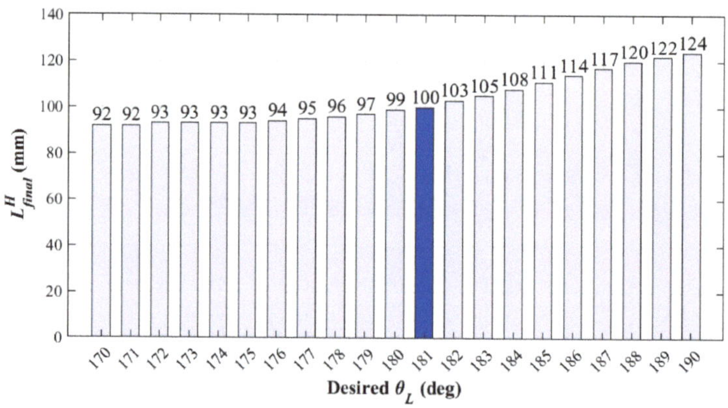

Figure 12. Parameter analysis of the hands-on phase where the upper limb posture represents a parallelogram if the target value of the left arm joint angle $\theta_L(t_f)$ is 181° (the one with deep blue bar).

Following a comparison with Figure 5, we find the phase switching conditions to achieve continuous transverse brachiation locomotion can be re-categorized as follows:

1. Swinging phase→Approaching-target phase: arm swing angle becomes $\theta_L^S = 28.8°$;
2. Approaching-target phase→Transition phase: vertical position of grasping gripper coincides with ledge position $y_R^A = 0$;
3. Transition phase→Hands-on phase: tail link swings back until joint angle $\theta_T^T = -21.1°$;
4. Hands-on phase→Swing phase: vertical position of returning gripper coincides with the ledge position $y_L^H = 0$.

4.2. *Locomotion for Transverse Ricochetal Brachiation*

The design proposed in Section 4.1 is based on postures with at least one hand holding onto the ledge. The energy accumulated during the swinging phase is lost to a certain degree under the constraints, so the grasping distance must allow for posture adjustment and mechanical flexibility when both grippers are gripping the ledge. For this reason, we propose incorporating a leap motion during the ledge gripping process. A leaping motion would more effectively utilize the energy accumulated during the swinging phase and increase the gait distance and mobility performance of the robot. In other words, we propose replacing the posture adjustment phase of the TLB gait with a two-hand-release phase to design a TLRB gait. The timing of switching from the approaching-target phase to both hands letting go for the flight phase is crucial. The wrong switching conditions will affect the subsequent landing posture, the grasping distance, and the success of the arms returning to their original position after landing.

Figure 13 presents a schematic diagram of the motions in the TLRB gait after the fore gripper has grasped the ledge. We applied the trajectory planning results of TLB as the basis of the gait design under the following design process:

1. The first step in the two-hand-release phase of TLRB is realizing the leap. However, as the left gripper releases the ledge in the instant before the right gripper reaches the ledge, we can assume that the arms begin to resume their original position once the right gripper has grasped the ledge. Based on the joint trajectory planning represented by Equation (11) for the TLB gait, the upper limbs almost form a parallelogram once the left arm resumes its original position. This can be referred to as the locomotion cycle of a TLRB gait.
2. As the execution of the flight stage is short, the various joint angle commands of the robot remain the same as they were at the end of the previous phase (i.e., the joint angles at the instant the left gripper releases the ledge).

3. We adopted the same parameter settings as those of the TLB gait for the initial robot posture, excitation frequency, and amplitude in tail joint trajectory planning.
4. Unlike the TLB gait, where the need for a posture adjustment phase causes the grasping distance to be restricted by mechanical flexibility, this gait mainly uses the energy accumulated in the swinging phase to lengthen the grasping distance as much as possible. The analysis results in Figure 9 show that as the TLRB gait does not require the posture adjustment phase, the constraint $L^{A}_{\text{Target}} = 170$ mm can be removed from Equation (8). In this way, the grasping distance d_g in the approaching-target phase is no longer restricted by mechanical flexibility. In other words, when the angle of the left arm in the swinging phase is $\theta_L = 5.84°$, the right gripper can release the ledge, and the inertia of the robot can be effectively utilized to achieve the maximum grasping distance $d_g = 101$ mm. Thus, we chose $\theta_L = 5.84°$ as the design condition for switching from the swinging phase to the approaching-target phase. Note that these are only the results derived based on the analysis of the dynamic models in the first two phases of the TLB gait. Later, the energy accumulated in previous phases can be further used to increase the total grasping distance during the leap.
5. As shown in Figure 13c, the robot switches to the hands-on phase once its fore (right) gripper grasps the ledge in the two-hand-release phase. The switching condition can be expressed using the following equation:

$$y^R_R = -l_R \cos\theta_R - l_{BW}\cos\theta_B - l_L\cos\theta_L + y^R_L = 0 \tag{13}$$

where y^R_R and y^R_L are the y coordinates of the left and right gripper joints in the two-hand-release phase.
6. As shown in Figure 13d, the hands-on phase mainly utilizes the horizontal inertia the robot obtained to leap. After the right gripper grasps the ledge, the left arm joint uses Equation (11) to bring the left gripper forward to grasp the ledge, thereby completing a TLRB locomotion cycle. At this point, the upper limbs of the robot again for a parallel four-bar linkage and the y coordinate of the left gripper joint $y^H_L = 0$ can also serve as the basis of judgment for the swing phase in the next locomotion cycle.

$$y^H_L = -l_R\cos\theta_R - l_{BW}\cos\theta_B - l_L\cos\theta_L = 0 \tag{14}$$

7. Figure 13a,b exhibit the schematics of the TLRB gait in the target-approaching phase. The focus of this movement phase is the timing of when the left gripper releases the ledge to perform the leap motion after the right gripper reaches out for a good grasping distance (d_g). Here, we used the right arm joint angle (θ_R) as the basis for phase switching, as shown in Figure 13c. Note that the d_g here is based on the ideal grasping distance derived from the dynamic analyses of the approaching-target phase and the flight phase and is different from the results obtained from the analysis in the previous section, which did not consider flight dynamics. Based on the above assumptions, we conducted a dynamic simulation of the entire gait using the switching conditions of the previous phases. As can be seen in Figure 14, there is a greater θ_R increase in the amount of time that the right gripper is airborne after the left gripper releases the ledge, and this increases the landing success rate. However, the grasping distance also decreases as the height of the right arm increases. Considering the mechanical restrictions of the body and arm links in the robot hardware, we chose to have the robot's left gripper release the ledge for the leap motion when the right arm raises to $\theta_R = 170°$ in the approaching-target phase. The resulting grasping distance is approximately 130 mm. Note that only when the robot is not airborne in the instant the right gripper releases the ledge is phase-switching considered successful. Simulations with this condition parameter revealed that L^{H}_{final} the distance between the two grippers in the hands-on phase is around 127 mm. To move closer to the upper limb ledge grip posture of parallel four-bar linkage, a smaller θ_R reference value can be used as the phase-switching condition. However, this may also cause

the flight duration of the right arm to be too short and result in gait failure during mechatronic system realization.

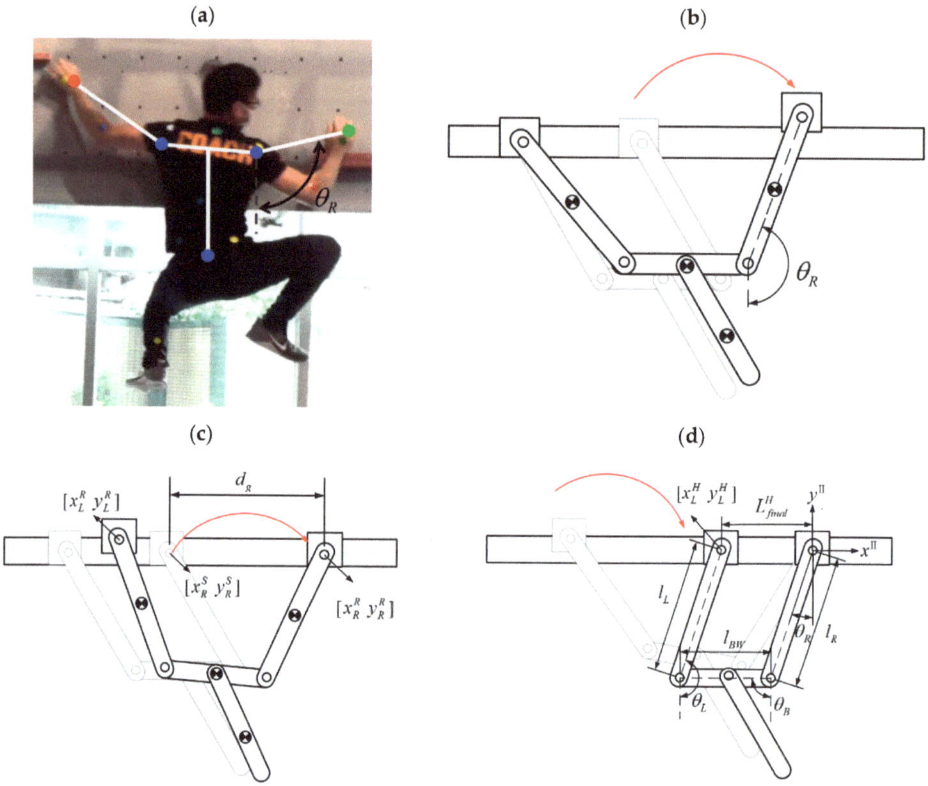

Figure 13. Schematic diagram of TLRB locomotion design: (**a**) athlete posture in approaching-target phase; (**b**) robot posture in approaching-target phase; (**c**) grasping distance d_g; (**d**) robot posture after finishing hands-on phase. Red arrow indicates moving direction.

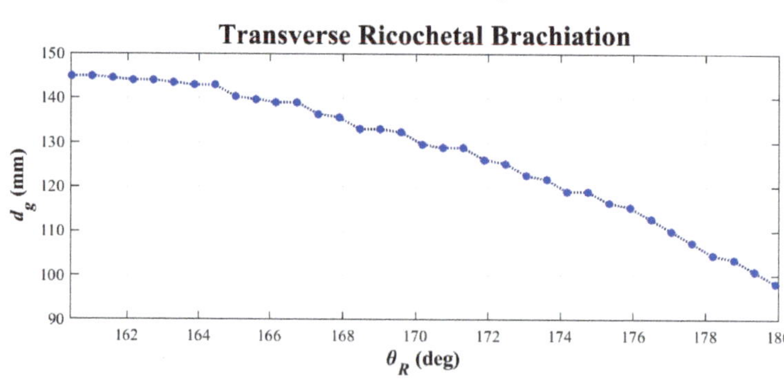

Figure 14. Analysis of phase switching condition θ_R and grasping distance d_g for TLRB (timing for the robot to release ledge-holding hand is related to posture of approaching-target arm).

The phase-switching conditions of the transverse brachiation locomotion gait proposed in this study can be organized as follows:

1. Swinging phase→Approaching-target phase: arm swing angle becomes $\theta_L^S = 5.8°$;

2. Approaching-target phase→Two-hand-release phase: robot releases the ledge-holding (left) hand when the joint angle of the approaching (right) arm reaches $\theta_R^A = 170°$;
3. Two-hand-release phase→Hands-on phase: vertical position of approaching hand coincides with ledge position $y_R^R = 0$;
4. Hands-on phase→End: vertical position of returning hand coincides with the ledge position $y_L^H = 0$.

5. Experiments and Discussion

In this section, we apply joint trajectory planning and phase-switching conditions derived from the previous analysis to a self-made transverse brachiation robot. We performed experiments with the two types of locomotion gaits to verify the feasibility of the proposed design and evaluate its locomotion performance.

5.1. Experiment Setup

Figure 15 shows a transverse ledge climbing scenario consisting of an aluminum extrusion and a long metal plate. Figure 15a shows the front view of the initial state of the robot grasping the ledge, and Figure 15b presents the top view. The grippers have a one-piece mechanical design similar to "Γ" in shape and use solenoid electromagnets to switch between gripping and releasing swiftly. We installed electromagnetic clutches between the shoulder joints and wrist joints of the robot to switch to the under-actuated mode for system excitation in the swinging phase. In the gripping and hands-on phases, it is switched to the servo mode to control gripper-end posture using belt transmission.

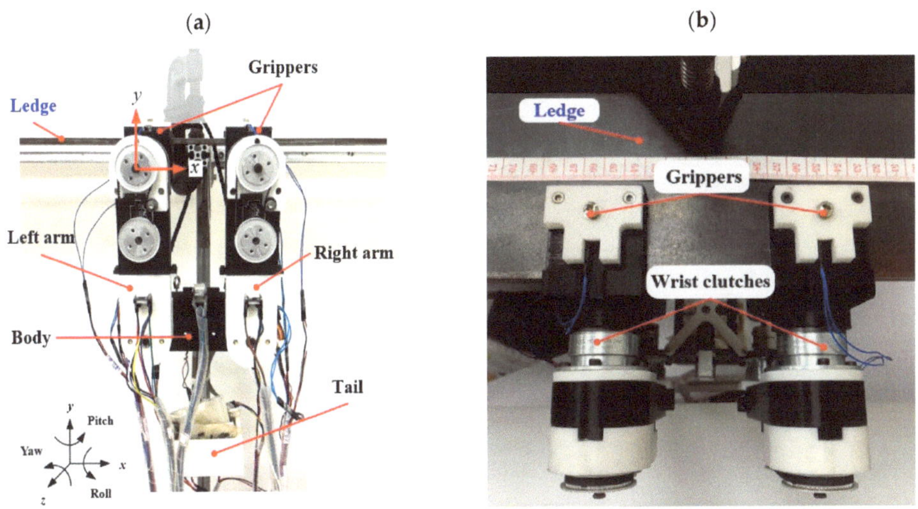

Figure 15. Test scenario for transverse brachiation experiments: (**a**) front view; (**b**) top view.

This study adopted a four-link robot consisting of two arms, a body, and a tail to realize anthropomorphic TLB. The shoulder and lower limb joints of the robot are equipped with DC motors with encoders, and we employed the STM32F407 Discovery board to realize the algorithms needed for joint motion control and the overall locomotion gait. Inertial measurement units are installed on the arm and body links, and the motor encoders transmit the various joint angle measurement results wirelessly to a personal computer via a Bluetooth module for further analysis. For a detailed description of the system architecture, please refer to [2].

5.2. Experiments for Transverse Brachiation

We first present experimental gait results for a single locomotion cycle and then add experimental results for two locomotion cycles to explain the possible problems and research challenges in the realization of continuous transverse brachiation gaits.

5.2.1. Single Locomotion Cycle

Figure 16 shows the results of one cycle of the gait. The top two plots show the robot joint trajectories obtained via simulation and experiment, whereas the bottom plot exhibits the changes in total energy throughout the experiment process. Figure 17 shows sequential images of the robot's motions during the experiment.

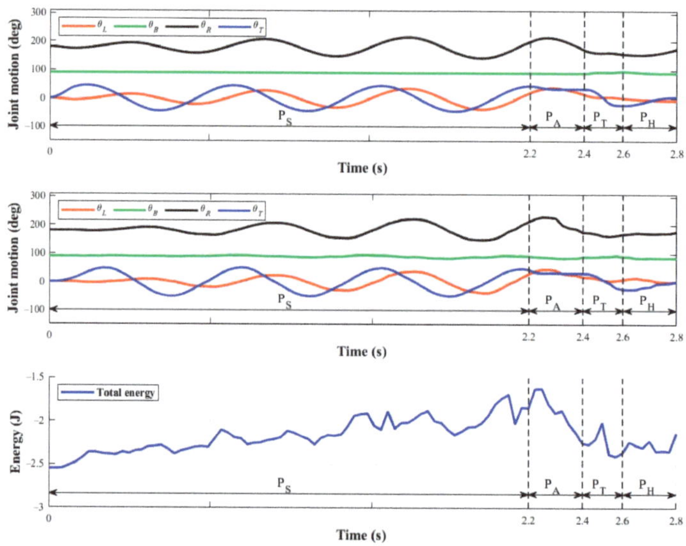

Figure 16. Single locomotion cycle results of transverse ledge brachiation: (**top**) joint trajectories from simulations; (**middle**) joint trajectories from experiments; (**bottom**) energy history from experiments.

Based on the switching conditions obtained in Section 4, the vertical black dashed lines in the graphs in Figure 16 separate the four different movement phases, marked P_S, P_A, P_T, and P_H. First, trends in experimental and simulation results were relatively similar. The body joint angle (green curve) remains at 90° (meaning the body remains in a vertical posture) during most of the process, and the tail joint (blue curve) swings back and forth until the arm joint angles are large enough for the robot to reach out and grasp the ledge. Before the hands-on phase, the left and right joint angles (red and black curves) are coordinated in sync with almost no phase difference, which demonstrates the feasibility and reliability of the proposed model. Next, we calculated the total energy of the entire gait using the following equation and presented the results in the bottom plot of Figure 16:

$$E_{Robot} = T_{Robot} + U_{Robot} \tag{15}$$

In the equation above, T_{Robot} denotes the kinetic energy of the robot, which can be derived from the total translational and rotational kinetic energy of each link; U_{Robot} represents the gravity potential energy of the robot, which can be derived using Equation (9). From Figure 16, we can see that excitation and kinetic energy accumulation in the swing phase (P_S) are achieved via the swinging of the lower limb. Thus, the energy in P_S increases gradually with the number of swings. After entering the approaching-target phase (P_A), the system continues the inertia accumulated in the previous phase to perform the transverse motion, so the energy increases at first, but then the overall potential energy swiftly declines due to gravity; thus, the total energy decreases as well. Next, in the posture transition phase (P_T), the swinging of the lower limb first increases the overall kinetic energy and potential energy of the system slightly. However, the range in which the upper limbs can adjust is limited; when the posture of the upper limbs is closer to their mechanical limit, the reduction in the kinetic energy of the robot is even more apparent. Although the potential energy of the system continues to increase, the kinetic energy drops at a greater rate. Thus, the plot shows that initially, the total energy of the system increases and then declines rapidly midway. Nevertheless, it does not return to the initial value when the robot is still. The approaching-target phase (P_H) mainly relies on the left arm returning to its original position to slightly increase the system energy, so the energy does not change significantly in this phase. The trends in the changes in energy during the locomotion process of the robot show that the designed gait can indeed maintain the energy in the various stages, meaning that after the robot begins to swing, its energy does not suddenly drop to the initial value. Thus, there are no sudden stops that cause incoordination in the gait or affect energy utilization efficiency.

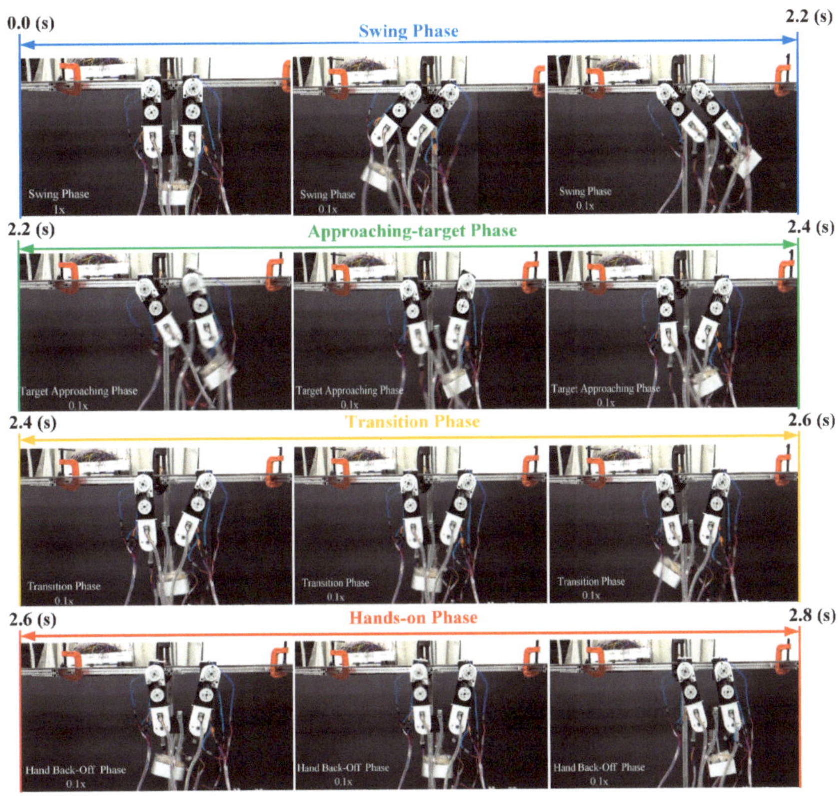

Figure 17. Sequential images of transverse ledge brachiation experiment: a single locomotion cycle.

To assess the mobility of the robot, we defined the total moving distance of the robot (M_d) under the premise that it is moving to the right, as the final moving distance of the end of the left arm, as shown in Equation (16), where $G^L_{initial}$ and G^L_{final}, respectively, denote the initial and final location of the left gripper. If the left arm of the robot cannot successfully move to the right in the current locomotion cycle, it means that the hands-on phase is not completed and that the moving distance of this locomotion cycle is zero.

$$M_d = G^L_{final} - G^L_{initial} \tag{16}$$

Based on this definition, the moving distances achieved by one locomotion cycle of the transverse brachiation gait were as shown in Table 3; simulation and experimental results were 69 mm and 63 mm, respectively. Aside from modeling errors, this difference may have been caused by the fact that the simulation was established on the assumption of perfect grasping and the fact that the result was obtained from a dynamic model based on a two-dimensional plane. In reality, the robot's grippers may slip slightly, causing three-dimensional deviations in posture and thereby reducing the moving distance. As it took 2.8 s to complete one locomotion cycle, the movement speed of the robot was 23 mm/s. Based on our experience, if the robot has to stop all motions during a phase in the entire moving process before it can continue to the next phase, then the movement speed achieved would be only about 3 mm/s. This was the case when the gait was not designed with energy continuation in mind. Thus, a design based on energy continuation increases the speed by seven or more times.

Table 3. Motion performance for transverse ledge brachiation.

	Simulation	Experiment
Grabbing distance at phase P_H (mm)	69	63
Time to finish a locomotion cycle (s)	2.8	2.8
Movement speed (mm/s)	24.6	22.5

5.2.2. Continuous Transverse Brachiation

The grasping target of transverse brachiation in this study is a rectangular ledge with boundary conditions in which it is not easy to attain a firm grip. Robots are easily affected by their three-dimensional dynamics, and once it releases the ledge and grabs hold again, the grip posture is generally different from that before releasing the ledge. Thus, after completing one locomotion cycle and beginning the next locomotion cycle, deviations in grip posture may be too great for the same locomotion gait to be successfully achieved. In other words, after the robot completes a few locomotion cycles, poor grip conditions may prevent it from commencing the swing and energy storage phase, or the grippers may gradually slip off the ledge, causing the robot to fall. This is one of the greatest challenges for TLB robots. To explore this interesting problem, we discuss the results of two cycles of transverse brachiation. Figure 18 presents the joint trajectory results. Again, the vertical black dashed lines separate the different movement phases.

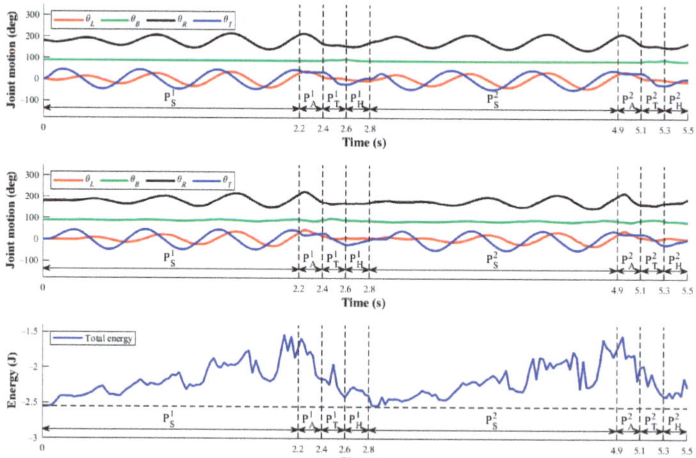

Figure 18. Continuous transverse brachiation for two locomotion cycles: (**top**) joint trajectories from simulations; (**middle**) joint trajectories from experiments; (**bottom**) system energy.

The middle plot in Figure 18 shows the second locomotion cycle beginning at 2.8 s, and the amplitudes of the robot's arm joint angles at the beginning of the swing phase are slightly smaller than those in the first locomotion cycle. It may be that as the arms return to their original positions at the end of the first locomotion cycle and begin the swing phase of the next cycle, the impact dynamics between the grippers and the ledge canceled out some of the system's energy and affected the robot's grip posture. This phenomenon can be verified with the total energy changes presented in the bottom plot of Figure 18. In the first cycle, the swinging phase lasted for 2.2 s before accumulating enough energy to begin the subsequent phases. In contrast, the swinging in the second cycle only lasted 2.1 s, but the moving distance resulting from an offset in the initial swinging posture in the second cycle decreased to 58 mm, which was 5 mm shorter than the 63 mm in the first cycle. Nevertheless, the system still retained some energy at the end of the first locomotion cycle that was not completely canceled out by the impact dynamics of the grippers, so there were still some energy continuation effects.

Experimental results revealed that the final posture of the arms at the end of the hands-on phase is key to whether the next cycle can be achieved successfully. The four-link robot is the most efficient in terms of inertia accumulation when the upper limbs form a parallel four-bar linkage with the ledge. When the posture of the robot's upper body at the end of the hands-on phase differs greatly from a parallel four-bar linkage, the swinging phase in the next cycle will be unable to accumulate energy successfully. Without this accumulation, the performance in subsequent phases will be negatively affected, and the gait might even fail. Based on the parallel four-bar linkage constraint, we used L^H_{final}, the distance between the two grippers of the robot at the end of the hands-on phase as an index to analyze the success rate of completing the next cycle. This index also helped us understand the influence of the robot's posture on continuous transverse brachiation. Figure 19 compiles the results of 14 experiments on two cycles of transverse brachiation, showing the distances between the grippers during the switch from the first to the second cycles. Two experiments failed (meaning the second locomotion cycle could not be executed) and are expressed using white bars; the remaining 12 successful experimental results are presented using blue bars. Theoretically, when L^H_{final} equals 10 mm, the width of the body, then the parallel four-bar linkage condition is satisfied, and the success rate of completing the next locomotion cycle will be higher. However, as shown in the figure, only Experiments 2, 5, and 8 came close to this condition at the end of the first locomotion cycle; in most of the experiments, the distance between the two grippers could not satisfy the condition needed for parallel four-bar linkage, but the second locomotion cycle could still be successfully achieved. However, the distance between the two grippers must be the same as the body width for good grip. If swinging or slippage causes any shifts (i.e., the robot's body is not parallel to the wall as originally assumed), then brachiation may fail. The experimental results showed that robots could complete the second cycle if the distance between the two grippers at the end of the hands-on phase fell between 8.9 cm and 11.56 cm. The mean and standard deviation of L^H_{final} were 10.1 cm and 0.7 cm, respectively. With limited degrees of freedom, it is difficult for the robot to maintain a good grip. In this case, the above data can be used to determine whether to execute the next cycle. Without parallel four-bar linkage, reaching for the ledge may result in failure, as in Experiments 3 and 14.

5.3. Experiments for Transverse Ricochetal Brachiation

Figures 20 and 21 display the simulation results of one TLRB cycle and sequential images of this gait. The sampling rate for simulations was 1 kHz, but due to the IMU readings, the sampling rate for experiments was only 100 Hz. For the latter, changes in total energy were measured by applying the image tracking analysis software Tracker [32] to 60 fps snapshots. For the convenience of presentation, we only show the results midway through the swinging phase in Figure 20.

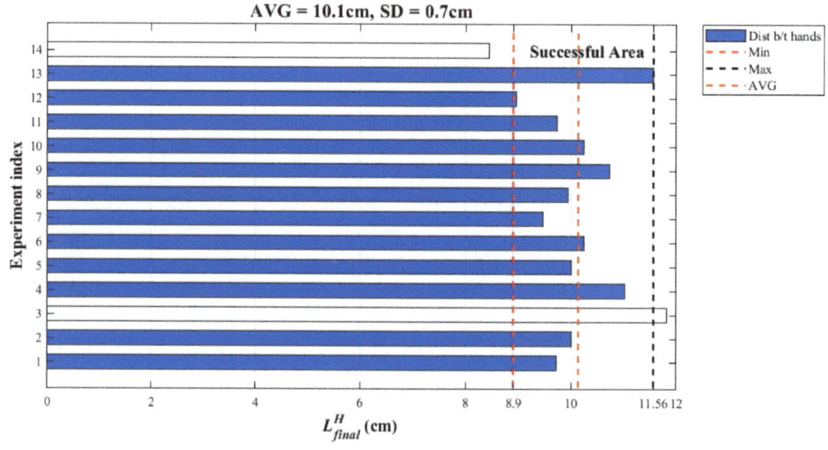

Figure 19. Continuous TLB experiments: the success of implementing the next cycle based on final hand posture in the hands-on phase. White bar indicates failure in experiments.

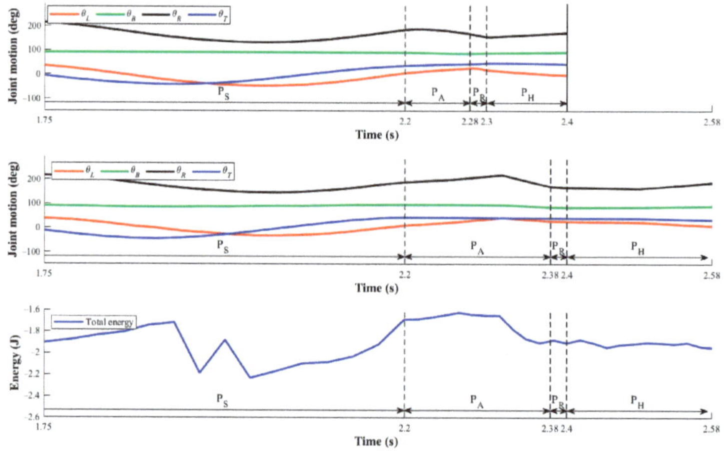

Figure 20. Results of transverse ricochetal brachiation: (**top**) joint trajectories from simulations; (**middle**) joint trajectories from experiments; (**bottom**) energy history from experiments.

As can be seen in the figure, simulation and experimental results for the joint trajectories are similar in the swinging phase but differ after the approaching-target phase begins. First, based on the original motion plans and phase-switching condition $\theta_R = 170°$, the simulation indicated that it would take 2.28 s − 2.2 s = 0.08 s to complete the approaching-target phase (P_A), after which the left gripper would release the ledge for the leap motion (i.e., execute P_R). However, in the experiment, it actually took 2.38 s − 2.2 s = 0.18 s to complete the approaching-target phase, thereby creating a delay of 0.18 s − 0.08 s = 0.1 s. In Section 5.2, the approaching-target phase in the TLB experiment was completed in approximately 2.4 s − 2.2 s = 0.2 s as expected. However, in the TLRB experiment, the larger arm motion in the approaching-target phase resulted in a later response. We, therefore, speculate that the dynamic responses of the robot arms may extend the amount of time needed to perform the entire gait due to mechatronic hardware limitations.

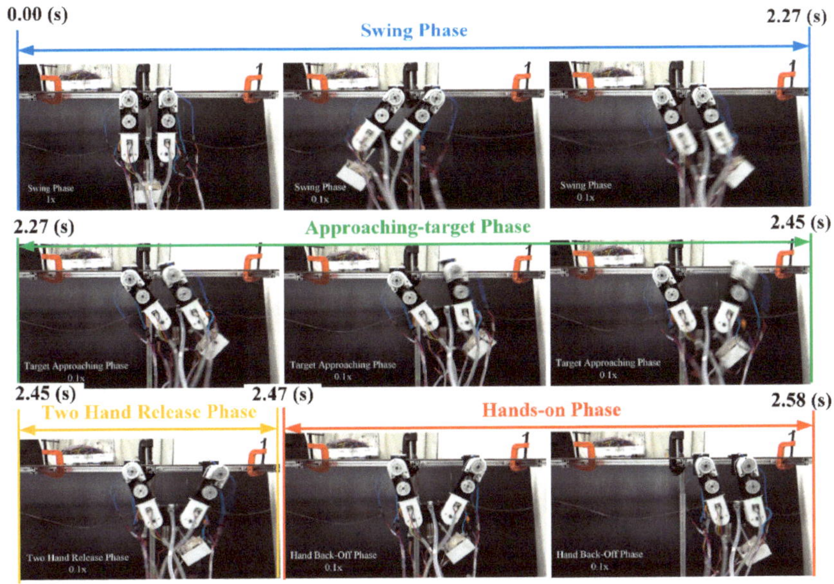

Figure 21. Sequential images of transverse ricochetal brachiation experiment.

It is not easy to see the leap of the robot in the images in Figure 21. This is because the two-hand-release phase (2.45 s~2.47 s) is closer to and even shorter than the response time of the solenoid electromagnet switches in the grippers. During this phase, the robot's grippers may still maintain a certain amount of contact with the ledge, which means that the robot's grippers may slide to the right along the surface of the ledge, thereby increasing the moving distance. Due to the short flight duration (about 0.02 s), we could not accurately determine the time at which the right gripper of the robot grasped the ledge based on the IMU readings. Using trial and error, we had the robot switch to the hands-on phase (P_H) at 2.4 s, and finally, the top view in Figure 22 shows that the robot had obviously turned by an angle of θ_{pitch} after re-grasping the ledge. With the grip posture of the two grippers already significantly different from the original posture, the robot could not execute the next cycle. Note that such posture deviations were also present within the TLB gait experiment in the previous section, although they were not as prominent. As long as θ_{pitch} is small enough, the robot can continue to perform swinging excitation. However, these deviations accumulate with each cycle, so only a finite number of locomotion cycles can be achieved. In another aspect, the energy plot shows that because the left gripper releases the ledge for the leap not long after the right gripper grasps the ledge, the reduction in total energy in P_H is only about half or less of the reduction in the P_H of the TLB gait. This demonstrates that during the flight phase, the transverse brachiation robot can more efficiently utilize the energy accumulated in the swing phase to achieve a longer moving distance.

The results above show that due to factors such as hardware limitations and environment dynamics, there are many difficulties in realizing the ideal TLRB gait. Even so, our TLRB experiment still achieved a grasping (moving) distance of 102 mm, which is at least 1.5 times that of the distance achieved by the TLB gait (63 mm). Next, we consider factors affecting the energy consumption of the actuators and analyze gait movement performance using the cost of transport (COT) [33] index, which is defined as follows:

$$\begin{gathered} \mathrm{COT} \triangleq \frac{E_a}{mgd} \\ E_a = \int |\tau| d\theta \end{gathered} \quad (17)$$

where E_a represents the total work performed by the various actuators in the joints of the robot; m is the total weight of the robot; and d denotes the moving distance of a locomotion cycle. Here, we make a simple comparison of the TLRB and TLB experimental results in Table 4. The TLRB gait indeed consumes more energy (by approximately 30%) but greatly increases the moving distance. Thus, the COT of the TLRB gait has better energy conversion efficiency and movement performance than the TLB gait does, and this is with limitations in the mechatronic hardware of the robot and much of the energy lost during the process due to contact dynamics. From the perspective of bionics, these two gaits are similar to the walking and running gaits of humans. Running consumes more energy than walking and tires the body more easily; however, it is a more efficient mode of locomotion. In another aspect, walking is the safer choice for continuous movement. Humans naturally switch between these two gaits depending on their physical capabilities and changes in the environment to move steadily, continuously, and efficiently. Thus, this study focuses on a means of enabling transverse brachiation robots to switch to a more suitable locomotion gait depending on the circumstances. Please refer to the following link for a video from our experiments: https://youtu.be/j11oPRQxvqA (accessed on 14 May 2023).

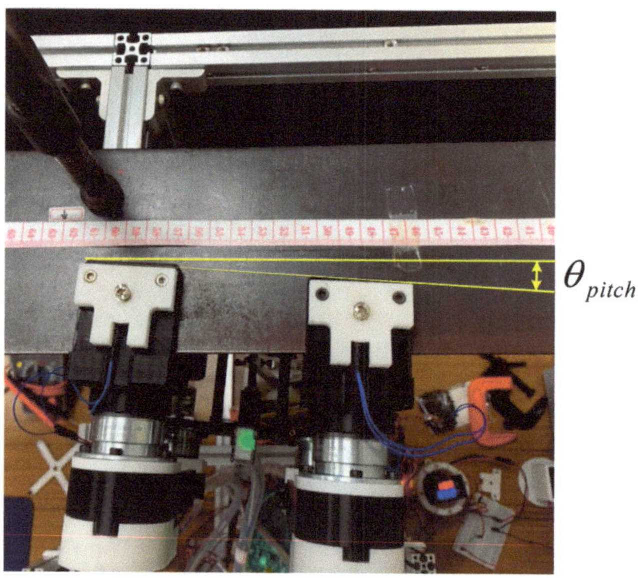

Figure 22. Substantial hand-holding posture deviation (θ_{pitch}) after finishing a locomotion cycle of transverse ricochetal brachiation. Colored lines are added to indicate the deviation angle.

Table 4. Comparison of proposed two robot locomotion styles.

	Transverse Ledge Brachiation	**Transverse Ledge Ricochetal Brachiation**
Grabbing distance (mm)	63	102
Time (s)	2.8	2.58
Movement speed (mm/s)	22.5	39.5
Robot weight (kg)	1.76	1.76
Actuator energy (J)	6.15	8.02
COT ($J \cdot kg^{-1} \cdot m^{-2} \cdot s^2$)	5.65	4.56

6. Conclusions

This study presents the locomotion design and implementation of an arm-body-tail ledge brachiation robot mimicking the transverse movements of climbing athletes. Under the constraint of two pairs of parallel links, we consider the cause-and-effect relationship between the designed phases to identify phase-switching conditions and desired joint motion. Experiments demonstrate that locomotion without leaping action can be successfully conducted for two consecutive cycles, and the locomotion of ricochetal brachiation represents a significant improvement in terms of grabbing distance and energy efficiency. However, the experiments all terminated during the locomotion process due to deviation in the hand-holding postures; this impeded energy accumulation in the next cycle and greatly increased the risk of falling. Future research could investigate landing posture dynamics using improved mechatronic design and three-dimensional modeling. Active posture compensation methods and advanced robot locomotion control could further improve the safety and robustness of transverse ledge brachiation.

Author Contributions: J.-M.L. and C.-Y.L. conceived and designed the robot locomotion and experiments; J.-M.L. developed the robot, performed the experiments, and analyzed the results; C.-Y.L. wrote the paper. All authors have read and agreed to the published version of the manuscript.

Funding: This work was supported by Ministry of Science and Technology, Taiwan, under grant number MOST-111-2628-E-011-011.

Data Availability Statement: Not applicable.

Conflicts of Interest: The authors declare no conflict of interest.

References

1. Lin, C.-Y.; Yang, Z.H. TRBR: Flight body posture compensation for transverse ricochetal brachiation robot. *Mechatronics* **2020**, *65*, 102307. [CrossRef]
2. Lin, C.-Y.; Tian, Y.-J. Design of Transverse Brachiation Robot and Motion Control System for Locomotion between Ledges at Different Elevations. *Sensors* **2022**, *22*, 4031. [CrossRef] [PubMed]
3. Official Website of the International Federation of Sport Climbing. Available online: https://www.ifsc-climbing.org/ (accessed on 22 March 2023).
4. Wang, S.; Jiang, H.; Myung Huh, T.; Sun, D.; Ruotolo, W.; Miller, M.; Roderick, W.R.T.; Stuart, H.S.; Cutkosky, M.R. Spinyhand: Contact load sharing for a human-scale climbing robot. *J. Mech. Robot.* **2019**, *11*, 031009. [CrossRef]
5. Wei, D.; Ge, W. Research on one bio-inspired jumping locomotion robot for search and rescue. *Int. J. Adv. Robot. Syst.* **2014**, *11*, 168. [CrossRef]
6. Patané, L. Bio-inspired robotic solutions for landslide monitoring. *Energies* **2019**, *12*, 1256. [CrossRef]
7. Seo, T.; Jeon, Y.; Park, C.; Kim, J. Survey on glass and façade-cleaning robots: Climbing mechanisms, cleaning methods, and applications. *Int. J. Precis. Eng. Manuf.* **2019**, *6*, 367–376. [CrossRef]
8. Schultz, J.T.; Beck, H.K.; Haagensen, T.; Proost, T.; Clemente, C.J. Using a biologically mimicking climbing robot to explore the performance landscape of climbing in lizards. *Proc. R. Soc. B* **2021**, *288*, 20202576. [CrossRef] [PubMed]
9. Zheng, Z.; Ding, N.; Chen, H.; Hu, X.; Zhu, Z.; Fu, X.; Zhang, W.; Zhang, L.; Hazken, S.; Wang, Z.; et al. CCRobot-V: A silkworm-like cooperative cable-climbing robotic system for cable inspection and maintenance. In Proceedings of the International Conference on Robotics and Automation, Philadelphia, PA, USA, 20–24 May 2022; IEEE: Piscataway, NJ, USA, 2022; pp. 164–170.
10. Lin, C.-Y.; Hu, T.-C. Locomotion control of ledge climbing robot using CPG control and infrared sensors. *Int. J. iRobotics* **2022**, *4*, 40–49.
11. Demonstration Video of Transverse Ledge Brachiation and Transverse Ledge Ricochetal Brachiation. Available online: https://youtu.be/l1WgzhDSBH8 (accessed on 22 March 2023).
12. Nakanishi, J.; Fukuda, T. A leaping maneuvre for a brachiating robot. In Proceedings of the IEEE International Conference on Robotics and Automation, San Francisco, CA, USA, 24–28 April 2000; IEEE: Piscataway, NJ, USA, 2000; pp. 2822–2827.
13. Cheng, H.; Rui, C.; Hao, L. Motion planning for ricochetal brachiation locomotion of bio-primitive robot. In Proceedings of the Annual International Conference on CYBER Technology in Automation, Control, and Intelligent Systems, Honolulu, HI, USA, 31 July–4 August 2017; IEEE: Piscataway, NJ, USA, 2017; pp. 259–264.
14. Andreuchetti, S.; Oliveira, V.M.; Fukuda, T. A survey on brachiation robots: An energy-based review. *Robotica* **2021**, *39*, 1588–1600. [CrossRef]
15. Nakanishi, J.; Fukuda, T.; Koditschek, D.E. A brachiating robot controller. *IEEE Trans. Rob. Autom.* **2000**, *16*, 109–123. [CrossRef]
16. Lu, Z.; Liu, G.; Zhao, H.; Wang, R.; Liu, C. Swing control for a three-link brachiation robot based on sliding-mode control on irregularly distributed bars. *Mech. Sci.* **2021**, *12*, 1073–1081. [CrossRef]
17. Davies, E.; Garlow, A.; Farzan, S.; Rogers, J.; Hu, A.P. Tarzan: Design, prototyping, and testing of a wire-borne brachiating robot. In Proceedings of the IEEE/RSJ International Conference on Intelligent Robots and Systems, Madrid, Spain, 1–5 October 2018; IEEE: Piscataway, NJ, USA, 2018; pp. 7609–7614.

18. Liu, C.; He, J.; Gao, B. Movement planning and control of an overhead power transmission line inspection bionic-robot. In Proceedings of the Annual International Conference on CYBER Technology in Automation, Control, and Intelligent Systems, Suzhou, China, 29 July–2 August 2019; IEEE: Piscataway, NJ, USA, 2019; pp. 25–29.
19. Saito, F.; Fukuda, T.; Arai, F. Swing and locomotion control for a two-link brachiation robot. *IEEE Control Syst. Mag.* **1994**, *14*, 5–12.
20. Kajima, H.; Hasegawa, Y.; Doi, M.; Fukuda, T. Energy-based swing-back control for continuous brachiation of a multilocomotion robot. *Int. J. Intell. Syst.* **2006**, *21*, 1025–1043. [CrossRef]
21. Yang, Z. Robot Brachiation with Energy Control. Master's Thesis, Carnegie Mellon University Pittsburgh, Pittsburgh, PA, USA, 2016.
22. Wu, W.; Huang, M.; Gu, X. Underactuated control of a bionic-ape robot based on the energy pumping method and big damping condition turn-back angle feedback. *Rob. Auton. Syst.* **2018**, *100*, 119–131. [CrossRef]
23. Nonoyama, K.; Shimizu, M.; Umedachi, T. Upside-down brachiation robot using elastic energy stored through soft body deformation. *IEEE Robot. Autom. Lett.* **2022**, *7*, 11291–11297. [CrossRef]
24. Hasegawa, Y.; Ito, Y.; Fukuda, T. Behavior coordination and its modification on brachiation-type mobile robot. In Proceedings of the IEEE International Conference on Robotics and Automation, San Francisco, CA, USA, 24–28 April 2000; IEEE: Piscataway, NJ, USA, 2000; pp. 3983–3988.
25. Rosa, N.; Barber, A.; Gregg, R.D.; Lynch, K.M. Stable open-loop brachiation on a vertical wall. In Proceedings of the IEEE International Conference on Robotics and Automation, Saint Paul, MN, USA, 14–18 May 2012; IEEE: Piscataway, NJ, USA, 2012; pp. 1193–1199.
26. Bretl, T. Motion planning of multi-limbed robots subject to equilibrium constraints: The free-climbing robot problem. *Int. J. Robot. Res.* **2006**, *25*, 317–342. [CrossRef]
27. Uno, K.; Takada, N.; Okawara, T.; Haji, K.; Candalot, A.; Ribeiro, W.F.R.; Nagaoka, K.; Yoshida, K. HubRobo: A lightweight multi-limbed climbing robot for exploration in challenging terrain. In Proceedings of the IEEE-RAS International Conference on Humanoid Robots, Munich, Germany, 19–21 July 2021; IEEE: Piscataway, NJ, USA, 2021; pp. 209–215.
28. Liu, C.; Liu, A.; Wang, R.; Zhao, H.; Lu, Z. Path planning algorithm for multi-locomotion robot based on multi-objective genetic algorithm with elitist strategy. *Micromachines* **2022**, *13*, 616. [CrossRef] [PubMed]
29. Arampatzis, A.; Brüggemann, G.P. Mechanical energetic processes during the giant swing before the Tkatchev exercise. *J. Biomech.* **2001**, *34*, 505–512. [CrossRef] [PubMed]
30. Gong, Y.; Hartley, R.; Da, X.; Hereid, A.; Harib, O.; Huang, J.K.; Grizzle, J. Feedback control of a cassie bipedal robot: Walking, standing, and riding a segway. In Proceedings of the American Control Conference, Philadelphia, PA, USA, 10–12 July 2019; IEEE: Piscataway, NJ, USA, 2019; pp. 4559–4566.
31. Hagh, Y.S.; Fekih, A.; Handroos, H. Robust finite non-singular terminal synergetic control for second order nonlinear systems subject to time-varying mismatched disturbances. In Proceedings of the IEEE Conference on Industrial Electronics and Applications, Chengdu, China, 1–4 August 2021; IEEE: Piscataway, NJ, USA, 2021; pp. 228–233.
32. Tracker—Video Analysis and Modeling Tool for Physics Education. Available online: https://physlets.org/tracker/ (accessed on 22 March 2023).
33. Chen, Y.; Grezmak, J.E.; Graf, N.M.; Daltorio, K.A. Sideways crab-walking is faster and more efficient than forward walking for a hexapod robot. *Bioinspir. Biomim.* **2022**, *17*, 46001. [CrossRef] [PubMed]

biomimetics

MDPI

Article

Low-Cost Angle Sensor for Robotics Applications Using Plastic Optical Fiber Based on Optical Loss Mechanism

Hyun-Woo Lee [1], Dae-Hyun Kim [2,*] and Sangwoo Shin [3]

1 Mechanical Engineering, Seoul National University of Science & Technology, 232 Gongneung-ro, Nowon-gu, Seoul 01811, Republic of Korea; leehyunwoo@seoultech.ac.kr
2 Department of Mechanical & Automotive Engineering, Seoul National University of Science & Technology, 232 Gongneung-ro, Nowon-gu, Seoul 01811, Republic of Korea
3 Department of Mechanical and Aerospace Engineering, University at Buffalo, The State University of New York, Buffalo, NY 14260, USA; sangwoos@buffalo.edu
* Correspondence: dkim@seoultech.ac.kr; Tel.: +82-2-970-6315

Abstract: Robotic systems and the human body consist of numerous joint structures, all of which require precise angle adjustments. At present, encoder, strain gauge, and electrical resistance-based sensors are commonly used for angle measurement. However, these sensors have limitations when used in underwater or in environments with strong electromagnetic waves. Therefore, we have developed an angle sensor based on step-index profile plastic optical fiber (SI-POF), which is cost-effective and highly durable, in this study in order to overcome the limitations of existing angle measurement sensors. To this end, the amount of light loss according to the gab and angle changes that occur when the POF angle sensor is applied to the robot arm was experimentally measured, and based on the results, a simulation of the amount of light loss when the two losses occurred at the same time was conducted. In addition, the performance of the POF angle sensor was evaluated by measuring sensitivity and resolution, and comparative verification with a commonly used encoder was conducted to verify the reliability of sensors in extreme environments, such as those with electromagnetic fields and those that are underwater. Through this, the reliability and practicality of the POF angle sensor were confirmed. The results obtained in this study suggest that POF-based angle sensors can contribute to the development of the biomimetic robot industry as well as ordinary robots, especially in environments where existing sensors are difficult to apply, such as areas with underwater or electromagnetic interference (EMI).

Keywords: step index profile plastic optical fiber (SI-POF); bending loss; coupling loss; robot arm joint; underwater; electromagnetic interference (EMI)

1. Introduction

Angle measurement is used in a variety of industries, from the traditional mechanical industry to clinical and motion capture, and related research is actively underway [1–4]. In particular, various robot systems including biomimetics, the use of which has recently been rapidly increasing, are composed of many joint structures, and the joint structures necessarily involve angle adjustment. An encoder, strain gauge, and electric resistance-based sensor are commonly used as conventional angle measurement methods [3–5]. However, these sensors have limitations in their use in extreme environments, such as those with strong electromagnetic waves or that are underwater, acting as an obstacle to expanding the application area of robotic systems. In contrast, optical fiber sensors have no effect on electromagnetic waves, have a wide operating temperature range, and are free from the effects of water or corrosion. In addition, it has many advantages, such as being able to transmit signals over long distances with little loss and being easy to embed due to its small size.

Fiber optic sensors can be classified by light intensity type [6,7], interference type [8], scattering type [9], and wavelength shift type [10–16] depending on the demodulation method. Among them, a wavelength shift type Fiber Bragg grating (FBG) sensor has been studied in depth for angle and shape estimation [15,16], but the optical system for FBG operation is very expensive, about USD 20,000 or more, which limits its application. In contrast, light intensity fiber optic sensors with easy demodulation techniques and low system prices are expanding their application areas in various industries. In particular, plastic optical fiber (POF), which is relatively durable compared to silica-based optical fiber, is highly applicable to the composition of light intensity sensors, and the system configuration price is as low as USD 30 or less [17,18]. In addition, POF is easy to handle because it has a large core diameter and numerical aperture (N.A.) [18], and has excellent flexibility, so related research has been expanding recently. Therefore, it can be a good option as a sensor for measuring angles when applied to joint parts such as robot arms in an extreme environment.

Therefore, in this study, we developed a cost-effective and easy-to-handle POF angle sensor to overcome the limitations of existing angle sensors. The process is illustrated in the flowchart presented in Figure 1. First, as described above, the development goal was set to create a POF angle sensor that can operate stably in an extreme environment. Second, for POF angle sensor development, the amount of light loss caused by the gab change and angle change generated when a POF angle sensor was applied to the robot arm was experimentally measured. Third, based on the previous results, a simulation of the amount of light loss when the two losses occurred at the same time was conducted. Fourth, the POF angle sensor proposed in this paper was applied to the robot arm joint structure and the performance of the sensor was verified by measuring sensitivity and resolution according to the change in the robot arm rotation angle. Also, comparative verification with a commonly used encoder was conducted to verify the reliability of the POF angle sensor proposed in this paper in extreme environments such as those with electromagnetic fields and that are underwater. Through this, the reliability and practicality of POF angle sensors were evaluated.

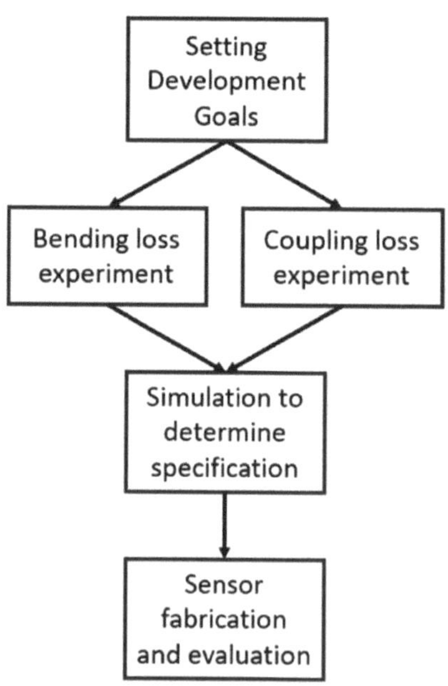

Figure 1. Research flow chart.

2. Light Loss Theory of POF

2.1. Bending Loss of POF

The theoretical analysis of the bending loss of multi-mode step-index profile plastic optical fiber (SI-POF) is based on geometric optical analysis. The light ray incident on the optical fiber is divided into a meridional ray that passes through the central axis of the core and a skew ray that does not pass through the central axis of the core. Most of the losses that occur when bending is applied to an optical fiber occur in meridional rays, so losses in skew rays are not considered [19]. When meridional rays traveling inside an optical fiber are incident on the interface between the core and cladding, two types of optical loss occur. The first is when the incident angle (θ_i) is smaller than the critical angle (θ_c). In this case, the light ray is refracted at the boundary and loss occurs. The second is when the incident angle (θ_i) is equal to or greater than the critical angle (θ_c), and the light ray leaks from the boundary and a relatively small loss occurs [19,20]. When curvature is applied to the optical fiber, the angle of incidence becomes smaller, as shown in Figure 2. Accordingly, through this principle, light loss due to refraction occurs in light rays whose incidence angle is smaller than the critical angle, which is expressed in Equation (1).

$$P_r = P_i(1 - T) \tag{1}$$

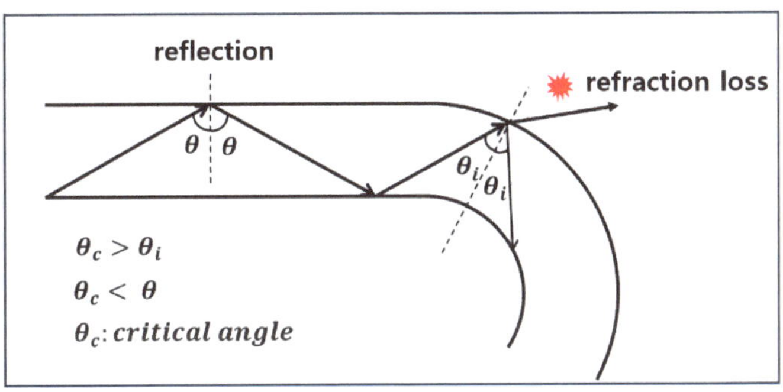

Figure 2. Refractive loss due to reduced angle of incidence.

Here, P_i and P_r are the incident and reflected rays, respectively, and T is the power transmission coefficient, calculated using Equation (2) [21].

$$T = \frac{4\cos\theta_i\left(\cos^2\theta_i - \cos^2\theta_c\right)^{1/2}}{\left[\cos\theta_i + \left(\cos^2\theta_i - \cos^2\theta_c\right)^{1/2}\right]^2} \tag{2}$$

In this way, the light loss resulting from curvature can be calculated through the sum of light losses due to refraction from all rays incident at an angle ($-\theta_c < \theta < +\theta_c$) smaller than the critical angle (θ_c) [19].

On the other hand, light rays with an incident angle (θ_i) greater than the critical angle (θ_c) continue to be reflected at the same incident angle within a curvature with a constant radius of curvature, as shown in Figure 3, and proceed along the curvature, and loss due to continuous leakage occurs in this process.

However, in the case of previous research [19], only the bending loss of optical fibers due to refraction, which occurs mostly at the beginning of curvature, is considered. On the other hand, cases where continuous loss due to leakage occurs, as shown in Figure 3, are not considered, so there is a limit to its application. In this paper, therefore, we overcome the limitations of existing theoretical analysis and experimentally measured the optical loss that occurs when optical fibers are used as angle sensors.

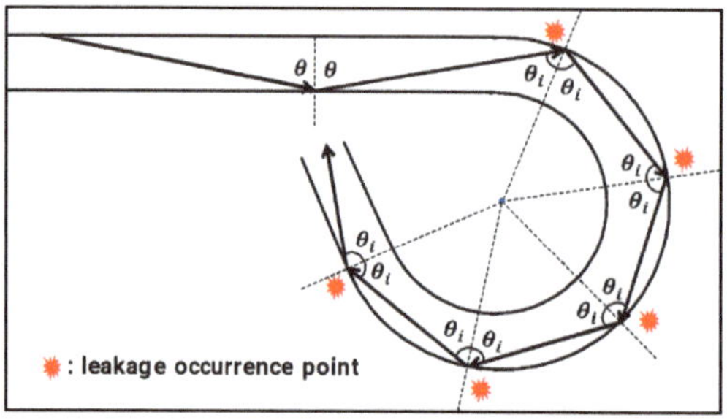

Figure 3. A beam of light continuing at the same angle of incidence within an arc.

2.2. Coupling Loss of POF

The coupling loss according to the gap distance between two aligned plastic optical fibers can be calculated through the illuminance distribution model of optical fibers based on photometric analysis [22]. In other words, in a light-transmitting fiber optical, as shown in Figure 4, the radiance of a small area $\Delta\alpha$ on the end surface and the size of the irradiance of a small area ΔA of the image plane that receives the light are shown for each position. Through Figure 5, the amount of the irradiance distribution for each area can be obtained, and the amount of the irradiance varies depending on the irradiation location (x_p, y_p). In Figure 5, the irradiance equations for areas (I), (II), and (III) can be obtained through Equations (3), (4), and (5), respectively.

$$E_{\text{I}} = M \tag{3}$$

$$E_{\text{II}} = \frac{L\pi}{2}[1 - \cos(\theta_2 - \theta_1)] \tag{4}$$

$$E_{\text{III}} = 2\left[\begin{array}{c} \int_{x_p - r_\alpha}^{x_0} \int_0^{\sqrt{r_\alpha^2 - (x - x_p)^2}} \frac{L y_p^2}{\left(y_p^2 + z^2 + (x_p - x)^2\right)^2} dz dx \\ + \int_{x_0}^{r_c} \int_0^{\sqrt{r_c^2 - x^2}} \frac{L y_p^2}{\left(y_p^2 + z^2 + (x_p - x)^2\right)^2} dz dx \end{array}\right] \tag{5}$$

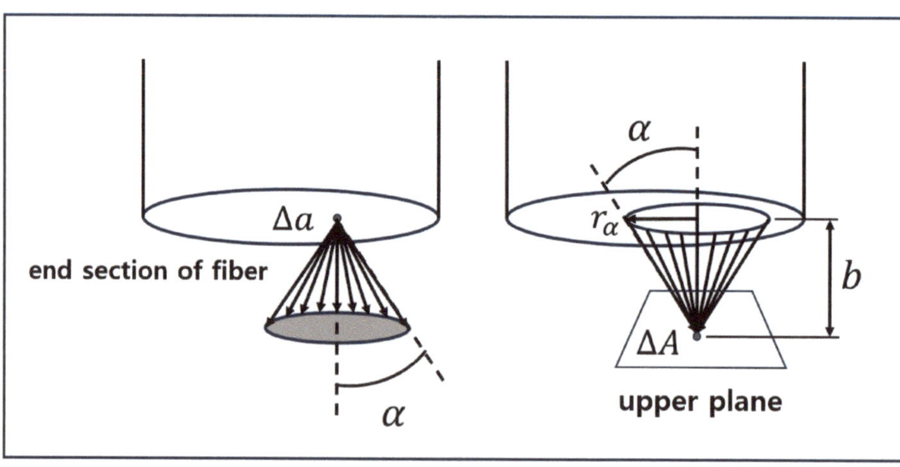

Figure 4. Basic concepts of light intensity distribution modeling [22].

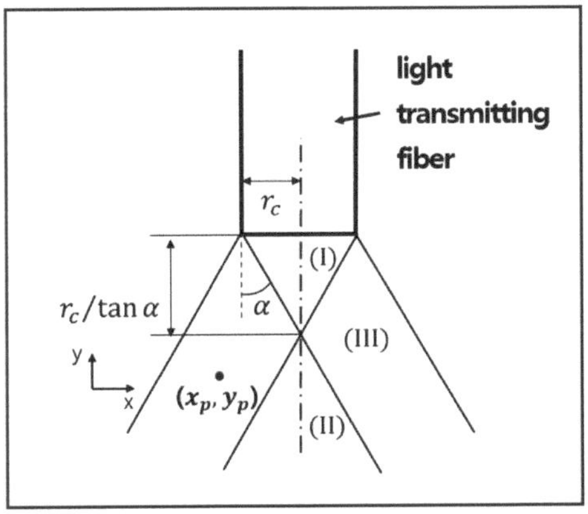

Figure 5. Illuminance distribution by plastic optical fiber area [22].

Here, L is the radiant luminance of a small area and M is the radiative divergence of the light-transmitting fiber, which has the relationship shown in Equation (6).

$$M = \frac{\pi L}{2}(1 - \cos 2\alpha) \tag{6}$$

Additionally, θ_1 and θ_2 in Equation (4) are calculated through Equations (7) and (8), and r_α and x_0 in Equation (5) are calculated through Equations (9) and (10), respectively.

$$\theta_1 = \sin^{-1}\frac{r_c - x_p}{\sqrt{y_p^2 + (r_c - x_p)^2}} \tag{7}$$

$$\theta_2 = \sin^{-1}\frac{r_c - x_p}{\sqrt{y_p^2 + (r_c + x_p)^2}} \tag{8}$$

$$r_\alpha = \sqrt{(x - x_p)^2 + z^2} \tag{9}$$

$$x_0 = \frac{r_c^2 - r_\alpha^2 + x_p^2}{2x_p} \tag{10}$$

3. Experiment for Optical Loss of POF

3.1. Experimental Apparatus and Method

In this section, a quantitative evaluation was conducted using experimental methods for the two types of light loss.

First, an experimental device, as shown in Figure 6, was prepared to measure bending loss as the rotation angle increases when the robot arm rotates. In addition, considering rotating parts with various radii, test specimens simulating a rotating part with four types of radii, including 3 mm, 6 mm, 9 mm, and 12 mm, were produced using a 3D printer.

The experiment was conducted by measuring the amount of light loss that occurs when an optical fiber is wound around each manufactured rotating part at 30-degree intervals from 0 degrees to 360 degrees. At this time, a constant load was applied to the optical fiber to ensure that there was no problem with the optical fiber's followability when the radius of curvature changed.

Second, an experimental device, as shown in Figure 7, was prepared to measure coupling loss according to the change in gap distance between two optical fibers. The experiment was conducted by aligning two strands of optical fibers on the same line using a 3-axis moving stage equipped with a micrometer and then changing the gap distance between the optical fibers. The amount of light loss occurring when the gap distance was changed from 0.0 to 39.0 mm at 1.0 mm intervals was measured.

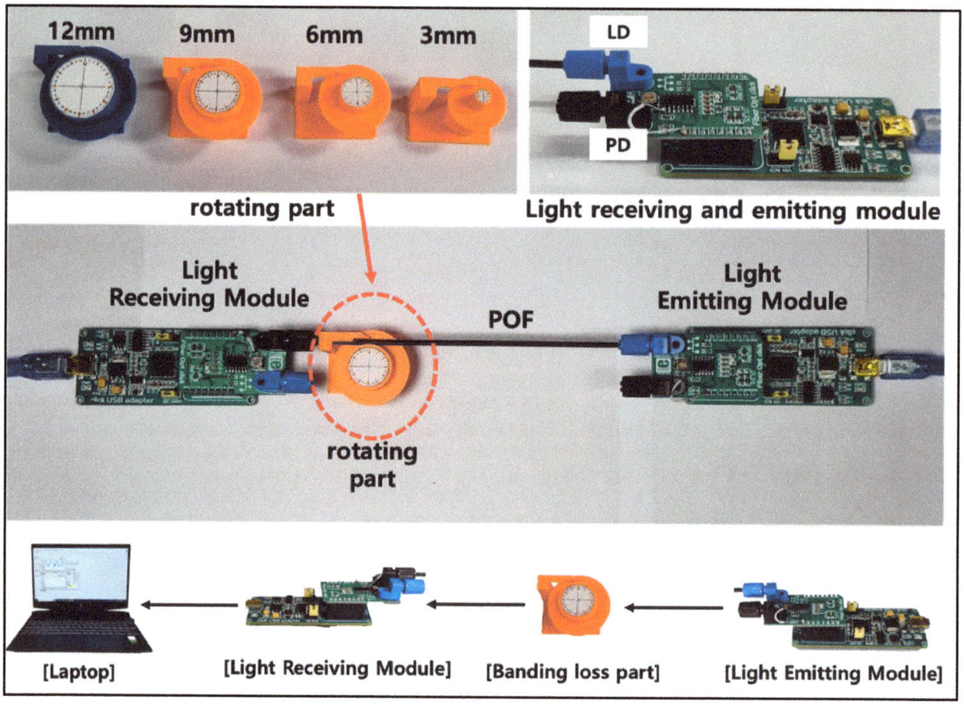

Figure 6. Experimental setup for a bending loss test.

In both types of experiments, GH4001, a multimode SI-POF from Eska™, was used as the optical fiber. The numerical aperture (N.A.) of the optical fiber used was 0.5, and the diameters of the core, cladding, and jacket were 980 μm, 1000 μm, and 2.2 mm, respectively, and the materials were polymethyl methacrylate, fluorinated polymer, and polyethylene, respectively. LD (IFE97, Industrial Fiber Optics) with a wavelength of approximately 660 nm was used as the light emitting module. In addition, as a light receiving module, a Fiber Opt click (Mikroe-1940, Mikro Elektronika d.o.o, Belgrade, Serbia) with a built-in photodetector (IF D91B, Industrial Fiber Optics) of the same wavelength was used.

Additionally, the light intensity data were connected to the measurement computer through the same company's click USB adapter (Mikroe-1433) and acquired by reading the output voltage through the LabVIEW software. All experiments were performed five times, and the measured values were averaged.

3.2. Experimental Results

First, the results of the bending loss measurement experiment according to the change in the angle of the rotating part are shown in Figure 8. The changes in result values according to four types of changes in the radius of the rotating part are also presented. In the figure, as the rotation angle increases from 0 degrees to 360 degrees, it shows an almost linear decrease until about 60 degrees, and then the nonlinearity rapidly increases. However, in the section after 150 degrees, there is a characteristic that shows linearity again.

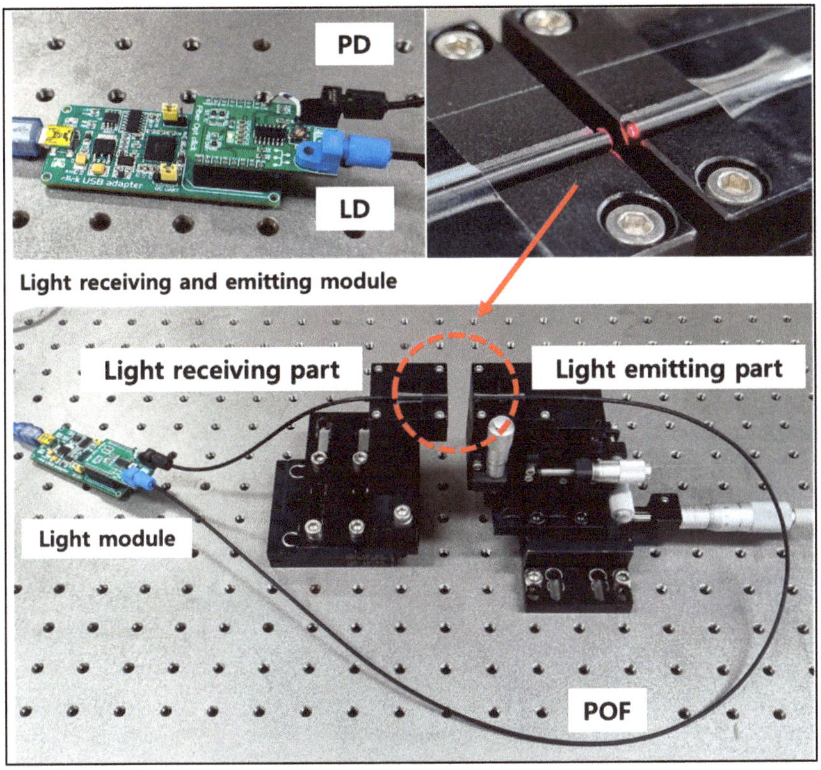

Figure 7. Experimental setup for a coupling loss test.

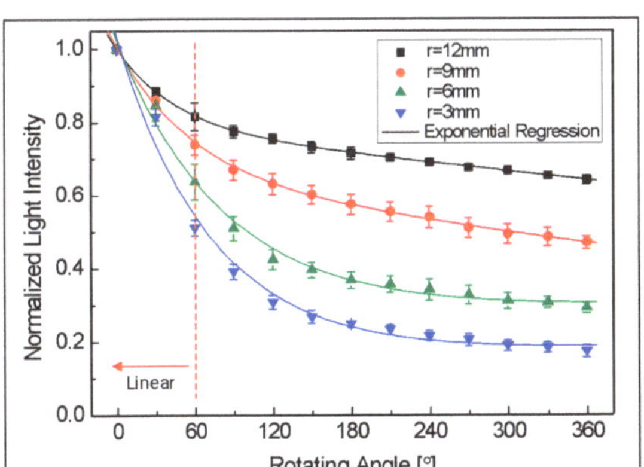

Figure 8. Normalized light intensity according to radius.

The sensitivity of banding loss is expressed in Figure 9. This indicated the average amount of light loss as a ratio of the initial amount of light as the angle changes by 1 degree within the 30-degree interval, and this result was approximated through exponential regression. In the range of 0 to 60 degrees, where linearity was good, sensitivity was also relatively excellent. From the geometrical optical theory in Section 2.1, it can be assumed that this is caused by loss due to refraction caused by a smaller angle of incidence at the beginning of the curvature. In addition, with respect to changes in the radius of the rotating part, if the radius is small, greater loss occurs depending on the angle change, so the sensor showed better results in terms of sensitivity. On the other hand, in the section after 60 degrees where nonlinearity increases, sensitivity is also gradually decreased. In particular, after about 150 degrees, the sensitivity shows a very low characteristic regardless of the change in the radius of the rotating part. This is because the influence of refraction loss is reduced in that section, and the influence of leakage loss occurring as it continues inside the curvature dominates. Therefore, as mentioned in a previous study [21], it was experimentally confirmed that at sufficient curvature, most of the refractive loss due to bending occurs in the section below about 120 degrees, and that leakage loss dominates in the section above 150 degrees.

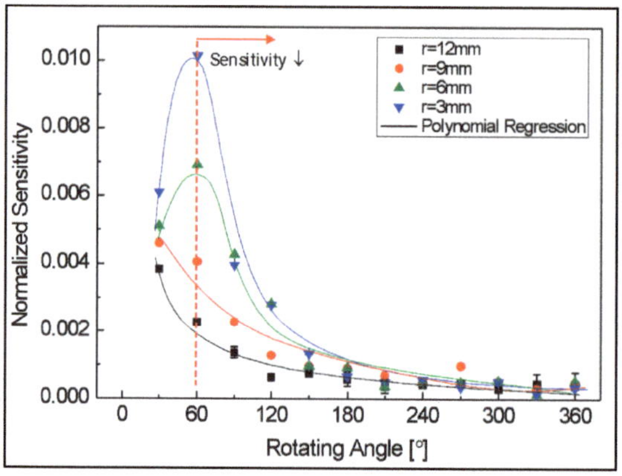

Figure 9. Normalized sensitivity of light intensity according to radius.

Second, the results of the coupling loss experiment according to the change in the gap between the two optical fibers are shown in Figure 10. Overall, the larger the gap, the greater the light loss. In particular, the degree of change rapidly decreased logarithmically as the gap increased. In addition, a high light loss of more than 90% occurred at a 10 mm gap distance.

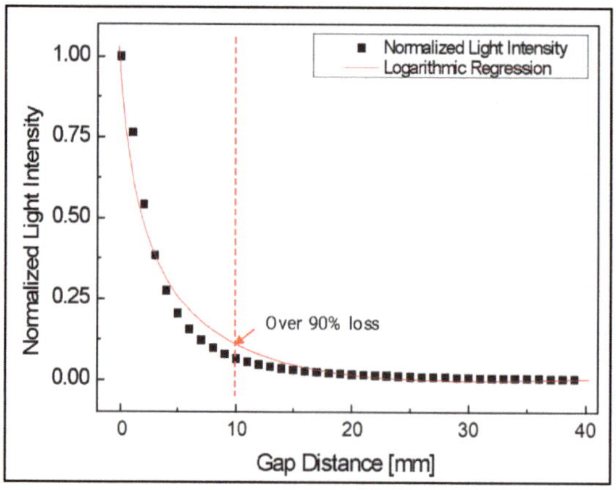

Figure 10. Normalized light intensity according to gap between two aligned optical fibers.

The sensitivity of coupling loss is expressed in Figure 11. This indicated the amount of light loss as a ratio of the initial amount of light as the gap changed by 1 mm, and this result was approximated through logarithmic regression. It showed a similar tendency to the results of Figure 10. The sensitivity at a gap distance of 10 mm was very low at less than 10% of the initial sensitivity.

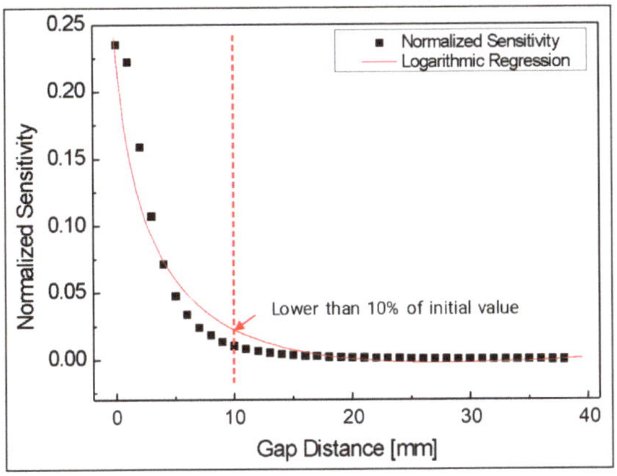

Figure 11. Normalized sensitivity of light intensity according to gap between two aligned optical fibers.

4. Robot Arm Device with POF Angle Sensor

4.1. POF Applied Robot Arm Device Design

In this chapter, a POF angle sensor applicable to a robot arm was manufactured using bending loss and coupling loss of POF, which were experimentally confirmed in Section 3.

First, the internal structure of the manufactured POF angle sensor was designed as shown in Figure 12. For easy application to robots, the POF angle sensor was designed based on the robot's arm and joint structure. The sensor is a form in which two arms are joined by a joint and can freely adjust the angle (α). Inside the tube installed on arm 1, two optical fibers are aligned in a straight line and designed to face each other. Based on the tube, the optical fiber on the left is fixed, and the optical fiber on the right is guided to pass through the curvature created in the joint as the device is given an angle, resulting in a distance from the left optical fiber inside the tube. As a result, as the gap between optical fibers inside the tube widens, light flows out, causing coupling loss, and the right optical fiber also causes bending loss due to curvature. Through this, a structure is formed in which bending loss and coupling loss occur simultaneously in the POF angle sensor when the angle of the robot arm rotating part changes.

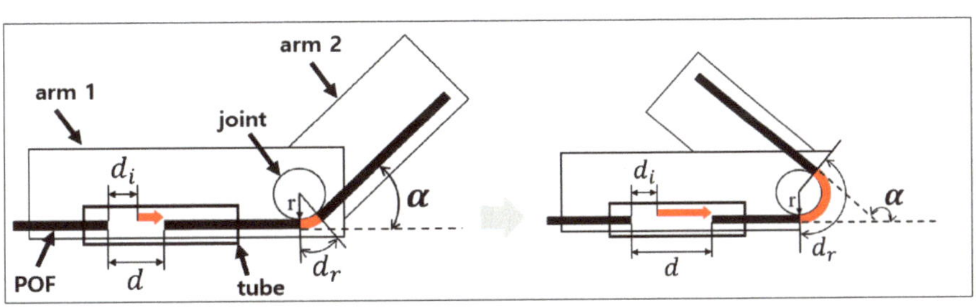

Figure 12. Conceptual approach for rotating angle detection mechanism.

Here, the gap distance (d) between two optical fibers is the sum of the length (d_r) of the optical fiber surrounding the curvature created by the angle (α) and the initial gap (d_i) between the two optical fibers, which is equal to Equation (11). In addition, the length

(d_r) of the optical fiber surrounding the curvature of the rotating part is calculated as in Equation (12), considering the radius of the POF of 1.1 mm.

$$d = d_i + d_r \tag{11}$$

$$d_r = \frac{\alpha}{360} \times 2\pi(r + 1.1) \tag{12}$$

Meanwhile, when bending loss and coupling loss act in combination, the light quantity (P) at each angle can be expressed as the product of the light quantity (P_b) when bending loss occurs and the light quantity (P_c) when coupling loss occurs, as shown in Equation (13).

$$P = P_b \times P_c \tag{13}$$

To determine the specifications of the joint to be applied to the robot arm, a simulation was conducted to determine the initial gap distance (d_i) between the two optical fibers and the radius of the joint (r). The values of Figures 8 and 10 obtained through the experiments in Section 3 were substituted into Equation (13) to analyze the characteristics according to the initial gap and the radius of rotation of the joint.

First, to determine the initial gap distance, the joint radius was fixed to 6 mm among the four types used in this study, and the change in light intensity according to the gap distance change was calculated. The initial gap distance was changed from 0 to 12 mm at 4 mm intervals, and the change in total light intensity was expressed as a ratio the initial value, as shown in Figure 13. In the figure, as the initial gap distance increases, the nonlinearity gradually decreases, which can be understood as the reason why the change in sensitivity is large at a gap distance of less than 10 mm in the results in Figure 11. In this study, therefore, the initial gap distance was decided to be 8 mm by comprehensively considering the decrease in the amount of light due to the increase in gap distance and the linearity issue required for the sensor.

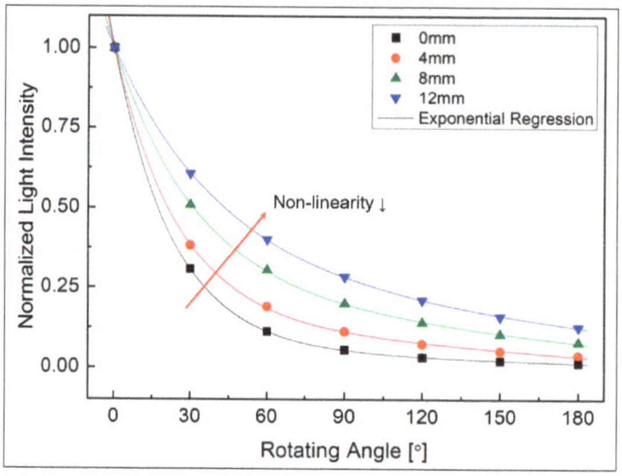

Figure 13. Normalized light intensity with initial gap of 0 mm, 4 mm, 8 mm and 12 mm (at r = 6 mm).

Second, to determine the radius of the joint, the gap distance was fixed to the previously determined 8 mm, and the change in light intensity was calculated according to the change in radius. The change in total light intensity according to the angle change for four types of radii of 3, 6, 9, and 12 mm was expressed as a ratio of the initial value, as shown in Figure 14. From the figure, it can be seen that as the radius becomes smaller, the amount of light changes linearly with respect to the angle change. This can be understood because the smaller the radius, the smaller the absolute movement amount due to rotation (d_r), and the increase in nonlinearity in Equation (13) also decreases. However, in order to manufacture a device, a radius of a certain size or more is required and the bending limit of the optical fiber itself (possibility of birefringence, etc.) need to be taken into consideration. Therefore, considering these points comprehensively, the minimum radius that can be manufactured was determined to be 7.5 mm.

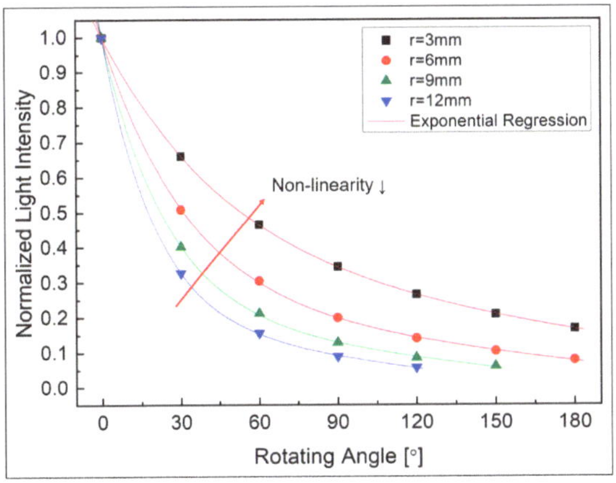

Figure 14. Normalized light intensity by radius change in rotating part (at d_i = 8 mm).

Based on the above results, a robot arm joint structure, as shown in Figure 15, was manufactured. In Figure 15, the device consists of a light receiver module, a light emitter module, a tube, and a joint, and each component was manufactured using a 3D printer. The joint is a structure in which two frames are joined by pins, allowing for free angle adjustment.

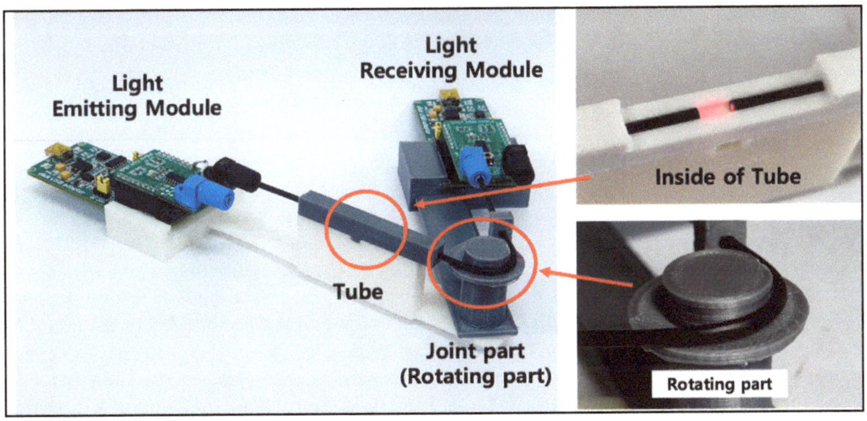

Figure 15. A robot arm part device with POF angle sensor.

4.2. Performance Evaluation of POF Angle Sensor

Figure 16 shows the results of an experiment measuring light intensity at each angle, repeated 5 times at 5-degree intervals from 0 to 165 degrees using the manufactured POF angle sensor. From the experimental results, the relationship between the change in rotation angle and the change in total light intensity can be derived as Equation (14).

$$y = 182\, e^{-\frac{x}{2}} + 51.78 - 57.67\, x \quad (14)$$

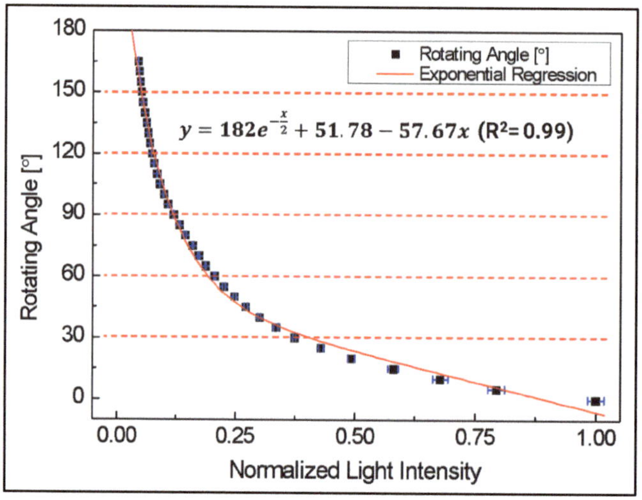

Figure 16. Performance test result of POF angle sensor.

Here, x is the amount of optical loss normalized to the initial amount of optical loss, and y is the rotation angle of the joint part.

From Equation (14), the rotation angle can be calculated by measuring the amount of light loss, but it can be seen that the overall nonlinearity is large. However, when divided into 30-degree sections, a linearity of at least $R^2 = 0.95$ or higher can be confirmed, as shown in Table 1. Therefore, by using this, it is possible to sufficiently confirm its applicability as an angle sensor.

Table 1. R square value for each 30-degree section in Figure 16.

Section	0~30°	30~60°	60~90°	90~120°	120~150°	150~165°
R^2	0.95	0.98	0.99	0.97	0.99	0.99

Meanwhile, Figure 17 shows the sensitivity to light measurement results by angle. This indicated the average amount of light loss as a ratio of the initial amount of light, as the angle changes by 1 degree within the 5-degree interval, and this result was approximated through exponential regression. In the figure, you can see that when the angle is small, the sensitivity is high, but at 60 degrees, the sensitivity decreases by more than 90% compared to the initial value. Additionally, in the section after 60 degrees, the change in light quantity can be assumed to be linear with respect to the angle change, so if it is used as a sensor using only that section, its usability as a sensor is good enough. However, in order to implement this, the optical fiber needs to be fixed for the initial 60-degree rotation amount, which requires additional work, such as the consideration of the fact that the operable angle range is limited and changes to the sensor structure.

In addition, the noise for each section was measured, and the resolution was obtained through this and displayed in Table 2. The resolution is a value obtained by dividing noise [mV] by the average sensitivity [mV/°] of each section, and the unit is a degree [°]. As the sensitivity decreases toward the rear section, the resolution tends to increase. Here, the noise was calculated as the width between the maximum and minimum values measured for 2 s at a sampling frequency of 50 Hz. It can be seen that noise exists almost uniformly regardless of the section.

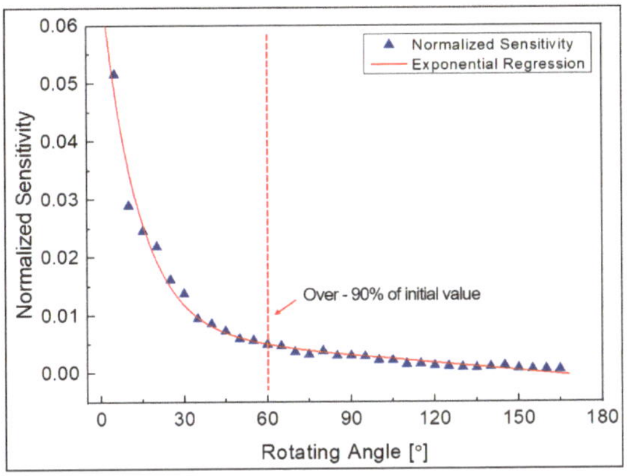

Figure 17. Normalized sensitivity of light intensity of POF angle sensor.

Table 2. Resolution and noise value for each 30-degree section.

Section	0~30°	30~60°	60~90°	90~120°	120~150°	150~165°
Noise [mV]	2.39	2.324	2.375	2.443	2.405	2.442
Resolution [°]	0.030	0.109	0.218	0.418	0.816	1.333

4.3. Comparative Evaluation of Sensors in Extreme Environments

In an additional experiment, the EMI influence of the POF angle sensor was compared with the encoder, a general angle sensor, to show the reliability of the POF angle sensor proposed in this study in an extreme environment. As shown in Figure 18, in order to artificially create an environment where a magnetic field exists, the experiment was conducted by repeatedly bringing the neodymium magnet with a maximum magnetic flux density of 0.412 Tesla close to the sensor within 10 mm and then placing it 150 mm away. Here, each sensor is fixed, and the magnet is prevented from touching the sensor. The encoder used in the experiment is the MSMF5AZL1S2 model from Panasonic Industry Co., Ltd. (Osaka, Japan).

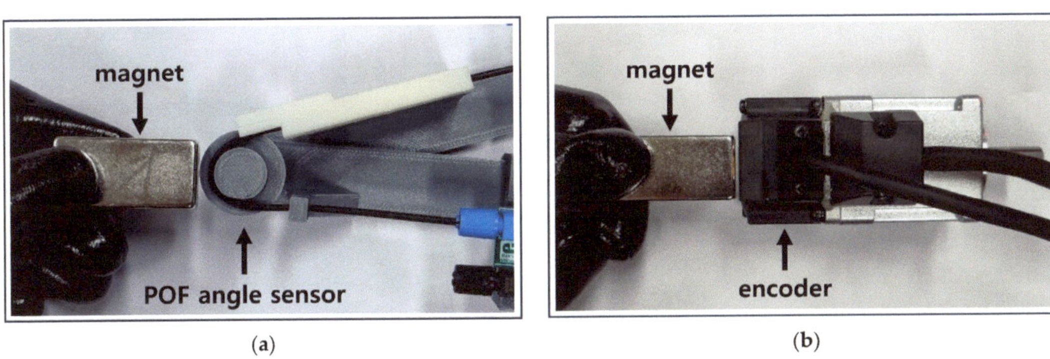

Figure 18. Exposing a magnetic field to each sensor: (**a**) POF angle sensor; (**b**) encoder.

For the POF angle sensor and the encoder, the noise in general environments and interference in environments exposed to magnetic fields are shown in the graphs of Figures 19 and 20, respectively. In addition, the relative ratio of the interference level in the environment exposed to the magnetic field and the noise level in the general environment for each sensor is shown in Table 3. Here, the noise signal and the interference were measured for 2 s at a sampling frequency of 50 Hz, and the noise level and interference level were calculated as the width between the maximum and minimum of the measured signal. In the case of POF angle sensor, there is no meaningful difference in interference level, even when exposed to magnetic fields, but in the case of the encoder, the interference level was increased by 9.67 times compared to the noise level. This result is predicted to be much greater if the magnet is brought closer to the encoder, or if it exposed to a stronger magnetic field. In addition, encoders cannot be used in water unless they are waterproofed, but POF angle sensors can be used regardless of whether they are waterproof or not. The above results confirm that the POF angle sensors proposed in this paper are reliable in magnetic fields and underwater environments, unlike existing sensors. The results obtained in this study suggest that POF angle sensors can be useful in environments where existing sensors are difficult to apply, such as areas where electromagnetic interference (EMI) occurs or underwater, which can contribute to expanding the application area of robot systems.

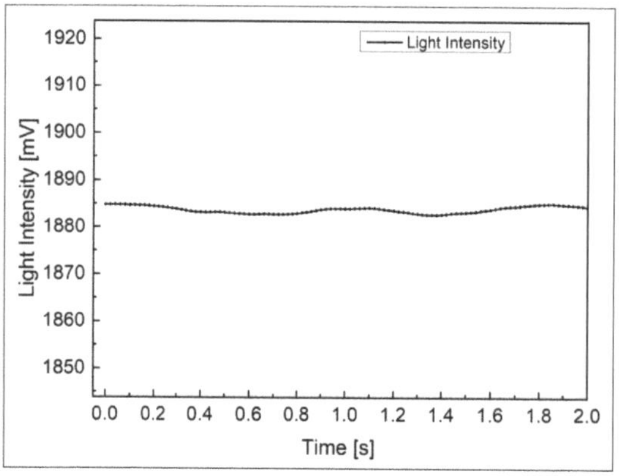

(a)

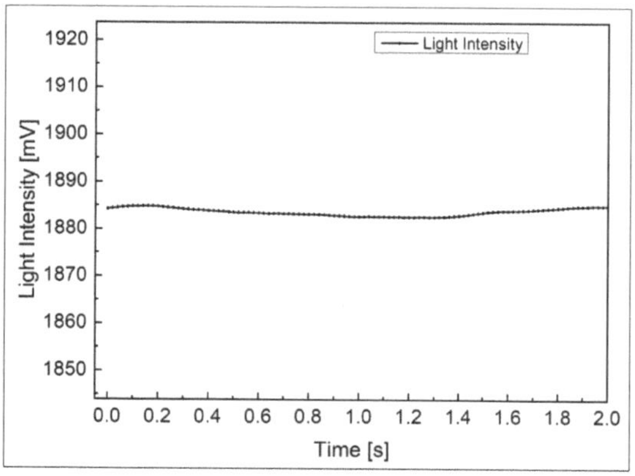

(b)

Figure 19. Signal of POF angle sensor: (**a**) noise in a general environment; (**b**) interference in the magnetic field environment.

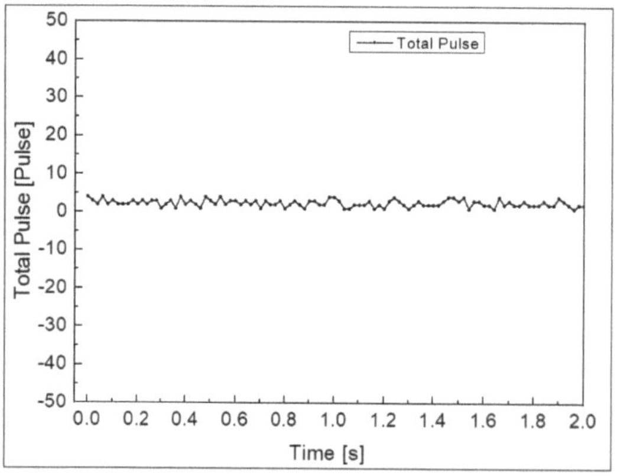

(a)

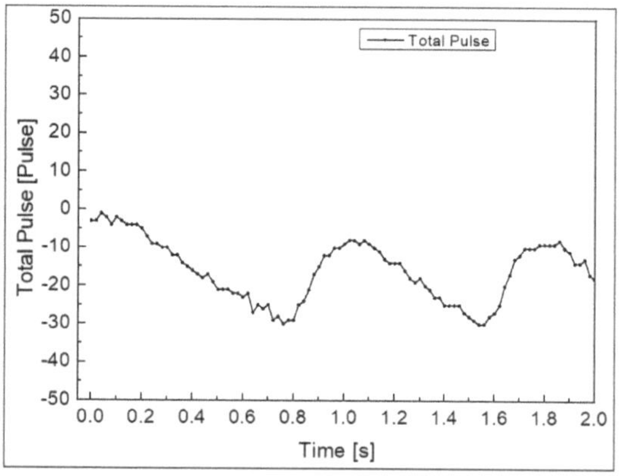

(b)

Figure 20. Signal of encoder: (**a**) noise in a general environment; (**b**) interference in the magnetic field environment.

Table 3. The relative ratio of the interference level in the environment exposed to the magnetic field, the noise level in the general environment, and usability in an underwater environment.

	Noise Level in a General Environment	Interference Level in Magnetic Field Environment	Relative Ratio of the Levels of Two Environments	Underwater Operation
Encoder	3 [Pulse]	29 [Pulse]	9.67	Inoperable
POF angle sensor	2.39 [mV]	2.40 [mV]	1.00	Operable

5. Conclusions

In this study, we developed a POF angle sensor applicable to robot arms by utilizing the optical loss characteristics of POF, which is inexpensive and highly durable. The results obtained through this study are as follows.

(1) The characteristics of bending loss and coupling loss that occur when POF is applied to the joints of the robot arm were confirmed experimentally, and suitability was confirmed through comparison with theoretical equations from previous studies.
(2) A POF angle sensor capable of measuring robotic joint angles using both bending and coupling losses has been successfully designed and fabricated.
(3) When manufacturing the POF angle sensor for a robot arm using POF, the appropriate initial gap distance (d_i) and joint radius (r) were determined to be 8 mm and 7.5 mm, respectively, taking comprehensive consideration of manufacturability and optical fiber characteristics.
(4) Through performance evaluation of the manufactured POF angle sensor, it was confirmed that angle measurement up to 165 degrees is possible. In addition, it was confirmed that the POF angle sensor has linearity in each section and a minimum resolution of 0.030 degree in the 0–30 degree section. In addition, a method of limiting the initial angle of the sensor to 60 degrees was also proposed to improve the linearity of the sensor.
(5) In order to confirm the reliability of each sensor in an extreme environment, an interference measurement experiment was conducted in an environment with a maximum magnetic flux density of 0.412 Tesla. The encoder confirmed that the interference level in the magnetic field environment increased 9.67 times compared to the noise in the general environment, whereas it was confirmed that there was no difference in the POF angle sensor.

Furthermore, the POF angle sensor developed in this study is expected to contribute to the development of the biomimetic robot industry as well as ordinary robots, especially in environments where existing electrical sensors are difficult to apply, such as areas with underwater or electromagnetic interference (EMI). In addition, this paper proposed a new sensor that can secure usability even in an environment where existing sensors are difficult to apply, but there are still limitations in linear section expansion and sensor resolution improvement. Therefore, it is judged that related improvement research is necessary in the future.

Author Contributions: Conceptualization, H.-W.L. and D.-H.K.; methodology, H.-W.L. and D.-H.K.; data curation, H.-W.L.; writing—original draft preparation, H.-W.L.; writing—review and editing, H.-W.L., D.-H.K. and S.S.; visualization, H.-W.L.; supervision, D.-H.K.; funding acquisition, D.-H.K. All authors have read and agreed to the published version of the manuscript.

Funding: This study was supported by the Research Program funded by the SeoulTech (Seoul National University of Science and Technology).

Institutional Review Board Statement: Not applicable.

Data Availability Statement: Data are included within this article.

Conflicts of Interest: The authors declare no conflict of interest.

References

1. Kim, S.G.; Shin, S.H.; Jeon, D.; Hong, S.H.; Sim, H.I.; Jang, K.W.; Yoo, W.J.; Lee, B. Fiber-optic Goniometer to Measure Knee Joint Angle for the Diagnosis of Gait Disturbance. *Trans. Korean Inst. Electr. Eng.* **2013**, *62*, 1009–1013. [CrossRef]
2. Joo, H.S.; Woo, J.H. Development of a Squat Angle Measurement System using an Inertial Sensor. *J. Korea Converg. Soc.* **2020**, *11*, 355–361.
3. Tesio, L.; Monzani, M.; Gatti, R.; Franchignoni, F. Flexible electro goniometers: Kinesiological advantages with respect to potentiometric goniometers. *Clin. Biomech.* **1995**, *10*, 275–277. [CrossRef] [PubMed]
4. Bini, R.R.; Hume, P.A.; Cerviri, A. A comparison of cycling SRM crank and strain gauge instrumented pedal measures of peak torque, crank angle at peak torque and power output. *Procedia Eng.* **2011**, *13*, 56–61. [CrossRef]
5. Just, A.; Krause, M.; Probst, R.; Bosse, H.; Haunerdinger, H.; Spaeth, C.; Metz, G.; Israel, W. Comparison of angle standards with the aid of a high-resolution angle encoder. *Precis. Eng.* **2009**, *33*, 530–533.
6. BJurado, P.; Calderón, A.G.; Cadavid, J.F.B.; Sucerquia, J.G. Intensity-modulated refractive index sensor based on optical fiber with slanted end. *Opt. Laser Technol.* **2023**, *157*, 108700.
7. DMontego, S.; Vázquez, C. Self-referenced optical networks for remote interrogation of quasi-distributed fiber-optic intensity sensors. *Opt. Fiber Technol.* **2020**, *58*, 102291.
8. Jiang, C.; Ren, A.; Dong, T.; Wu, H.; Xu, K.; Chen, P. An optical fiber voltage sensor based on self-mixing interference. *Opt. Fiber Technol.* **2023**, *75*, 103201. [CrossRef]
9. Toal, D.J.F.; Flanagan, C.; Lyons, W.B.; Nolan, S.; Lewis, F. Proximal object and hazard detection for autonomous underwater vehicle with optical fibre sensors. *Robot. Auton. Syst.* **2005**, *53*, 214–229. [CrossRef]
10. Kim, H.Y.; Lee, J.H.; Kim, D.H. Muscular Condition Monitoring System Using Fiber Bragg Grating Sensors. *J. Korean Soc. Nondestruc. Test* **2014**, *34*, 362–368. [CrossRef]
11. Kang, D. Analysis of the effects of concentrating electromagnetic waves during carbon composite curing using microwaves. *J. Korean Soc. Nondestruc. Test* **2022**, *42*, 136–143. [CrossRef]
12. Kang, D.; Kim, H.-Y.; Kim, D.-H.; Park, S. Thermal characteristics of FBG sensors at cryogenic temperatures for structural health monitoring. *Int. J. Precis. Eng. Manuf.* **2016**, *17*, 5–9. [CrossRef]
13. Kang, D.; Kim, H.-Y.; Kim, D.-H. Enhancing thermal reliability of fiber-optic sensors for bio-inspired applications at ultra-high temperatures. *Smart Mater. Struct.* **2014**, *23*, 074012. [CrossRef]
14. Kim, H.; Kang, D.; Kim, M.; Jung, M.H. Microwave curing characteristics of CFRP composite depending on thickness variation using FBG temperature sensors. *Materials* **2020**, *13*, 1720. [CrossRef] [PubMed]
15. Umesh, S.; Padma, S.; Srinivas, T.; Asokan, S. Fiber Bragg Grating Goniometer for Joint Angle Measurement. *IEEE Sensors J.* **2018**, *18*, 216–222. [CrossRef]
16. Zhang, L.; Qian, J.; Shen, L.; Zhang, Y. FBG sensor devices for spatial shape detection of intelligent colonoscope. *IEEE Int. Conf. Robot. Autom.* **2004**, *1*, 834–840.
17. Park, H.J.; Kim, D.H. A Study on the Bend Loss and Light intensity Distribution of Plastic Optical Fiber. *J. Korean Soc. Nondestruc. Test* **2021**, *41*, 183–190. [CrossRef]
18. Bilro, L.; Alberto, N.; Pinto, J.L.; Nogueira, R. Optical Sensors Based on Plastic Fiber. *Sensors* **2012**, *12*, 12184–12207. [CrossRef]
19. Arrue, J.; Zubia, J. Analysis of the decreases in attenuation achieved by properly bending plastic optical fibers. *IEEE Proc. Optoelectron.* **1996**, *143*, 135–138. [CrossRef]
20. Snydre, A.W.; Love, J. *Optical Waveguide Theory*; Champman and Hall: London, UK, 1983; pp. 135–152.
21. Arrue, J.; Zubia, J.; Fuster, G.; Kalymnios, D. Light power behaviour when bending plastic optical fibers. *IEE Proc. Optoelectron.* **1998**, *145*, 313–318. [CrossRef]
22. Shin, W.C.; Hong, J.H. Modeling of Transmitting Light Irradiance Distribution of Step-index Multimode Optical Fiber. *Korean J. Opt. Photonics* **2006**, *17*, 136–142.

biomimetics

Article

Bioinspired Whisker Sensor for 3D Mapping of Underground Mining Environments

Virgilio Gomez [1,*], Walid Remmas [2], Miguel Hernando [1], Asko Ristolainen [2] and Claudio Rossi [1]

1 Centre for Automation and Robotics, Universidad Politécnica de Madrid—CSIC, 28006 Madrid, Spain; miguel.hernando@upm.es (M.H.); claudio.rossi@upm.es (C.R.)
2 Centre for Biorobotics, Department of Computer Systems, School of Information Technologies, Tallinn University of Technology, 12618 Tallinn, Estonia; walid.remmas@taltech.ee (W.R.); asko.ristolainen@taltech.ee (A.R.)
* Corrrespondence: virgilio.gomez.lambo@upm.es

Abstract: Traversing through challenging, unstructured environments, particularly in mining scenarios characterized by dust concentration, darkness, and lack of communication presents formidable obstacles for traditional sensing technologies. Drawing inspiration from naked mole rats, characterized as being skilled subterranean navigators that depend heavily on touch to navigate their environment, this study introduces a new whisker-sensing disk designed for 3D mapping in unstructured environments. The disk comprises a circular array of 32 whisker sensors, each featuring a slender flexible plastic rod attached to a compliant base housing a 3D Hall-effect sensor. The whisker sensor is modeled using both analytical and data-driven approaches to predict rotation angles based on magnetic field measurements. The validation and comparison of both models are performed by evaluating data from other whisker sensors. Additionally, a series of experiments demonstrates the whisker disk's capability in performing 3D-mapping tasks, along with successful implementation on diverse robotic platforms, highlighting its future potential for effective 3D mapping in complex and unstructured subterranean environments.

Keywords: whisker sensor; 3D mapping; bioinspired design

1. Introduction

Navigating unstructured environments poses formidable challenges for autonomous robots, as they struggle with complexities, such as clutter, unpredictable terrain, and the limitations of traditional mapping technologies. This is particularly prevalent in mining environments, known for their extreme and unstructured nature, where dust concentrations and darkness create additional hurdles for traditional localization, sensing, and mapping methods that rely on Global Positioning Systems (GPS), cameras, and Light Detection and Ranging (LiDAR) systems [1]. Recognizing the need for innovative solutions, the development of alternative sensory approaches becomes critical for robots tasked with mapping and navigating these cluttered and difficult-to-access environments [2]. A promising approach involves the use of touch-based sensors to map the surface of the mine, with bioinspired mammalian sensors emerging as a viable alternative due to their demonstrated robustness, simplicity, and precision [3]. The versatility of such sensors is evident in their application for various purposes, ranging from object localization [4–6], contour and texture recognition [3,7,8], and even fluid flow direction and viscosity [9–11]. With particular emphasis on robot navigation and mapping, the integration of active and passive whiskers into the structure of mobile robots and robot manipulators [12–16] has shown great potential for obstacle avoidance and Simultaneous Localization and Mapping (SLAM) tasks [17,18]. However, not much effort has been put into point cloud generation for environment reconstruction. Mapping 3D spaces presents additional complications, as whisker-based sensors need to navigate and interpret spatial information effectively. Factors such as environmental variability, surface textures, and the need for real-time adaptability contribute to the intricacies faced in developing robust models for whisker-based sensors in 3D-mapping applications.

The ROBOMINERS project [1] strategically focuses on introducing innovative robotic concepts to overcome these challenges and improve mineral access in Europe, thus reshaping the mining ecosystem. A crucial objective is to enable robots to navigate and execute selective mining procedures by equipping them with the capability to map mining environments and dynamically adjust their paths for optimized mineral extraction. To address this challenge, prototypes that incorporate various alternative sensor technologies are currently in development and are undergoing testing. These prototypes explore a spectrum of options, from spectrometry-based sensors [19] to contact-based sensors. Drawing inspiration from the remarkable tactile abilities of the naked mole rat, a notable rodent species renowned for its navigation and contact detection capabilities in subterranean environments (Figure 1a) [20], an autonomous robotic miner (Figure 1c) could navigate challenging environments, addressing complex issues such as executing mining procedures and precise 3D-mapping tasks and navigating in subterranean areas [21]. Similar to the naked mole rat, which is heavily dependent on its somatosensory systems (touch) [22] and uses a series of sensitive sensory hairs to orient itself and navigate subterranean landscapes (Figure 1b), the robotic miner strives to replicate the efficiency of nature in overcoming obstacles in its quest for optimal performance and adaptability.

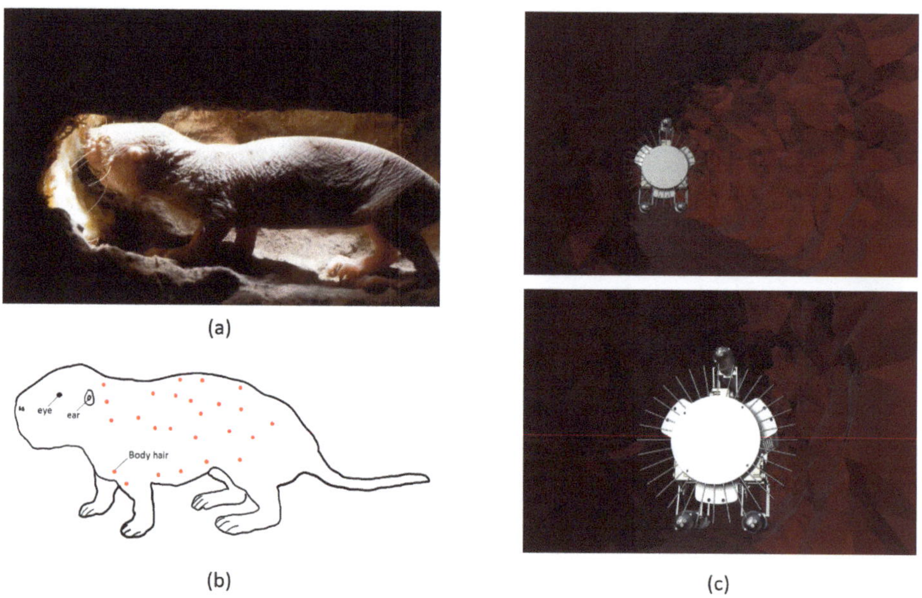

Figure 1. (**a**) Naked mole rat in subterranean environments (Picture taken by Neil Bromhall, Shutterstock). (**b**) Body hair location for an adult naked mole rat. Each body hair is represented by a small red dot. This rodent presents a systematic array of sensitive sensory hairs on its body, as described in [23], that help it navigate and orient itself in lightless subterranean environments. (**c**) Concept of robotic miner equipped with touch based sensor for navigation and mapping of unstructured 3D environments.

This study introduces a novel whisker-based disk that uses 3D Hall-effect sensors to map 3D unstructured environments. The disk design tries to approximate a section of the mole rat body, where the sensory hairs are approximately distributed in a ring-like pattern across the whole body. Compared to other designs, the whisker is housed in a non-permanent manner to the silicone rings, and the magnet holder features a spherical joint that serves as the pivot point to transmit the rotational movement of the whisker. The whisker is modeled following both analytical and data-driven approaches to determine the rotational angles of the whiskers, given the magnetic field felt by the sensor when the whisker makes contact with an object. In this paper, the whisker disk is installed on a serial manipulator to demonstrate the mapping of 3D surfaces. Furthermore, the design can be attached to a scaled-down robotic miner to give the robot the capability to map 3D environments during operation.

This article is structured as follows. In Section 2, we introduce the design and manufacturing process of the sensor. In Section 3, we present an approximation of an analytical model of the phenomena and propose an alternative sensor model based on a data-driven approach. In Section 4, we perform different experiments to compare and validate the rotational models presented. In Section 5, we carry out 3D-mapping experiments using the serial manipulator and showcase the adaptability of the presented solution to other robotic platforms. Finally, in Section 6, we draw conclusions from the results obtained in the study.

2. Sensor Design and Manufacture Process

The prototype presented (Figure 2a) is based on a circular array of 32 whiskers that can be attached to different robots to map non-structured environments. The main component of the disk is the whisker sensor, composed of a slender flexible polystyrene rod that is inserted at one end into a 3D-printed part called a "magnet holder", where a neodymium permanent magnet is affixed to the opposite end of this piece. The magnet possesses axial magnetization, aligning its magnetic field with the Hall effect sensor's x-axis. The magnet holder features a spherical contact that is housed inside a silicone structure composed of two rings that are attached to the disk, as shown in Figure 2b. The dimensions of the sphere are slightly larger with respect to the silicone structure to ensure the restriction of the magnet holder inside the rings. The silicone structure allows the magnet holder to rotate in two planes as the whisker makes contact with a surface.

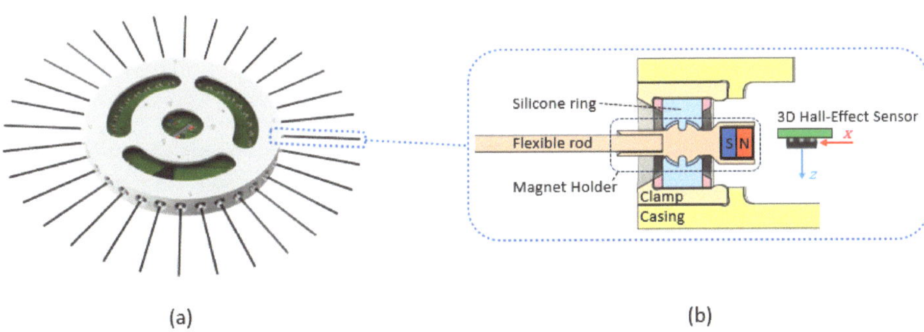

Figure 2. Whisker disk prototype. (**a**) The sensor comprises a circular array of 32 whiskers. (**b**) Detailed section view of a single whisker sensor.

Sensor Manufacturing Process

The whisker disk pieces were manufactured using a combination of fused deposition modeling (FDM) and stereolithography (SLA) printing. The disk structure was printed using an industrial grade FDM printer (Stratasys, Dimension Elite), while the molds for the inner cavity of the silicone structure and the magnet holder were printed with a high resolution SLA printer (Formlabs Form 3B) with a layer thickness of 50 µm.

To manufacture the silicone rings, we first prepared the mold pieces by applying a silicone release agent and secured them with a set of screws. During the mixing procedure, the silicone mixture (Zhermack Elite double 22) underwent a two-minute vacuum treatment to eliminate any potential bubble formation. Following degassing, the mixture was carefully injected into the mold. Finally, to guarantee the absence of bubbles within the silicone structure, the mold was reintroduced into the vacuum chamber and pressurized throughout the rest of the curing process. This step was repeated for the upper ring, ensuring uniformity and structural integrity.

Finally, to assemble the whisker disk (Figure 3), we first clamped the two rings with their respective silicone structure and secured the structure inside the casing that comprises the disk with a set of screws. To assemble the magnet holder, we glued the magnets to the rear part and attached the rods to the front part. The magnets used were N52 grade 4mm cylindrical neodymium magnets. The rods selected were 2 mm polystyrene rods. The magnet holder was then inserted into the silicone rings until the spherical housing occupied the silicone cavity. Lastly, a Printed Circuit Board (PCB) disk was installed where the Hall-effect sensors are mounted, ensuring that the corresponding sensor's X-axis was collinear to each whisker. It should be noted that since the presented prototype is made up of 32 whiskers and the selected Hall-effect sensors do not have enough configurable addresses, it was necessary to include four Inter-Integrated Circuit (I2C) multiplexors (PCA9548AD) in the PCB.

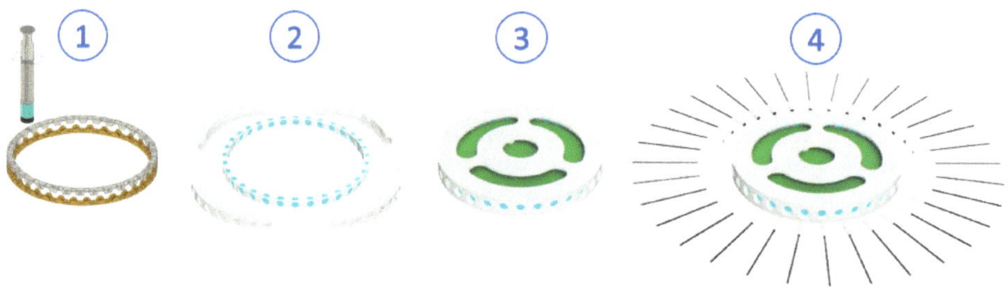

Figure 3. Whisker disk assembly procedure. First, the silicone base is prepared by injecting the mixture into the plastic rings. Next, the silicone rings are clamped and assembled into the disk structure with a set of screws. Finally, the PCB is installed, and the whiskers are inserted through the silicone ring until the spherical contact reaches its position.

3. Sensor Modeling

In order to determine the sensor model that correctly maps sensor measurements, (B_x, B_y, B_z) to rotation angles (θ_y, θ_z), we performed a comprehensive modeling procedure employing analytical and data-driven methodologies.

To collect the necessary data to train and validate both models, we designed and built a test platform featuring a Universal Robot (UR) UR3e manipulator. The manipulator gripped a single whisker at its tip and performed randomized movements within its workspace. The setup is presented in Figure 4, where the sequence of movements is based on performing pure rotations with respect to the pivot point of the whisker for both angles. Throughout this process, both the robot's rotation angles with respect to the whisker and the sensor measurements for each position were recorded. A demonstration of the data capture process is available in the Supplementary Video S1.

The whisker workspace was determined by establishing rotational limits for both possible angles, followed by generating a meshgrid that encompassed all attainable positions. To facilitate movement to these positions, a Python script was written using UR's RTDE python client library [24]. To read the sensor output, a microcontroller was utilized by running an Arduino routine that averaged the first 16 measurements and, upon request from the Python script, printed the averaged sensor data.

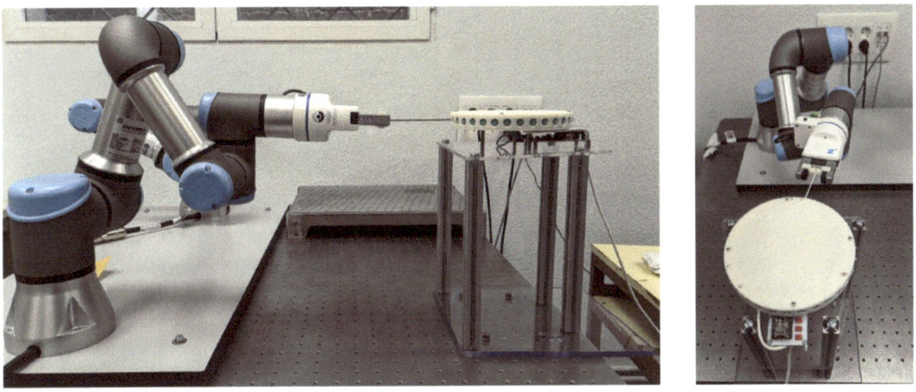

Figure 4. Testbed to capture whisker sensor data featuring a UR3e serial manipulator.

3.1. Whisker Rotation Model

To model the expected rotation of the whisker, given the magnetic field measurements, we considered the relationships provided by the manufacturer to estimate the linear and rotational movements of the magnet, considering an approximation of the relationship between the magnetic fluxes of each axis and the orientation of the magnet [25]. The description of the proposed model and references is presented in Figure 5, where the angles θ_y and θ_z represent the rotation on the y and z axes, O represents the pivot point, M is the center point of the magnet, and S is the center of the sensitive area of the sensor.

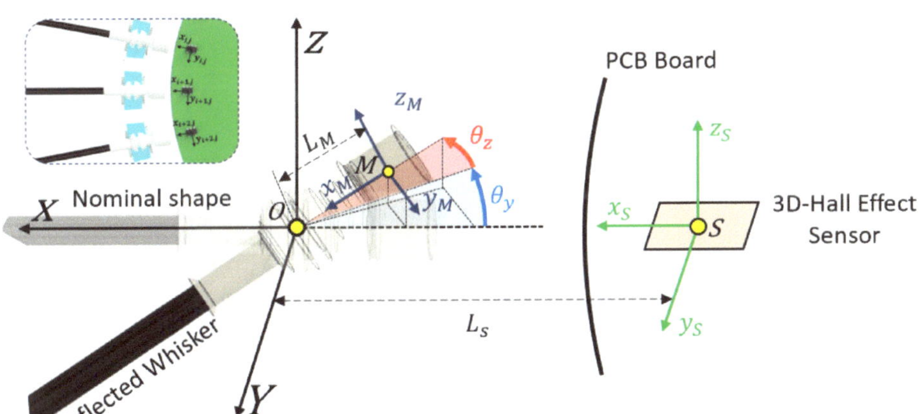

Figure 5. Whisker sensor rotational model description. The whisker can rotate in two different angles from the pivotal point generating the change in the position and orientation of the magnet and therefore of the magnetic field felt by the Hall-effect sensor.

The magnet position can be estimated considering the approximation presented in (1–3), where $\vec{B}$ represents the magnetic flux vector measured by the sensor, and sc_{θ_y}, b_{θ_z}, b_{θ_y} and b_{θ_z} represent the scaling and bias factors for both bending angles, respectively. Both scaling factors and biases are adjusted based on experimental data. It is worth noting that both the earth's magnetic field and neighboring magnets can affect the magnetic flux of the sensor, therefore changing the estimation of the magnet position. However, these effects resulted in a relatively small impact and were excluded from the model.

$$\vec{B} = \{B_x \quad B_y \quad B_z\} \tag{1}$$

$$\theta_z = \frac{atan2(B_y, B_x)}{sc_{\theta_z}} - b_{\theta_z} \tag{2}$$

$$\theta_y = -\frac{atan2\left(B_z, \sqrt{B_x^2 + B_y^2}\right)}{sc_{\theta_y}} - b_{\theta_y} \tag{3}$$

Once the rotation angles are calculated, assuming that the point of contact between the whisker and the surface occurs at the tip of the whisker, we can determine an approximate location of the surface point. It should be noted that the error resulting from this consideration is acceptable for a mining scenario, where the whisker length with respect to the overall size of the robot is relatively small.

Data-Driven Approach

We decided to implement a data-driven approach based on machine learning algorithms to account for the nonlinearity, variability, and inconsistencies derived from the whisker sensor manufacturing process. This approach allows us to provide a more adaptable solution, with the drawback of reducing the explainability of the model.

To select the algorithm that best combines computational cost and accuracy, we carried out a comparative study. Four state-of-the-art algorithms, Extreme Learning Machine (ELM), Multilayer Perceptron (MLP), Support Vector Regressor (SVR), and K-Nearest Neighbors Regressor (KNN-R) were considered based on previous experience. ELM and MLP are both neural network architectures, but ELM is known for its simplicity and rapid training [26], while MLP typically involves deeper architectures and backpropagation [27]. SVR is based on the principles of support vector machines, where the model learns the importance of a variable to characterize the relationship between input and output [28]. The aim is to find a hyperplane that best fits the data, considering a margin around the predicted values. KNN-R is a non-neural network method that relies on instance similarity [29].

To compare the algorithms, a sensitivity analysis was performed first to identify the models that perform the best for each algorithm. The prediction time and precision were evaluated using the mean absolute error (MAE) and the root mean square error (RMSE) as the main comparison metrics. A total of 900 points were recorded for a single whisker, where the rotations for both angles varied between −20° and 20°. The data were divided in an 85/15 split for training/testing sets, and the algorithms were run on a 1.8 GHz AMD Ryzen 7 5700U computer with 16 GB RAM, running a Python script.

Observing the results in Table 1, it can be seen that the best balance between the computational cost and precision is obtained by the ELM. The four models are capable of predicting the behavior of the whisker given the sensor input. However, the ELM is ranked the highest in speed and the second in terms of precision, only surpassed by the KNN-R algorithm. Recognizing the importance of a rapid and lightweight model is essential to maintain a satisfactory sensing frequency.

Table 1. Machine learning algorithms comparison.

Whisker Rotation	ELM (N_hidden_neurons = 30)			SVR (C = 8, Epsilon = 0.11)			MLP Single Layer (N_hidden_neurons = 100)			KNN-R (N_neighbors = 3)		
	MAE [rad]	RMSE [rad]	Time [ms]	MAE [rad]	RMSE [rad]	Time [ms]	MAE [rad]	RMSE [rad]	Time [ms]	MAE [rad]	RMSE [rad]	Time [ms]
θ_z	0.0221	0.0273	0.1	0.0453	0.0529	1.14	0.0286	0.0364	0.37	0.0155	0.0205	1.05
θ_y	0.0364	0.0465		0.0512	0.0609		0.0447	0.0609		0.0230	0.0315	

4. Results and Discussion

Six whiskers were randomly selected and evaluated in terms of the accuracy and computational cost for both models. The analytical model was fine-tuned with respect to scaling and bias factors through an optimization procedure using the nonlinear least squares method, while the data trained a specific ELM neural network algorithm. For each whisker, a test set of 200 points each was evaluated.

Although both models exhibited highly accurate performance across nearly all whiskers considered, the data-driven approach outperformed the analytical method in terms of efficiency. Analyzing Tables 2 and 3, we observe that the predictions of the ELM model are not only more precise but also faster than those of the analytical model. This efficiency gain can be explained based on the approximations of the analytical model, where there are no considerations for imperfection, misalignment, and tolerances in the sensor's mechanical design, which introduces challenging-to-model nonlinearities.

Table 2. Calibration for the analytical whisker rotational model.

Whisker	Model Parameters				θ_z			θ_y			Time [ms]
	sc_{θ_z}	b_{θ_z} [rad]	sc_{θ_y}	b_{θ_y} [rad]	MAE [rad]	RMSE [rad]	R^2	MAE [rad]	RMSE [rad]	R^2	
w_1	2.24	−0.0533	1.66	−0.0228	0.036	0.0489	0.945	0.0514	0.0723	0.8801	0.85
w_2	2.44	0.001	1.9	−0.04	0.0352	0.0469	0.9513	0.0577	0.0736	0.8681	0.72
w_3	2.27	−0.0321	1.56	0.01	0.0355	0.0477	0.949	0.052	0.0676	0.8973	0.64
w_4	2.65	0.054	2.12	0.05	0.0507	0.0702	0.8914	0.0818	0.1057	0.765	0.72
w_5	2.54	0.02	2	−0.01	0.0457	0.063	0.9157	0.0593	0.0829	0.8464	0.73
w_6	1.97	0.0476	1.5	−0.0834	0.036	0.0489	0.9468	0.0417	0.0557	0.931	0.73

Table 3. Calibration for the data driven whisker rotational model.

Whisker	θ_z			θ_y			Time [ms]
	MAE [rad]	RMSE [rad]	R^2	MAE [rad]	RMSE [rad]	R^2	
w_1	0.040	0.055	0.93	0.024	0.033	0.97	0.46
w_2	0.037	0.048	0.94	0.028	0.036	0.97	0.28
w_3	0.034	0.042	0.96	0.024	0.030	0.98	0.40
w_4	0.048	0.060	0.92	0.028	0.039	0.97	0.47
w_5	0.049	0.064	0.91	0.029	0.038	0.97	0.35
w_6	0.035	0.046	0.95	0.026	0.033	0.98	0.39

As shown in Figure 6, a comparison is drawn between a finely-tuned analytical model and a specific ELM applied to a single-whisker dataset. The figure highlights the precision of these models within the robot's captured workspace, presenting MAE and RMSE values for each cell. Upon observation, it becomes evident that the ELM performs well across the entire range of the whisker sensor, while the analytical model exhibits higher errors at higher rotational values. These findings affirm that the analytical model encounters challenges in predicting higher rotational angles, as nonlinearities are more likely to impact precision in this context.

In order to extend the model to cover the remaining whiskers, we opted to create a comprehensive general model for simplicity. This involved combining all the recorded whisker data into a single dataset. We then tuned and evaluated both analytical and data-driven models using these unified models. For the analytical model, we determined both bias and scaling factors by averaging the results presented in Table 2. Regarding the ELM model, we trained the new model using the combined whisker dataset. This time, we adopted a 70/30 split for training the ELM, where the total size of the dataset comprises 2094 recordings.

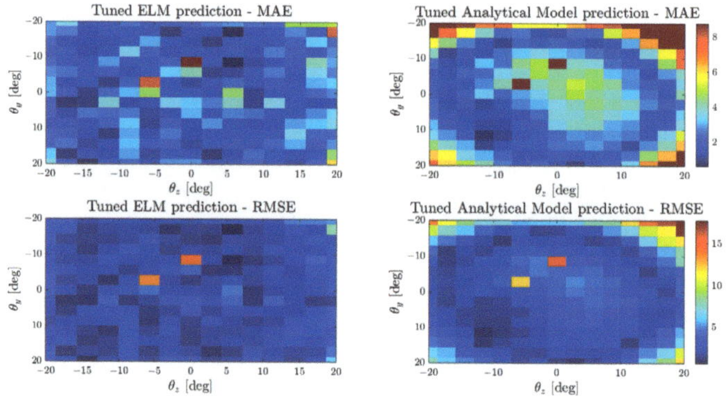

Figure 6. Mean average error and root mean square error comparison between the ELM and analytical models tuned for a single whisker.

To validate these models, we again compared the prediction and speed of both models in the test set to ensure that the ELM was not trained with them. Table 4 and Figure 7 present the comparison of both models for the representation of the whisker workspace. As expected, the ELM model results in a more suitable model, since it presents lower errors and deviation across the combined whiskers workspace.

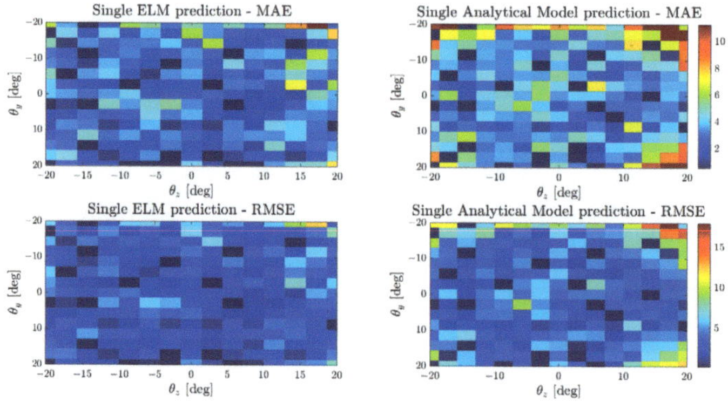

Figure 7. Mean average error and root mean square error comparison between the ELM and analytical models for the combined whisker dataset.

Consequently, the data-driven approach was chosen to ensure a faster sensor response given its superior precision and speed. The use of the same neural network trained with data from the collection of whiskers, along with appropriately scaled input values from other whiskers, was determined to yield relatively accurate results. However, to improve the accuracy and reliability of the model, each whisker sensor must undergo a specific tuning process to address imperfections that arise from the manufacturing process.

Table 4. Whisker rotation model comparison for general models for the combined whisker dataset.

Whisker Rotation Angles	Single ELM (N_hidden_neurons = 30)				Single Analytical Model ($sc_{\theta_z} = 2.37$, $sc_{\theta_y} = 1.75$ $b_{\theta_z} = -0.01$, $b_{\theta_y} = -0.02$)			
	MAE [rad]	RMSE [rad]	R^2	Time [ms]	MAE [rad]	RMSE [rad]	R^2	Time [ms]
θ_z	0.0352	0.0464	0.95	0.18	0.0505	0.0665	0.90	0.87
θ_y	0.0388	0.0537	0.94		0.0589	0.0833	0.84	

5. 3D-Mapping Implementation

After successfully validating the whisker model, we proceeded to implement it for a 3D-mapping scenario. In this occasion, the test platform, shown in Figure 8, incorporates the whisker disk as the end effector of the Universal Robots UR3e manipulator, while a rectangular box structure is securely attached to a technical table to mimic the basic geometry of a mine.

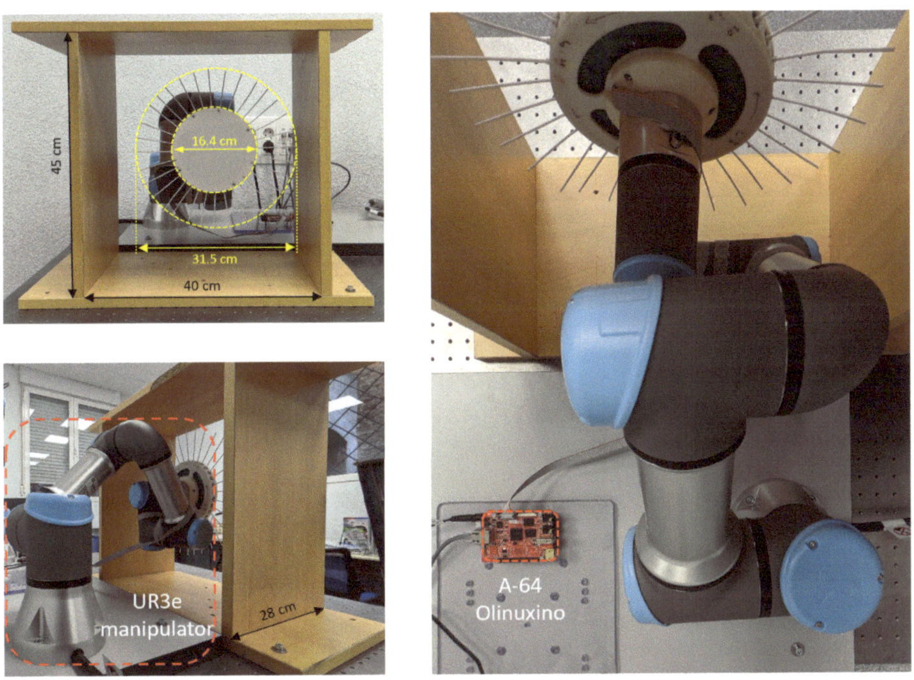

Figure 8. 3D mapping testbed for environment reconstruction. The wood structure represents a simple section of a mining environment. The surfaces to be mapped are the inner faces of the wooden box.

In order to visualize the point cloud produced by the whisker disk, we developed an application utilizing UR's ROS2 driver package [30] and Rviz for visualization purposes. The disk was attached to the UR3e arm to determine its position and orientation in space. The whiskers were represented using a rigid bar featuring two rotational joints at the pivot point of the whisker, with each rotation angle determined by the whisker rotational model.

For reading the 3D Hall-effect sensors, an Olimex A64 computer running an Ubuntu image was programed to execute a routine that verified the proper performance of each sensor and then published the readings in an ROS2 topic. A custom message was created to transmit all 32 magnetic field measurements in a single packet. The topic publishing rate was fixed at 10 Hz.

To generate the point cloud, a node was written to receive the custom whisker message and convert it into the spatial position of the whisker tip. Before publishing the point cloud, a condition to filter out non-deflected whiskers was implemented. The condition specified that the sum of the squares of the rotation angles of the whisker must exceed a threshold to be included in the point cloud. The threshold was empirically determined as 0.09, which yielded the best results.

3D-Mapping Experiment

The mapping experiment was carried out by teleoperating the robot arm to ensure that the whiskers were in contact with the surface being mapped. Figure 9 and Video S1 highlight the generation of the point cloud based on deflected whiskers. This approach demonstrated good performance in determining the shape and approximate geometry of the environment.

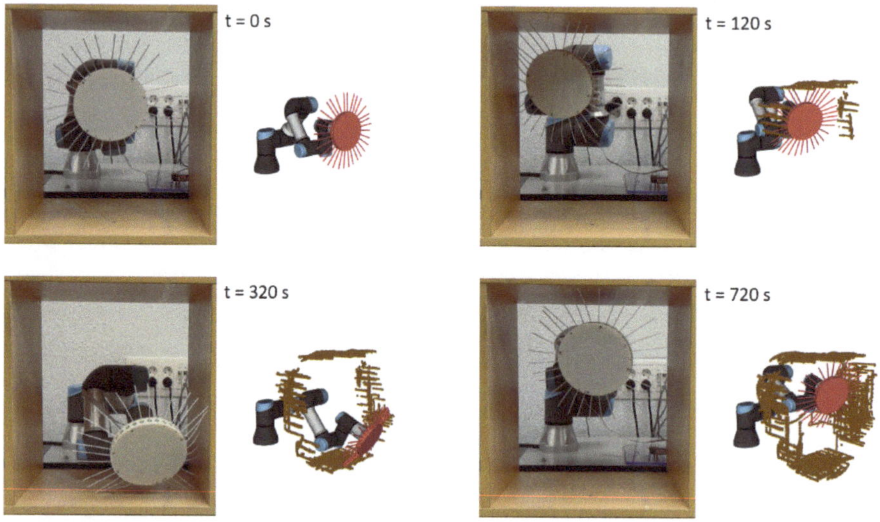

Figure 9. Environment reconstruction from the 3D-mapping experiment.

The generated point cloud is shown in Figure 10. It reveals an accurate representation of the box geometry, even though it presents minor errors observed in the central section of the box. These errors may arise from deflections in the whiskers, causing them to bend to such a degree that they do not fully return to their nominal shape and are captured by the point cloud.

To quantify the accuracy of the 3D geometry prediction, we performed additional mapping experiments and exported and processed the resulting point cloud by applying a clustering algorithm and implementing a filter based on the Mahalanobis distance to eliminate outliers. The results of the refined point cloud are presented in Figure 11 and Table 5, which confirms the reliability of the proposed method with average errors close to millimeter precision on almost all surfaces. To obtain the distance difference from each surface, we projected the captured points onto the closest neighboring surface of the box and measured the Euclidean distance for each pair of points. Our analysis revealed that the predicted dimensions of the box measure 42.74 cm $\times$ 46.42 cm $\times$ 29.11 cm, showing a good approximation from the actual dimensions of the box (40 cm $\times$ 45 cm $\times$ 28 cm).

Table 5. Maximum and average errors derived from the post-processed point cloud and the projected points onto the adjacent box surface.

Box Inner Surface	Max Error [m]	Mean Error [m]
Left	0.0339	0.0149
Right	0.0394	0.0132
Top	0.022	0.0052
Bottom	0.0593	0.0123

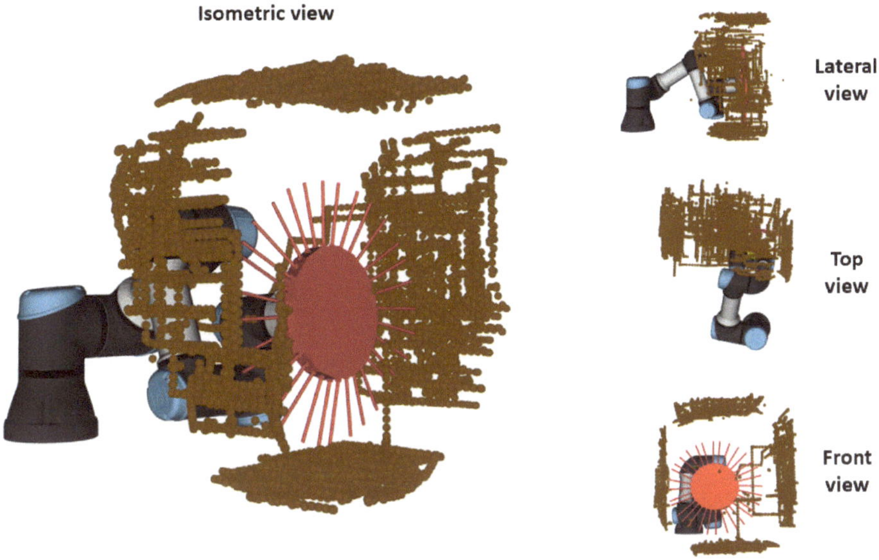

Figure 10. Orthographic views of the generated 3D pointcloud resulting from the mapping experiment.

Finally, it should be noted that the presented solution has been designed to work as a robotic module that can be adapted to the scaled-down version of the robominer platform [21], to give the robot the ability to map unstructured environments while operating. However, this concept can be easily adapted to larger mine-ready platforms.

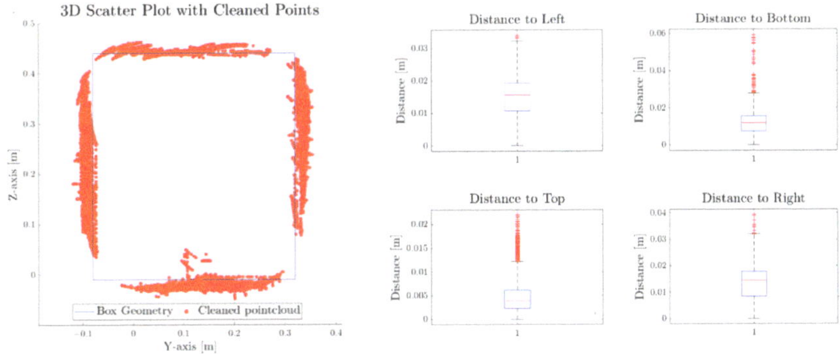

Figure 11. Three-dimensional point cloud comparison and error estimation between the ground truth and the prediction of each surface.

6. Conclusions

This study presented the design, modeling, and implementation of a sensor disk composed of a circular array of 32 whisker sensors for 3D mapping in unstructured environments. The design of the whisker comprises a plastic piece called a magnet holder, where a polystyrene rod and N52 permanent magnet are attached, and a 3D Hall-effect sensor that measures the magnetic flux generated by the magnet. The magnet holder features a spherical contact that acts as a ball joint. Each magnet holder is housed inside a silicone structure, and the magnetic sensors are installed on a single PCB.

The whisker was modeled following both analytical and data-driven approaches. For the analytical model, a series of relationships between the rotation angles and the magnetic flux measurements were derived based on the sensor manufacturer recommendations. To account for misalignment and variability from whisker to whisker, scaling and bias factors were included into the model and tuned based on experimental data, following a nonlinear least squares method. For the data collection process, a testbed featuring a UR3e robot was built to record the rotations of the whisker and the sensor measurements of the magnetic sensor.

For the data-driven approach, first a model selection process determined which algorithm resulted in the best performance considering both the precision and prediction time. Four widely used algorithms were considered, and we finally selected the ELM, as it presented the best balance between speed and accuracy.

To evaluate the performance of the models on the different sensors, six different whiskers were evaluated for both models. When comparing the results of the models, it was determined that the data-driven approach resulted in a more robust and faster performance than the analytical model, since it can account for nonlinearities that are not present in the analytical model. To account for the remaining sensors, a general model was developed and tested for both approaches, where the data captured from the recorded whiskers were merged into a single dataset. After tuning and evaluating the model, it was determined that the data-driven approach resulted in a more precise and robust solution. In addition, it was observed that the analytical model had difficulties estimating the rotational angles for large deflections of the whisker sensor, due to the appearance of difficult-to-model nonlinearities that are not considered. It is worth noting that to achieve optimal performance, each whisker must undergo an individual tuning process.

Once the whisker model was validated, the sensor disk was mounted in a serial manipulator to perform 3D-mapping experiments. A second testbed included a UR3e manipulator and a rectangular wood box that resembles a simple section of a mine geometry. The robot arm was teleoperated to map the inner surface of the rectangular box, resulting in a successful mapping of the geometry, thus validating the sensor capacity to reconstruct 3D environments. Although the sensing device was tested on a robot manipulator, the simple and easily adaptable design can be mounted on different robotic platforms, including mobile robots.

Recommendations for future developments include improvements to the mechanical design of the whisker to ensure its return to its nominal shape after bending as well as the design of new shapes of the whisker sensor to mimic the mole rat snout to improve the mapping performance in tight corners. Further advancements should also focus on refining the model to accurately identify contact points along the whisker, particularly in scenarios involving complex surface interactions, thus enhancing the overall accuracy and applicability of the system. Other studies can be carried out to generate a smart autonomous mapping algorithm that optimizes the mapping procedure.

Supplementary Materials: The following supporting information can be downloaded at: https://zenodo.org/records/10447310 (accessed on 31 December 2023), Video S1: whisker_demo.mp4.

Author Contributions: Conceptualization, V.G. and M.H.; Formal analysis, W.R. and M.H.; Funding acquisition, C.R.; Investigation, V.G. and M.H.; Methodology, V.G., W.R., and A.R.; Project administration, C.R.; Software, V.G. and W.R.; Supervision, M.H., A.R., and C.R.; Visualization, W.R.;

Writing—original draft, V.G.; Writing—review and editing, W.R., M.H., A.R., and C.R.; All authors have read and agreed to the published version of the manuscript.

Funding: This research was funded by the European Commission's Horizon 2020 research and innovation program (ROBOMINERS, grant no. 820971).

Data Availability Statement: The data captured for the whisker sensors are available in the following repo: https://github.com/virgilio96upm/whisker-modeling/tree/main (accessed on 31 December 2023).

Acknowledgments: We would like to thank Oscar Rudek, Roza Gkliva, and Jaan Rebane for their collaboration in the project.

Conflicts of Interest: The authors declare no conflicts of interest. Funding institutions had no role in the design of the study; in the collection, analyses, or interpretation of data; in the writing of the manuscript; or in the decision to publish the results.

References

1. Lopes, L.; Bodo, B.; Rossi, C.; Henley, S.; Žibret, G.; Kot-Niewiadomska, A.; Correia, V. ROBOMINERS—Developing a bio-inspired modular robot-miner for difficult to access mineral deposits. *Adv. Geosci.* **2020**, *54*, 99–108. [CrossRef]
2. Pearson, M.J.; Mitchinson, B.; Sullivan, J.C.; Pipe, A.G.; Prescott, T.J. Biomimetic vibrissal sensing for robots. *Philos. Trans. R. Soc. B Biol. Sci.* **2011**, *366*, 3085–3096. [CrossRef]
3. Solomon, J.H.; Hartmann, M.J.Z. Extracting Object Contours with the Sweep of a Robotic Whisker Using Torque Information. *Int. J. Robot. Res.* **2010**, *29*, 1233–1245. [CrossRef]
4. Zhao, C.; Zhang, S.; Xie, T.; Zeng, L. A novel whisker sensor with variable detection range for object positioning. *Rev. Sci. Instruments* **2022**, *93*, 035007. [CrossRef] [PubMed]
5. Lin, M.A.; Reyes, E.; Bohg, J.; Cutkosky, M.R. Whisker-Inspired Tactile Sensing for Contact Localization on Robot Manipulators. In Proceedings of the 2022 IEEE/RSJ International Conference on Intelligent Robots and Systems (IROS), Kyoto, Japan, 23–27 October 2022; pp. 7817–7824. [CrossRef]
6. Pearson, M.J.; Salman, M. Active Whisker Placement and Exploration For Rapid Object Recognition. In Proceedings of the 2019 IEEE/RSJ International Conference on Intelligent Robots and Systems (IROS), Macau, China, 3–8 November 2019; pp. 672–677. [CrossRef]
7. Zeng, L.; Zhang, S.; Xie, T.; Zhao, C. Contour recognition method of Hall-effect-based whisker sensor. *Meas. Sci. Technol.* **2022**, *33*, 065104. [CrossRef]
8. Jiang, Q.; Wei, G.; Zhao, C. Fiber Bragg grating-based biomimetic whisker for shape and texture recognition. *J. Instrum.* **2018**, *13*, P11013. [CrossRef]
9. Miersch, L.; Hanke, W.; Wieskotten, S.; Hanke, F.D.; Oeffner, J.; Leder, A.; Brede, M.; Witte, M.; Dehnhardt, G. Flow Sensing by Pinniped Whiskers. *Philos. Trans. R. Soc. Lond. B Biol. Sci.* **2011**, *366*, 3077–3084. [CrossRef] [PubMed]
10. Ju, F.; Ling, S.F. Sensing fluid viscosity and density through mechanical impedance measurement using a whisker transducer. *Meas. Sci. Technol.* **2013**, *24*, 055105. [CrossRef]
11. Zheng, X.; Kamat, A.M.; Harish, V.S.; Cao, M.; Kottapalli, A.G.P. Optimizing Harbor Seal Whisker Morphology for Developing 3D-Printed Flow Sensor. In Proceedings of the 2021 21st International Conference on Solid-State Sensors, Actuators and Microsystems (Transducers), Orlando, FL, USA, 20–24 June 2021; pp. 1271–1274. [CrossRef]
12. Fox, C.; Evans, M.; Pearson, M.; Prescott, T. Tactile SLAM with a biomimetic whiskered robot. In Proceedings of the 2012 IEEE International Conference on Robotics and Automation, Saint Paul, MN, USA, 14–18 May 2012; pp. 4925–4930. [CrossRef]
13. Xu, P.; Wang, X.; Wang, S.; Chen, T.; Liu, J.; Zheng, J.; Li, W.; Xu, M.; Tao, J.; Xie, G. A Triboelectric-Based Artificial Whisker for Reactive Obstacle Avoidance and Local Mapping. *Research* **2021**, *2021*, 9864967. [CrossRef]
14. Jung, D.; Zelinsky, A. Whisker based mobile robot navigation. In Proceedings of the IEEE/RSJ International Conference on Intelligent Robots and Systems. IROS'96, Osaka, Japan, 8 November 1996; Volume 2, pp. 497–504. [CrossRef]
15. Brown, L.; Carrasco, J.; Watson, S.; Lennox, B. Elbow Detection in Pipes for Autonomous Navigation of Inspection Robots. *J. Intell. Robot. Syst.* **2019**, *95*, 527–541. [CrossRef]
16. Wei, Z.; Shi, Q.; Li, C.; Yan, S.; Jia, G.; Zeng, Z.; Huang, Q.; Fukuda, T. Development of an MEMS based biomimetic whisker sensor for tactile sensing. In Proceedings of the 2019 IEEE International Conference on Cyborg and Bionic Systems (CBS), Munich, Germany, 18–20 September 2019; pp. 222–227. [CrossRef]
17. Milford, M.; Wyeth, G.; Prasser, D. RatSLAM: A hippocampal model for simultaneous localization and mapping. In Proceedings of the IEEE International Conference on Robotics and Automation, 2004. Proceedings. ICRA'04. 2004, New Orleans, LA, USA, 26 April–1 May 2004; Volume 1, pp. 403–408. [CrossRef]
18. Salman, M.; Pearson, M.J. Advancing whisker based navigation through the implementation of Bio-Inspired whisking strategies. In Proceedings of the 2016 IEEE International Conference on Robotics and Biomimetics (ROBIO), Qingdao, China, 3–7 December 2016; pp. 767–773. [CrossRef]

19. Burlet, C.; Stasi, G.; Godon, S.; Gkliva, R.; Piho, L.; Ristolainen, A. ROBOMINERS resilient reflectance/fluorescence spectrometers. In Proceedings of the EGU General Assembly 2023, Vienna, Austria, 24–28 April 2023. [CrossRef]
20. Boublil, B.L.; Diebold, C.A.; Moss, C.F. Mechanosensory Hairs and Hair-like Structures in the Animal Kingdom: Specializations and Shared Functions Serve to Inspire Technology Applications. *Sensors* **2021**, *21*, 6375. [CrossRef]
21. Gomez, V.; Hernando, M.; Aguado, E.; Sanz, R.; Rossi, C. ROBOMINER: Development of a Highly Configurable and Modular Scaled-Down Prototype of a Mining Robot. *Machines* **2023**, *11*, 809. [CrossRef]
22. Park, T.; Lewin, G.; Buffenstein, R. Naked Mole Rats: Their Extraordinary Sensory World. In *Encyclopedia of Animal Behavior*; Breed, M.D., Moore, J., Eds.; Academic Press: Oxford, UK, 2010; pp. 505–512. [CrossRef]
23. Crish, S.D.; Rice, F.L.; Park, T.J.; Comer, C.M. Somatosensory organization and behavior in naked mole-rats I: Vibrissa-like body hairs comprise a sensory array that mediates orientation to tactile stimuli. *Brain. Behav. Evol.* **2003**, *62*, 141–151. [CrossRef] [PubMed]
24. Universal Robots. RTDE Client Library-Python. Available online: https://github.com/UniversalRobots/RTDE_Python_Client_Library (accessed on 31 December 2023).
25. Infineon Technologies. TLV493D-A1B6. Available online: https://www.infineon.com/cms/en/product/sensor/magnetic-sensors/magnetic-position-sensors/3d-magnetics/tlv493d-a1b6/ (accessed on 25 November 2023).
26. Huang, G.B.; Zhu, Q.Y.; Siew, C.K. Extreme learning machine: A new learning scheme of feedforward neural networks. In Proceedings of the 2004 IEEE International Joint Conference on Neural Networks (IEEE Cat. No.04CH37541), Budapest, Hungary, 25–29 July 2004; Volume 2, pp. 985–990. [CrossRef]
27. Murtagh, F. Multilayer perceptrons for classification and regression. *Neurocomputing* **1991**, *2*, 183–197. [CrossRef]
28. Zhang, F.; O'Donnell, L.J. Chapter 7—Support vector regression. In *Machine Learning*; Mechelli, A., Vieira, S., Eds.; Academic Press: Cambridge, MA, USA, 2020; pp. 123–140. [CrossRef]
29. Kramer, O., K-Nearest Neighbors. In *Dimensionality Reduction with Unsupervised Nearest Neighbors*; Springer: Berlin/Heidelberg, Germany, 2013; pp. 13–23. [CrossRef]
30. Universal Robots. Universal Robots ROS2 Driver. Available online: https://github.com/UniversalRobots/Universal_Robots_ROS2_Driver (accessed on 31 December 2023).

biomimetics

MDPI

Article

Bio-Inspired Space Robotic Control Compared to Alternatives

Timothy Sands [1,2]

[1] Department of Mechanical and Aerospace Engineering, Cornell University, Ithaca, NY 14853, USA; dr.timsands@alumni.stanford.edu

[2] Department of Mechanical and Aerospace Engineering, Naval Postgraduate School, Monterey, CA 93943, USA

Abstract: Controlling robots in space with necessarily low material and structural stiffness is quite challenging at least in part due to the resulting very low structural resonant frequencies or natural vibration. The frequencies are sometimes so low that the very act of controlling the robot with medium or high bandwidth controllers leads to excitation of resonant vibrations in the robot appendages. Biomimetics or biomimicry emulates models, systems, and elements of nature for solving such complex problems. Recent seminal publications have re-introduced the viability of optimal command shaping, and one recent instantiation mimics baseball pitching to propose control of highly flexible space robots. The readership will find a perhaps dizzying array of thirteen decently performing alternatives in the literature but could be left bereft selecting a method(s) deemed to be best suited for a particular application. Bio-inspired control of space robotics is presented in a quite substantial (perhaps not comprehensive) comparison, and the conclusions of this study indicate the three top performing methods based on minimizing control effort (i.e., fuel) usage, tracking error mean, and tracking error deviation, where 96%, 119%, and 80% performance improvement, respectively, are achieved.

Keywords: bioinspiration; biomimetics; robotics; control; bio-inspired locomotion; bio-robotics; bio-inspired robots; biomechanics

1. Introduction

To accumulate energy, baseball pitchers (Figure 1) "wind up", initially moving the ball in the opposite direction to the desired destination. The general shape of space robots is not dissimilar to baseball pitchers, and this study evaluates the efficacy of trajectory shaping for space robots inspired by the biomimicry of baseball pitching Box 1.

(a)

(b)

Figure 1. Space robots may be represented as cylindrical center rigid bodies and highly flexible appendages. (**a**) A U.S. Naval Academy pitcher throws to home plate at a baseball tournament (image credit: Technical Sergeant David W. Carbajal) [1,2]. (**b**) NASA's first humanoid space robot (image credit: NASA) [3,4].

Box 1. Problem Statement

Problem Statement: Amongst the many available options for autonomously controlling deep space robots, is one method better than another at simultaneously seeking exact precision, vibration elimination, fuel minimization, and robustness?

1.1. Broad Context and Why This Study Is Important

Despite recent demonstrations of operations in space of capabilities at the system level of several autonomous functions [5,6] including robotics, contemporary space operations assessment and action planning rely upon pre-scripted sequences of commands from ground operations personnel [7]. Considering challenging and distant robotics missions (e.g., to Mars), ground operators are unlikely to predict all the likely encounters and physical interactions occurring in parts of space seldomly experienced before [8]. Limited knowledge in complex situations demands autonomy, perceivably establishing a new frontier for space exploration. Another obvious example is the utilization of very small spacecraft in cislunar orbits to refuel, repair, and replenish earth-orbiting spacecraft at lower altitudes, necessitating grappling potentially unknown masses of uncooperative spacecraft. Extreme fuel efficiency seems mandatory, especially due to the low mass and volume of underactuated robotic repair spacecraft [8] like those depicted in Figure 2, leading naturally to control minimization as a primary figure of merit [9].

(**a**) (**b**)

Figure 2. (**a**) NASA mission to repair and refuel satellites on orbit (image credit: NASA [4,10]. (**b**) Satellite servicing mission of the Defense Advanced Research Projects Agency (DARPA). Image credit: DARPA [11,12].

1.1.1. Cislunar Space

A recent primer on cislunar space published by the U.S. Air Force [13] intended to aid the development of expertise, capabilities, plans, and operational concepts. The importance of the orbits manifested in the December 2019 creation of a new branch of the military charged with the duties to defend and protect American space interests, especially since highly perturbed orbits depart predictable locations; intercept, rendezvous, and proximity operations (depicted in Figure 3) become quite complicated and potentially unpredictable. Cislunar orbits are generally no longer planar and no longer elliptical (certainly not circular), and spacecraft positions are no longer easy to articulate geometrically.

(a) (b)

Figure 3. Operations in cislunar orbits **[14]**. (**a**) Schematic defining features of cislunar space. Image credit NASA [15]. (**b**) Cislunar satellite inspector. Image credit: Air Force Research Laboratory and National Aeronautics and Space Administration [2,4].

1.1.2. Cislunar Robotic Operations

Actuators for space robots were recently reviewed in [16], highlighting time delays as a key limiting issue for successful operations, some of which are displayed in Figure 4. A disparate review [17] assembled over the last three years subdivided the trends in research development achievements. The reviews focus on two different treatments of modeling uncertainties in hopes of not needing to increase design margins: either increasing the accuracy of parameter discrimination or alternatively developing methods with inherent robustness. The later review [17] emphasized the fact that task performance execution success correlated with onboard computing power.

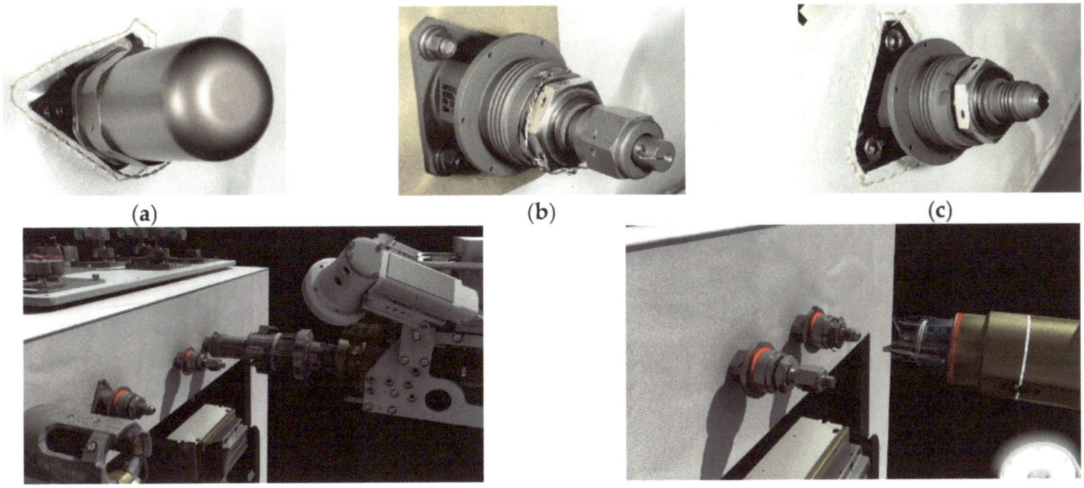

(a) (b) (c)
(d) (e)

Figure 4. NASA Robotic Refueling Mission (RRM) task: refueling. Individual pieces of hardware show the seals that typical satellite fuel valves have. (**a**) A tertiary cap with a "lock wire" visible underneath; (**b**) a safety cap/actuation nut with a securing lock wire; (**c**) an exposed fuel valve; (**d**) a safety cap tool removing a safety cap; and (**e**) a nozzle tool being connected to the now exposed fuel valve, enabling fuel transfer (image credit: NASA [4,18]).

High-speed maneuvering necessitating the real-time solution of inverse kinematics was proposed in [18,19] for the capture of space debris with safety from possible collisions for dual-arm continuum manipulators with input saturation. Real-time trajectory planning provides an alternative when paired with an adaptive controller using deep reinforcement learning proposed in [19]. Meanwhile, trajectory tracking for attitude maneuvers with vibration suppression was proposed in [20] in the presence of actuator and disturbance uncertainties. Disturbances and actuator saturation was investigated in [21], while compliance control was emphasized in [22], while so–called "soft robotics" was modeled in [23]. Modeling is key, since the robot hardware needs a mathematical model to facilitate experimentation in computers.

In the broadest sense, the space robot (like those depicted in Figures 1, 3 and 5) may be considered generically as a cylindrical main body with a robotic appendage attached, leading to both rigid-body and flexible-body treatments. Figure 5 displays two variations on laboratory hardware replicating space robots. Subfigure (b) is the Navy space robot laboratory hardware that serves as the baseline system analyzed in subsequent sections of this manuscript. The terminology is defined in Table 1 which displays proximal variable definitions and nomenclature. Such convenience to the readership is repeatedly provided (e.g., Tables 2–5) throughout the manuscript.

Table 1. Table of proximal variables and nomenclature [1].

Variable/Acronym	Definition	Variable/Acronym	Definition
$\hat{y}$	Centerline unit vector	k	Appendage stiffness
T_C	Control torque	$m_i, I_i \forall i = 2 \ldots 5$	Flexible masses and inertias
J_1	Main body inertia mass moment	J_2	Flexible inertia mass moment
θ_1	Main body rotation angle	$\theta_i, W_i \forall i = 2 \ldots 5$	Translation and rotation angles

[1] Such tables are offered throughout the manuscript to aid readability.

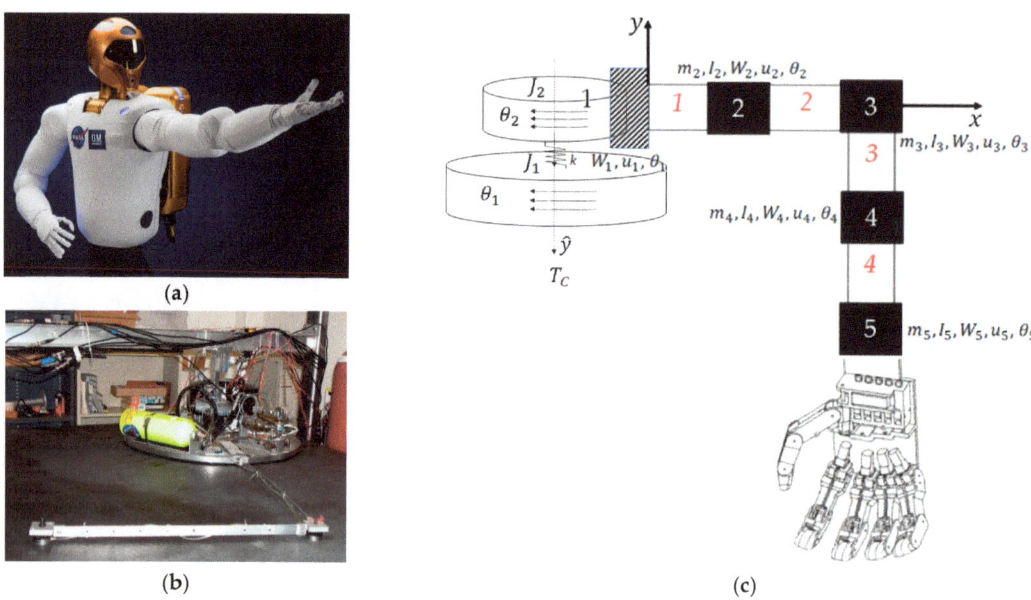

Figure 5. Space robots with cylindrical center rigid bodies and highly flexible appendages. (**a**) NASA's first humanoid space robot. Image credit: NASA [3]. (**b**) Laboratory flexible rotational spacecraft hub with a free-floating, planar air-bearing, very light robotic arm, the schematic of which is displayed in subfigure (**c**).

The literature reveals many mathematical approaches [13] for modeling and also for control, while the readership might be bereft and confused about which method or combinations of methods should be considered for their particular application. This manuscript compares a recently proposed bio-inspired approach to many of the contemporary alternatives.

1.2. Broad Review of Modeling and Control from First Principles to Modern Instantiations

Modeling can include external and internal constraints on every particle expressed in a Euclidean space [24] or where Newton's Laws [25] and Euler's equation [26] may be combined using Chasle's theorem [27] to define space robot motion states in six degrees of freedom [28]. Control options include both feedback [29] and feedforward (sometimes necessitating system identification) [30] in addition to shaping of the commanded trajectory to ameliorate the deleterious effects of interactions between the control and robot modes that reduce appendage pointing and positioning accuracy. Feedback control is arguably the best understood to adjust robot performance by designing the closed loop system [31]. Well-established, classical techniques include proportional plus integral plus derivative (PID) control [32], emphasizing oscillation design and maintenance of stability following a commanded movement trajectory. A common modern control method used to minimize a performance measure is the linear quadratic regulator [33] ubiquitously applied to linear time-invariant systems, while the presumption for robotic space missions used for refueling and resupply particularly must assume time variance, since fuels are being relocated and hardware may be removed and replaced. A less frequently first-used method is feedforward control, elaborated in [34] to be particularly useful if the movements of the space robot are predictable. Feedforward control is also referred to as open-loop control. The recent resurgence of artificial intelligence has not overlooked space robotics, or medical purposes [35,36]. Robots can be trained by machine learning algorithms to compensate for varying tasks and environments. The internet of things (IoT) is another newly conceived possibility for remote robot control and monitoring [37].

The vibration of flexible bodies is strongly driven by the nature of the (impact) excitation or rate of application of external forces. Accordingly, the dynamics of physical contact are important, leading to a field of study known as contact dynamics. Such dynamics are modeled using a so-called Hertz model in [38], which studied gripping a non-cooperative spacecraft focusing on contact compliance control. The study highlighted the key nature of the vibration of flexible spacecraft parts, which could lead to repeated continuous collisions. The high flexibility of the robot arms is a key focus of attention for capturing non-cooperative spacecraft. This study presented simulation experiments indicating that compliance control seems key to successful performance.

The natural vibrational frequency of the robotic arms is dominated by the arms' masses and structural stiffness (resistance to motion, either translational or rotational). Tracking control of variable stiffness actuators was studied in [39], highlighting robot arm link motor disturbances and variability in actuator stiffnesses (naturally), which may render tracking control schemes ineffective. A mechanism for learning was proposed to compensate for disturbance uncertainties, leading to a novel framework for designing controls, starting with back-stepping tuning of feedback control, then parameter tuning with finite switching. Validation was offered using computer simulations.

A well-known method called input shaping from the 1990s has recently been hybridized in [40,41], seeking to address vibration residuals in flexible, multi-mode systems. Modal analysis is used to decouple the system by transforming to the reference frame defined by the modal (eigen) vectors. The command is shaped in the diagonalized modal reference frame, where the three contending alternatives were versine (sinusoidal), ramp, or cycloid plus ramped sinusoidal. The methods were compared in simulation experiments which validated high robustness to parameter uncertainty. *The general approach is duplicated in this present manuscript comparing bio-inspired whiplash options.*

The study in [42] establishes the present study's comparative benchmark offered by classical control methods augmented with signal processing filters, and the prequel's benchmark space robot is the same U.S. Navy system, with natural frequencies of vibration that are so low as to be excited by even low-bandwidth feedback systems. The utilization of versine shaping proved superior to step commands when the structural filters were designed to achieve system stability margins (classical gain margin and phase margin). The study in [43] is an intermediate sequel investigation on the same Navy space robot system, where so-called systems theory methods of Lev Pontryagin were used to develop an open-loop optimal control to minimize maneuver time with quiescent final conditions. The surprising results indicated a "whiplash"-shaped command (initially in a direction opposite to the commanded terminal state) minimized maneuver time. Results were produced in a commercial, pseudospectral optimization software, and then validation was performed analytically using six necessary conditions of optimization: (1) the Hamiltonian minimization condition; (2) adjoint equations; (3) the terminal transversality condition; (4) the Hamiltonian final value condition; (5) the Hamiltonian evolution equation; and (6) Bellman's principle. Importantly, the biomimicking "whiplash" shaping failed to validate one of the six necessary conditions motivating continuing research (including this present manuscript as a new sequel). The biomimicry whiplash control [43] behavior (depicted earlier in this manuscript in Figure 1) is evaluated in this present study applied to shaping the commanded input rather than as a feedforward control.

In addition to high structural flexibility, motor torque limits and joint flexibility are additional considerations emphasized in [44] which presented a model-based autonomous generation method for trajectories assuming base excitation and large time delays for any communications with earth. Computer simulations for verification were presented alongside validation by laboratory experiments on an air-bearing table.

Bio-inspired techniques were specifically highlighted in [45] including the adhesion mechanism and its locomotion system mimicking geckos, spider-inspired actuators, joints inspired by human knees, particle transport by peristaltic motion, and locus-inspired digging mechanisms. Nearly a decade later, NASA presented [46], introducing all-terrain legged rovers inspired by mountain goats, human-like planetary exploration robots, deep drilling inspired by gophers, artificial muscles as actuators, snake-like robotics for traversing narrow openings and passages. Just a few years later, Ellery offered a tutorial review of bio-inspired approaches to robotic manipulation for space debris salvage [47], highlighting sensorimotor control mimicking the human brain for several strategies including (i) sensorimotor planning, learning and control, (ii) optimal feedback control, (iii) impedance control, (iv) predictive control, and (v) Bayesian inferencing. That same year, Ellery also illustrated how to build a biological machine using engineering materials and methods [48]. Just last year, the study in [49] offered an alternative approach applied to a disparate system relative to the research in the present study presented in this manuscript: A biomimetic approach for control of a seven-degrees-of-freedom robotic arm. Namely, central nervous system-based motor control (neural networks with deep learning) is presented as a more directly competing deterministic alternative with the motor control methods presented by Menezes for underwater robotics [50]. Another aspect of the present study is vibration elimination, while an alternative was included in [51], where through slowing the seed and thereby reducing the negative effects of hitting the ground, many species of nuts have evolved differently by producing a rigid layer of protective shell around their seed.

Operationally, highly flexible space robots need to move around large loads of heavy items, potentially leading to large tracking errors. The study in [52] proposed damping–stiffness control including joint dynamics in a comprehensive model utilizing Luenberger observers for unmeasurable quantities. Damping was treated as a feedforward (plus gain), while the feedback is used to suppress perturbations. Verification in simulations hinted at a potential 98% percent improvement, while laboratory experiments merely validated an 88% improvement.

The dynamics (mathematical models) were the focus of [53], seeking the proper element number for inclusion in appendage models, where the end effector trajectory was controlled in the feedforward. Sliding mode control including gravity effects was proposed in [54] including nonlinear dynamics typically decomposed into separate flexible and rigid subsystems (as was in the present manuscript's study), but modelling accuracy strongly drove performance. As the control of highly flexible robotic systems becomes more commonplace, a MATLAB®/Simulink® toolbox evolved and was presented in [55], effectively reducing the coding burden of future investigations.

1.3. The Current State of the Research Field and Key References

The study in [38] describes substantial on-going proof-of-concept investigations including space debris removal, life-extension services, on-orbit assembly, and manufacturing, while identifying remaining challenges, particularly simulation of the true six-degrees-of-freedom dynamics of large-scale microgravity operations, especially for robotic systems. *The fourth option, so-called "whiplash shaping", is the bio-inspired method of shaping the input commanded trajectories for the space robot.*

1. Gain stabilization [56]: Tuning of gain to achieve stability of the rigid-body mode. Advantages: simplicity and based on well-known mathematics. Disadvantages: imprecise and uses effort (or equivalently fuel) wastefully compared to more modern methods.
2. Classical second-order structural filtering [56]: Second-order filters designed for each chosen resonance and anti-resonance, usually of the lowest mode or the lowest two modes to ensure stability. Advantages: aids fuel usage of gain stabilization approaches. Disadvantages: mathematic model must have precision and remain not time varying.
3. Input shaping [40,57–59]: Modification of open-control frequency content using time-delayed control application. Advantages: existing proofs of mathematical optimality. Disadvantages: lacks robustness.
4. *Whiplash compensation* [60,61]: Initially commanding maneuver in the opposite direction of desired end-state. Advantages: existing proofs of mathematical optimality. Disadvantages: lacks robustness.
5. Rigid-body, minimum-fuel input trajectory shaping: Apply control analytically derived from constrained control-minimization boundary value problem solutions. Advantages: existing proofs of mathematical optimality. Disadvantages: lacks robustness.
6. Single-frequency trajectory shaping: The fashion commanded trajectory from a single sinusoid chosen to avoid mode frequencies of the flexible robot. Advantages: simplicity. Disadvantages: still need accurate mathematical models to properly pick the single frequency.
7. Flatten the curve to improve stability: Use option #2 to compensate for all structural modes seeking to create a magnitude response curve resembling such a curve for a second-order rigid-body system (primary motivation remains increased system stability). Advantages: aids fuel usage of gain stabilization approaches. Disadvantages: mathematic model must have precision and remain not time varying.
8. Flatten the curve to improve trajectory tracking: This option is like option #7, except choosing parts of modes (resonance or anti-resonance) to minimize trajectory tracking errors (proposed in this manuscript). Advantages: aids fuel usage of gain stabilization approaches. Disadvantages: mathematic model must have precision and remain not time varying.
9. Deterministic artificial intelligence: Use physics to define robot self-awareness, while adapting or learning time-varying physical system parameters (e.g., mass, mass moments, stiffness, and damping). Advantages: proofs of optimality, robustness and simple algorithms. Disadvantages: relatively unknown compared to peer methods.
    - 9.1 Self-awareness statements [62]: Use governing equations from physics to exclusively define robot self-awareness, while prescribing necessary trajectories to be tracked (currently only sinusoidal trajectories and control-minimizing trajectories are in the literature).
    - 9.2 Adaption or optimal learning [63]: Use classical control methods (e.g., the "M.I.T. Rule" [64] to adapt system parameters to minimize tracking errors or alternatively use least squares estimation methods (e.g., batch, recursive, and extended).

1.4. Controversial and Diverging Hypotheses—Literature Gaps

Two disparate paradigms are evident regarding response magnitude curve shaping: flattening the curve to improve stability [42] versus flattening the curve to improve trajectory tracking (to be addressed in this manuscript).

1.5. Main Aim of the Work and Highlighting of Principal Conclusions

The goal is to provide the readership an extensive study comparing performances of available options based on multiple figures of merit: necessary fuel expenditure, mean tracking errors, and tracking error deviations.

1.6. Novelties Presented

1. Commanded trajectory-shaping options are compared using control effort and tracking accuracy, and recommendations are offered.
2. Feedforward controls are compared using control effort and tracking accuracy, and recommendations are offered.
3. Commanded trajectories are compared with filtered feedback and no feedforward using least control effort tracking accuracy, and recommendations are offered.
4. Mode 1 filtering options are compared using control effort tracking accuracy, and recommendations are offered.
5. Mode 3 filtering options are compared using control effort tracking accuracy, and recommendations are offered.
6. Mode 4 filtering options are compared using control effort tracking accuracy, and recommendations are offered.
7. Overall recommendations are made for selection of commanded trajectories, feedforward controls, and filtered versus unfiltered feedback.
8. The least control effort was achieved with step trajectories, rigid-body optimal feedforward control and unfiltered feedback, while recommendations are offered based on tracking accuracy and control effort.

2. Materials and Methods

Section 2.1 introduces modeling of the highly flexible space robot using as a benchmark the free-floating (on an air-bearing) flexible spacecraft simulator at the Naval Postgraduate School depicted in Figure 6, which also depicts system diagrams including both rigid-body and flexible robotic appendages. Development of system models (equations) follows next, where the resulting detailed equations are provided in Appendix A. Section 2.2 introduces competing control methodologies and depicts simulation topologies, where detailed simulation codes are provided in Appendix B.

Selectable options include commanded trajectories, feedforward controls, feedback controls, and structural filtering.

2.1. Space Robot Modeling

The highly flexible space robot laboratory hardware depicted in Figure 6, subfigure (a) is decomposed into a rigid-body base (simulating the near-rigid spacecraft) depicted in subfigure (b) with highly flexible, lightweight robotic appendages modeled using the finite element method in subfigure (c). The first node of the finite element representation is attached to the spacecraft, while flexible elements (mass and stiffnesses) are lumped at subsequent nodes, where Table 2 conveniently defines variables and nomenclature.

Equations of motion may be derived using at least several disparate methods commonly understood in kinetics: Hamilton's Principle, Lagrange's equations, and Chasle's theorem combining Newton's Law and Euler's equations. Since prequel [43] elaborated Hamilton's Principle, while prequel [42] elaborated Lagrange's equations, this present manuscript elaborates Chasle's Theorem. Newton's Law for translational motion is expressed in Equation (1), where displacements are expressed in coordinates of an inertial reference frame, while similarly Euler's moment equations are displayed in Equation (2). Expressing motion in coordinates of a non-inertial reference frame modifies Equations (1) and (2) to Equations (3) and (4), respectively. A representative two-node system of equations including internal flexible "spring forces" that resist motion is displayed in Equations (5) and (6), respectively, for each node, and then assembled into matrix-vector notation in Equation (7), where Table 2 conveniently defines variables and nomenclature.

$$F|_{inertial} = ma|_{inertial} = m\dot{v}|_{inertial} = m\ddot{x}|_{inertial} \quad (1)$$

$$T|_{inertial} = J\alpha|_{inertial} = m\dot{\omega}|_{inertial} \quad (2)$$

$$F = m\ddot{x} + m\frac{d\omega}{dt} \times r' + 2m\omega \times v' + m\omega \times (\omega \times r') \quad (3)$$

$$T = J\dot{\omega} + \omega \times J\omega \quad (4)$$

$$m_1\ddot{x}_1 = -k_1x_1 + k_2(x_2 - x_1) \quad (5)$$

$$m_2\ddot{x}_2 = -k_2(x_2 - x_1) + F \quad (6)$$

$$\underbrace{\begin{bmatrix} m_1 & 0 \\ 0 & m_2 \end{bmatrix}}_{[M]}\underbrace{\begin{Bmatrix} \ddot{x}_1 \\ \ddot{x}_2 \end{Bmatrix}}_{\{\ddot{x}\}} + \underbrace{\begin{bmatrix} k_1 + k_2 & -k_2 \\ -k_2 & k_2 \end{bmatrix}}_{[K]}\underbrace{\begin{Bmatrix} x_1 \\ x_2 \end{Bmatrix}}_{\{x\}} = \underbrace{\begin{Bmatrix} 0 \\ 1 \end{Bmatrix}F}_{\{F\}} \quad (7)$$

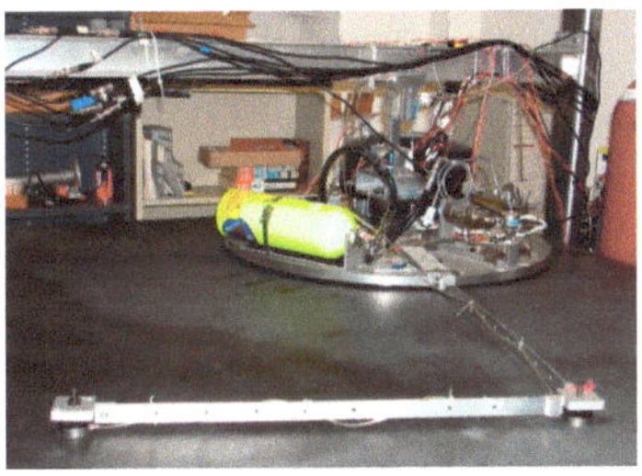

(a)

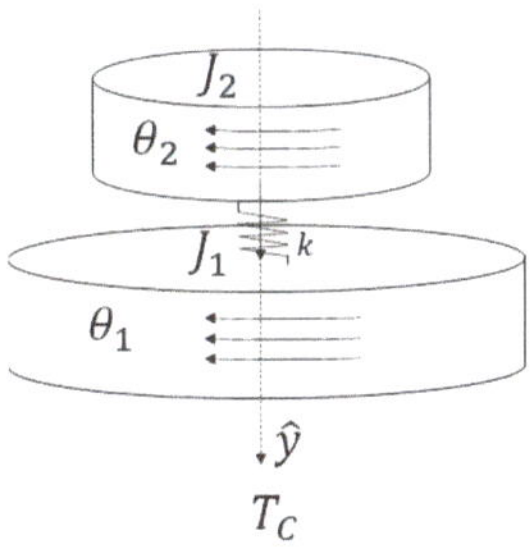

(b)

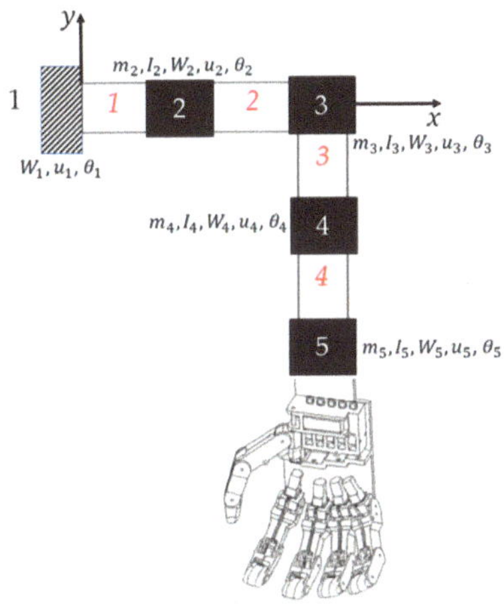

(c)

Figure 6. (**a**) Laboratory flexible spacecraft robotic arm (very lightweight) attached to a free-floating planar air-bearing rotational hub. Image used in compliance with image use policy [2], "U.S. Department of Defense photographs and imagery, unless otherwise noted, are in the public domain". (**b**) Schematic of center-lined, cylindrical, rigid spacecraft. (**c**) The flexible arm is modeled using the lumped-mass technique, where arm mass is distributed to discretized nodes. Images taken from [60] in compliance with respective image use policies [65].

Euler's moment equations in Equation (2) are elaborated for the flexible space robot in Equation (8) and resembled in Equation (9) to more closely resemble the basic expression of Newton's Law, where variables and nomenclature are conveniently defined in Table 3.

$$I_{zz}\ddot{\theta} + \sum_{i=1}^{n} D_i\ddot{q}_i + I_w\ddot{\theta}_w = T_D \quad (8)$$

$$I_{zz}\ddot{\theta} + \sum_{i=1}^{n} D_i\ddot{q}_i = \sum T \quad (9)$$

Table 2. Table of proximal variables and nomenclature [1].

Variable/Acronym	Definition	Variable/Acronym	Definition
$\hat{y}$	Centerline unit vector	k	Appendage stiffness
T_C	Control torque	$m_i, I_i \forall i = 2 \ldots 5$	Flexible masses and inertias
J_1	Main body inertia mass moment	J_2	Flexible inertia mass moment
θ_1	Main body rotation angle	$\theta_i, W_i \forall i = 2 \ldots 5$	Translation and rotation angles

[1] Such tables are offered throughout the manuscript to aid readability.

Table 3. Table of proximal variables and nomenclature [1].

Variable/Acronym	Definition
$F\vert_{inertial}, T\vert_{inertial}$	Externally applied force and torque expressed in inertial coordinates
F, T	Externally applied force and torque expressed in non-inertial coordinates
m, J	Body's mass and mass moment of inertia
$a\vert_{inertial} = \dot{v}\vert_{inertial} = \ddot{x}\vert_{inertial}$	Resulting accelerations expressed in inertial coordinates
$\omega, \dot{\omega}$	Angular velocity and acceleration vectors
$x_1, x_2; \ddot{x}_1, \ddot{x}_2$	Translational velocity and acceleration vectors
k_1, k_2	Flexible member stiffnesses
$[M], [K]$	Assembled matrices of masses and stiffnesses

[1] Such tables are offered throughout the manuscript to aid readability.

Table 4. Table of proximal variables and nomenclature [1].

Variable/Acronym	Definition
I_{zz}	Body principal moment of inertia with respect to Z-axis
$\ddot{\theta}$	Angular acceleration of the system rotation angle, θ
D	Rigid–elastic coupling term
$\ddot{q}$	Acceleration in generalized displacement coordinates
I_w	Reaction wheel principal moment of inertia with respect to C, Z axis
$\ddot{\theta}_W$	Angular acceleration of the reaction wheel rotation angle, θ_W
T	Control torque of the spacecraft reaction wheel
T_D	Disturbance torques

[1] Such tables are offered throughout the manuscript to aid readability.

Isolating the first term of Equation (9) leads to Equation (10), and slight arithmetic leads to Equation (11).

$$\ddot{\theta} + \frac{\sum_{i=1}^{n} D_i}{I_{zz}} \ddot{q}_i = \frac{\sum T}{I_{zz}} \tag{10}$$

$$\ddot{\theta} = \frac{\sum T}{I_{zz}} - \frac{\sum_{i=1}^{n} D_i}{I_{zz}} \ddot{q}_i \tag{11}$$

Detailed implementation on the flexible space robot depicted in Figure 6 is included in the appendix to aid repeatability. Mode shapes and (constant) natural vibrational frequencies are obtained by spectral decomposition (i.e., the eigenvalue problem). Meanwhile, mass and moments of inertia (locations of mass) vary, leading to time-invariant frequenices and shapes, thus motivating the proposals in this manuscript with the anticipation of future research into deterministic artificial intelligence.

2.2. Competing Control Design Methodologies

Many options are available in the literature to the robot designer to control the highly flexible system in space. The immediate prequels elaborated gain stabilization, classical second-order structural filtering, input shaping, and whiplash compensation (where whiplash trajectory shaping is implemented in this study). Meanwhile, this present manuscript iterates several remaining trajectory-shaping options: rigid-body minimum-fuel trajectory shaping, single-frequency sinusoid, and options for flattening the magnitude response plot, while establishing some groundwork to prepare for future efforts with deterministic artificial intelligence.

1. Gain stabilization,
2. Classical second-order structural filtering,
3. Input shaping,
4. Whiplash compensation,
5. Rigid-body minimum-fuel input trajectory shaping,
6. Single-frequency trajectory shaping,
7. Flatten the curve to improve stability,
8. Flatten the curve to improve trajectory tracking,
9. Deterministic artificial intelligence:
    9.1 Self-awareness statements, and
    9.2 Adaption or optimal learning.

2.3. Selectable Options: Trajectories, Feedforward, Feedback, and Filtering

Figure 7 depicts simulations created in SIMULINK® including subsystems for the selectable commanded trajectory, selectable feedforward controls, feedback controller, structural filters, and the selection subsystem to activate feedforward, feedback, and structural filtering. Those subsystems are fed to control the flexible space robot's subsystem in a unit-feedback loop resulting in a displayable rotation angle.

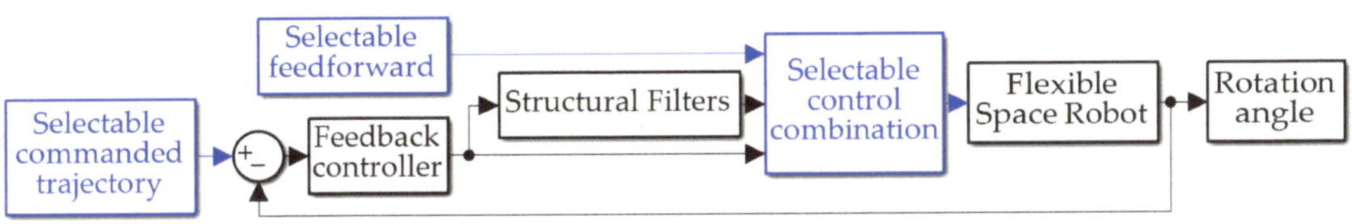

Figure 7. Topology of simulation created in Simulink®.

2.3.1. Commanded Trajectories

The Simulink® subsystem used to select between commanded trajectories is displayed in Figure 8a including ubiquitous step commands, rigid-body control-minimizing optimal commands, whiplash compensation, time-delay input-shaped trajectories, and single sinusoidal commanded trajectories. Figure 8b displays a subsystem used to formulate trajectories that are non-zero only when maneuvering. Meanwhile, Figure 8c,d display notion subsystem outputs.

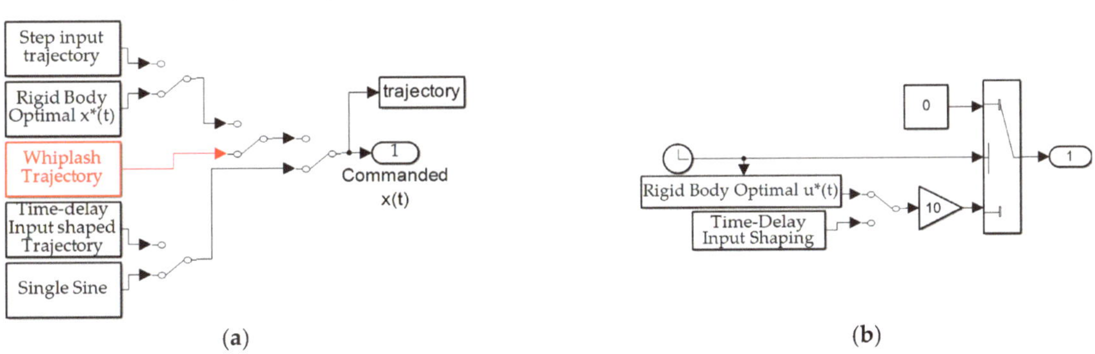

Figure 8. *Cont.*

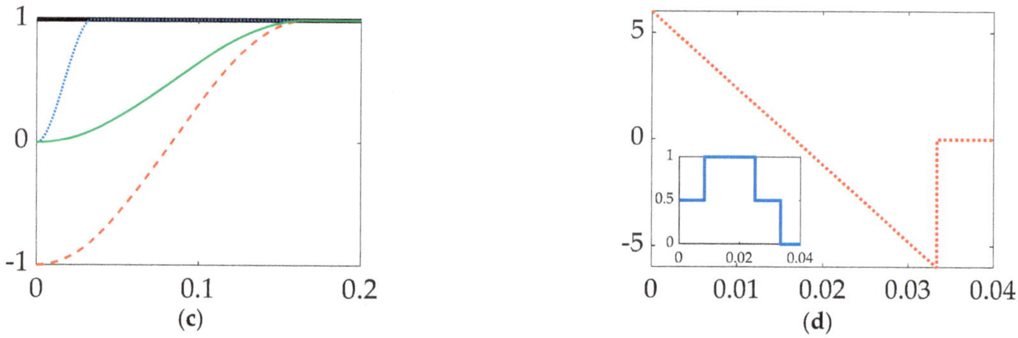

Figure 8. Topology of simulation created in Simulink®: (**a**) selectable command trajectory; (**b**) selectable feedforward; (**c**) selectable commanded trajectories: the thick solid black line is the unit step function, the rigid-body minimum fuel is the thin solid green line, ***bio–inspired whiplash trajectory*** is the red dashed line, and the single sinusoidal commanded trajectory is the dotted blue line. (**d**) Feedforward controls: minimum fuel optimal feedforward is the dashed red line, while time-delay input-shaped feedforward is the solid blue line in the inset plot.

2.3.2. Feedback Filtering

Filtering feedback controls is simulated as depicted in Figure 9a, where the iteration of controls is achieved by a subsystem of manual switches depicted in Figure 9b. Activation of all structural filtering (four bandpass filters placed at the spectral location of four anti-resonances plus four notch filters placed at the spectral location of four resonance frequencies) results in modifying the frequency response magnitude plot depicted in Figure 10a,b, where the spikes and dips of resonance and anti-resonance, respectively, have been smoothed by the structural filters. Former analysis designed these filters primarily using stability as motivation, while this present research evaluates several figures of merit including trajectory tracking and control usage. Figure 10c,d, respectively, depict notion impacts of notch and bandpass filters on the frequency response magnitude plot (i.e., Bode plot's magnitude), where Table 4 provides convenient definitions of proximal variables and nomenclature.

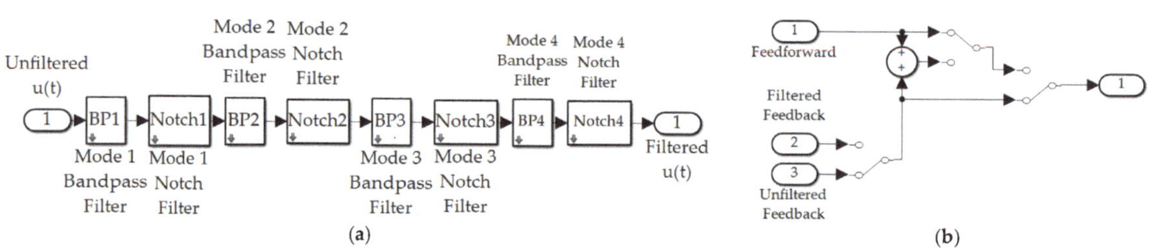

Figure 9. Topology of simulation created in Simulink®: (**a**) structural filters; (**b**) selectable control combination.

$$\frac{Output(s)}{Input(s)} = \frac{\frac{s^2}{\omega_z^2} + \frac{2\zeta_z}{\omega_z^2}s + 1}{\frac{s^2}{\omega_p^2} + \frac{2\zeta_p}{\omega_p^2}s + 1} \tag{12}$$

$$K_\infty = 40log_{10}(\omega_p/\omega_z) \tag{13}$$

$$\phi_{max} = cos^{-1}\left[\frac{\left(2\zeta_c\sqrt{\omega_p/\omega_z}\right)^2 - (\omega_p/\omega_z - 1)^2}{\left(2\zeta_c\sqrt{\omega_p/\omega_z}\right)^2 + (\omega_p/\omega_z - 1)^2}\right] \tag{14}$$

$$K_{max} = 20log_{10}(\zeta_z/\zeta_p)\ dB \tag{15}$$

$$\omega_1/\omega_c = \sqrt{2\zeta_z\zeta_p + 1 - \sqrt{(2\zeta_z\zeta_p + 1)^2 - 1}} \tag{16}$$

$$\omega_2/\omega_c = \sqrt{2\zeta_z\zeta_p + 1 + \sqrt{(2\zeta_z\zeta_p + 1)^2 - 1}} \tag{17}$$

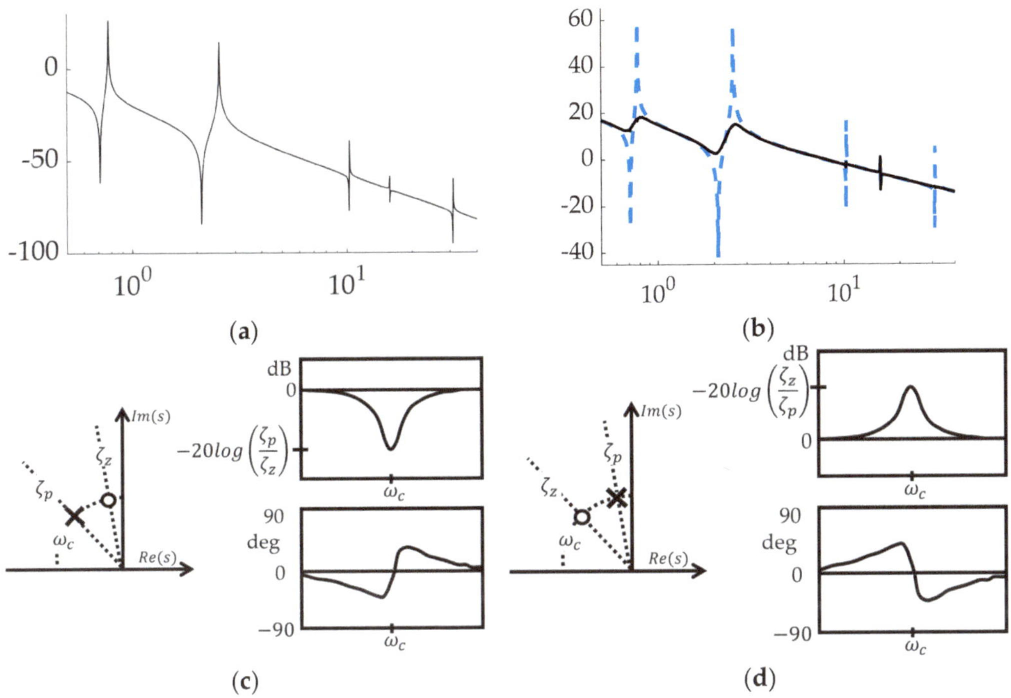

Figure 10. Feedback filtering for structural resonances and anti-resonances: (**a**) frequency response plot of unfiltered, PID controlled space robot with decibel frequency on the abscissa and response magnitude on the ordinant; (**b**) flattened curve, frequency response plot when all four modes are filtered with both bandpass and notch filters with decibel frequency on the abscissa and response magnitude on the ordinant; (**c**) second-order notch filters optionally applied at resonances with decibel frequency on the abscissa and response magnitude on the ordinant of the right-hand subplot with Real (Re) and Imaginary (Im) parts displayed in the left-hand plot; (**d**) second-order bandpass filters optionally applied to anti-resonances with decibel frequency on the abscissa and response magnitude on the ordinant of the right-hand subplot with Real (Re) and Imaginary (Im) parts displayed in the left-hand plot.

Table 5. Table of proximal variables and nomenclature [1].

Variable/ Acronym	Definition
$Im(s)$	Imaginary component of transient response
$Re(s)$	Real component of transient response
ζ_p	Damping ratio of pole in denominator of Equation (12)
ζ_z	Damping ratio of zero in numerator of Equation (12)
ω_c	Center frequency of filter placement
ω_p	Center frequency of filter pole placement in denominator of Equation (12)
ω_z	Center frequency of filter zero placement in numerator of Equation (12)
dB	Decibels

Table 5. *Cont.*

Variable/ Acronym	Definition
log	Base-10 logarithm
$Output(s)$	Displacement or rotation expressed in Laplace domain
$Input(s)$	Control force or torque expressed in Laplace domain
K_∞	Steady state gain
ϕ_{max}	Maximum phase lead occurring at frequencies determined by ζ_z and ζ_p
K_{max}	Maximum gain occurring when $\omega_p = \omega_p$

[1] Such tables are offered throughout the manuscript to aid readability.

2.3.3. Feedforward Controls

Feedforward controls are strictly taken as the self-awareness statements in deterministic artificial intelligence, where the intension is to use this manuscript as benchmarks for the seminal development of deterministic artificial intelligence for flexible space robotics in the sequel. For example, presently Equation (8) is modified to Equation (18) by prescribing the motion states (using the chosen commanded trajectories) and using time-invariant estimates of the physical parameters (e.g., mass, mass moments, stiffnesses, and eventually damping). The sequel will modify Equation (9) is modified to Equation (19) by prescribing the motion states (using the chosen commanded trajectories) and using time-invariant estimates of the physical parameters (e.g., mass, mass moments, stiffnesses, and eventually damping). develop adaption and learning methods for the estimates making them time varying. Follow-on work will combine the transport theorem in Equation (18) with the rigid–elastic coupled system in Equation (19) which should permit deterministic artificial intelligence to learn time-varying natural frequencies stemming from time-varying mass, mass moments, stiffnesses, and damping.

$$T \equiv \hat{J}\dot{\omega}_d + \omega_d \times \hat{J}\omega_d \tag{18}$$

$$\hat{I}_{zz}\ddot{\theta}_d + \sum_{i=1}^{n} \hat{D}_i (\ddot{q}_i)_d \equiv \sum \boldsymbol{T} \tag{19}$$

Selectable options for designing commanded trajectories, feedforward and feedback controls, and structural filtering were elaborated in Section 2.3. Presently concluding the Methods and Materials, Section 3 next presents the parameters used in the simulation experiments, and then presents the results of many experiments.

3. Results

This section provides a concise and precise description of the experimental results favoring multi-plots with accompanying tables of comparative figures of merit to aid the readership's efforts ascertaining the relative efficacy of the approaches presented. Presented next is the data interpretation, as well as the experimental conclusions that can be drawn, culminating in a very large table of thirteen of the best approaches (of the twenty-six approaches iterated), displayed with comparative figures of merit. Mini summaries are provided, allowing the reader to discard thoughts of relatively inferior methods, while continuing to the next set of comparisons, eventually narrowing to a grouping of the best available options of those surveyed. The simulation parameters are provided in Table 5 to aid repeatability of the presented results.

3.1. *Comparing Commanded Trajectories with Unfiltered Feedback*

The first options experiments compared disparate options for comparing commanded trajectories with unfiltered, classical PID feedback control. Space robot rotation angles are displayed versus time (scaled to unity) in Figure 11 with corresponding figures of merit in Table 6 revealing obviously superior tracking performance of using single-sinusoidally shaped commanded trajectories without feedforward control and unfiltered classical PID feedback control. Orders-of-magnitude improvements in trajectory tracking mean error and deviations is achievable with more than three-fold increase in fuel utilization (i.e., control effort).

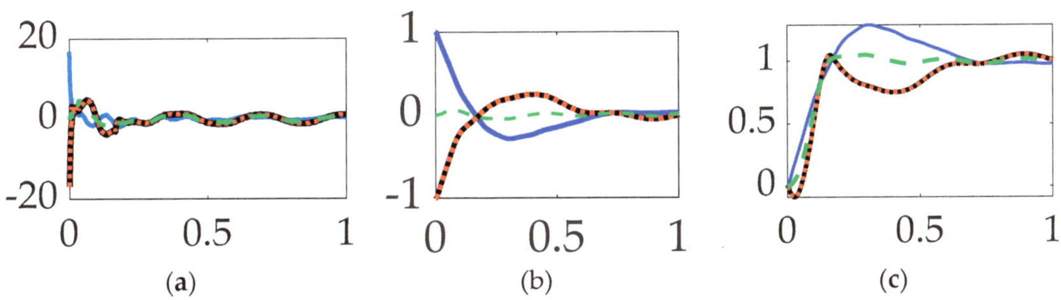

Figure 11. Trajectory command comparative simulation experiments performed in Simulink® with normalized time on the abscissae. The solid blue line indicates step trajectory, no feed-forward, unfiltered; the solid black line indicates whiplash trajectory, no feedforward, unfiltered; the red dotted line indicates time-delayed input-shaped trajectory, no feedforward, unfiltered, where the ordinants display: (**a**) control in [Newton meters], (**b**) tracking error in [degrees], and (**c**) rotation angle in [degrees]. Qualitative results correspond to quantitative figures of merit in Table 6. Rigid-body minimum-fuel input trajectory shaping performed so poorly as to not be presentable.

Table 6. Trajectory command comparative simulation experiments performed in Simulink®. Quantitative figures of merit correspond to qualitative results in Figure 11.

Control Methods [1]	Control Effort	Tracking Error Mean	Tracking Error Deviation
Step trajectory, no feedforward, unfiltered	**0.27662**	0.025967	0.29883
Bio-inspired whiplash trajectory, no feedforward, unfiltered	1.0997	−0.026376	0.2936
Time-delayed input-shaped trajectory, no feedforward, unfiltered	1.0997	−0.026376	−0.27936
Single-sinusoid trajectory, no feedforward, unfiltered	0.69228	**0.00052658**	**0.025702**

[1] Rigid-body minimum-fuel input trajectory shaping performed so poorly as to not be presentable.

Interim summary. *When comparing commanded trajectories, step trajectories surprisingly led to the least control effort, while single-sinusoid trajectories produce the most accurate tracking, with 150% more control effort.* ***Bio-inspired whiplash compensation performed essentially as well as time-delayed input shaping.***

3.2. Comparing Feedforward Controls with Unfiltered Feedback

Having compared trajectory-shaping options in Section 3.1, this section includes the results of direct comparison of disparate options for feedforward control in the presence of unfiltered, classical PID feedback control. Figure 12 reveals that relatively superlative performance is obtained by using single-sinusoidally shaped, commanded trajectories with time-delay input-shaped feedforward control in the presence of unfiltered classical feedback controls, where figures of merit in Table 7 reveal the increased trajectory tracking performance necessitates a non-trivial increase in fuel expenditure (control effort).

Interim summary. When comparing feedforward controls, rigid-body optimal feedforward with step trajectory command surprisingly led to the least control effort, while time-delay input-shaped feedforward with single-sinusoid trajectories commanded produced the most accurate tracking, with 280% more control effort.

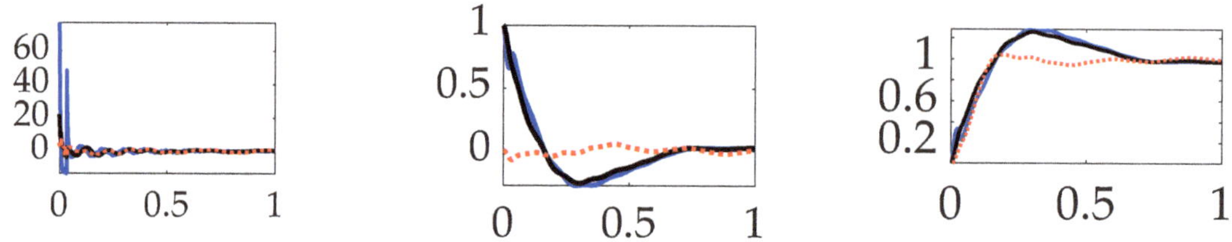

(**a**) (**b**) (**c**)

Figure 12. Selectable feedforward comparative simulation experiments performed in Simulink® with normalized time on the abscissae. The solid blue line indicates step trajectory, rigid-body optimal feed-forward, unfiltered; the solid black line indicates step trajectory, time-delay input shaping feedforward, unfiltered; the red dotted line indicates single-sine trajectory, time-delay input shaping feedforward, unfiltered, where the ordinants display: (**a**) control in [Newton meters], (**b**) tracking error in [degrees], and (**c**) rotation angle in [degrees]. Qualitative results correspond to quantitative figures of merit in Table 7. Single-sinusoid trajectory, rigid-body optimal feedforward, unfiltered performed so poorly as to not be presentable.

Table 7. Selectable feedforward comparative simulation experiments performed in Simulink®. Quantitative figures of merit correspond to qualitative results in [degrees].

Control Methods [1]	Control Effort	Tracking Error Mean	Tracking Error Deviation
Step trajectory, rigid-body optimal feedforward, unfiltered	**0.15916**	0.030193	0.29752
Step trajectory, time-delay input-shaped feedforward, unfiltered	0.211539	0.026717	0.27847
Single-sinusoid trajectory, rigid-body optimal feedforward, unfiltered	294.3845	6.1359	3.9368
Single-sinusoid trajectory, time-delay input-shaped feedforward, unfiltered	0.61639	**−0.0014197**	**0.026812**

[1] Rigid-body minimum-fuel input trajectory shaping performed so poorly as to not be presentable.

3.3. Comparing Commanded Trajectories with Filtered Feedback

Having discerned the advantages of single-sinusoidally shaped, commanded trajectories and time-delay input-shaped feedforward control, this section repeats the comparison in Section 3.1 (which included only unfiltered feedback), but this time iterates the options when structural filters are use. Figure 13 reveals the lowest control effort (i.e., fuel usage) with nominal tracking performance using unshaped step commands, no feedforward controls with structurally filtered feedback; however, substantially improved target tracking performance is achievable at using higher control efforts by commanding optimal (fuel-minimizing) trajectories that are constrained by rigid-body dynamics equations.

Interim summary. When comparing commanded trajectories with filtered feedback and no feedforward, step trajectory commands surprisingly led to the least control effort, while rigid-body optimal trajectories achieved an order of magnitude higher accuracy with 2345% more control effort.

Table 8. Selectable commanded trajectories with filtered feedback (and no feedforward) comparative simulation experiments performed in Simulink®. Quantitative figures of merit correspond to qualitative results in Table 8.

Control Methods [1]	Control Effort	Tracking Error Mean	Tracking Error Deviation
Step trajectory, no feedforward, filtered feedback	**0.028957**	0.026908	0.30033
Rigid-body optimal trajectory, no feedforward, filtered feedback	3.0251	**0.0020571**	**0.040348**
Bio-inspired whiplash, no feedforward, filtered feedback	0.70804	−0.023355	0.27816
Single-sine trajectory, no feedforward, filtered feedback	0.70804	−0.023355	0.27816

[1] Rigid-body minimum-fuel input trajectory shaping performed so poorly as to not be presentable.

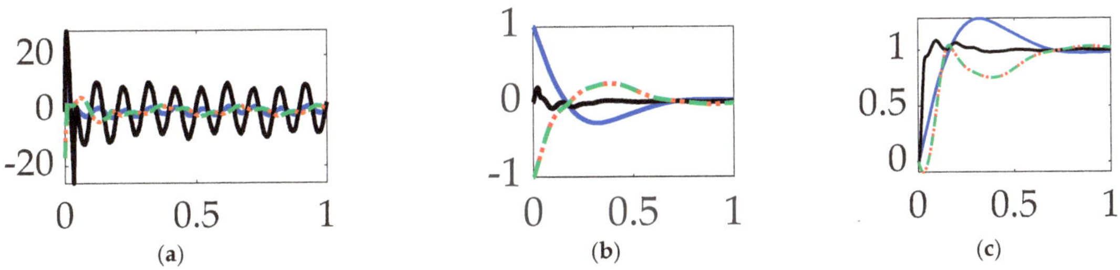

Figure 13. Selectable commanded trajectory comparative simulation experiments performed in Simulink® with normalized time on the abscissae. The solid blue line indicates step trajectory, no feedforward, filtered feedback; the solid black line indicates rigid-body optimal trajectory, no feedforward, filtered feedback; the red dotted line indicates whiplash trajectory, no feedforward, filtered feedback; and the green dashed line indicates single-sine trajectory, no feedforward, filtered feedback, where the ordinants display: (**a**) control in [Newton meters], (**b**) tracking error in [degrees], and (**c**) rotation angle in [degrees]. Qualitative results correspond to quantitative figures of merit in Table 8.

3.4. Comparing Mode 1 Filtering with Single-Sinusoidal Trajectories and No Feedforward

The final sections of the manuscript present experiments with iterated structural filtering: mode 1, mode 2, mode 3, mode 4, and then all of modes 1–4. Single-sinusoidally commanded trajectories are carried over from Sections 3.1–3.3, identified as a burgeoning best practice. This section iterates compensation of the lowest (first) flexible mode by sequentially compensating for the anti-resonance, the resonance, and then both modal features. The prequel literature predominantly designs structural filters for the sake of stability, often leading to emphasizing notch filtering the first resonant peak, while the results here (designed for target tracking performance rather than stability) indicate superior performance is obtainable by only compensating for the first anti-resonance frequency. This assertion is buttressed by the qualitative results in Figure 14, the quantitative companion of which is displayed in Table 9. Selectable mode 1 feedback filtering comparative simulation experiments performed in Simulink®. Quantitative figures of merit correspond to qualitative results in Figure 14′s display of meaningful performance figures of merit.

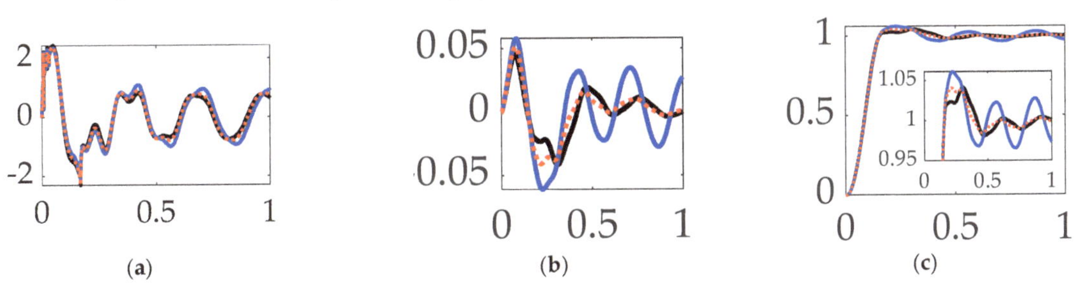

Figure 14. Selectable mode 1 feedback filtering comparative simulation experiments performed in Simulink® with normalized time on the abscissae. The solid black line indicates single-sine trajectory, no feedforward, mode 1 bandpass filtered; the solid blue line indicates single-sine trajectory, no feedforward, mode 1 notch filtered; the red dotted line indicates single-sine trajectory, no feedforward, mode 1 bandpass and notch filtered, where the ordinants display: (**a**) control in [Newton meters], (**b**) tracking error in [degrees], and (**c**) rotation angle in [degrees]. Qualitative results correspond to quantitative figures of merit in Table 8.

Interim summary. *When comparing mode 1 filtering options bandpass only not surprisingly led to the least control effort, but surprisingly also produced the most accurate tracking.*

Table 9. Selectable mode 1 feedback filtering comparative simulation experiments performed in Simulink®. Quantitative figures of merit correspond to qualitative results in Figure 14.

Single-Sine Trajectory, No Feedforward, Iterated Feedback Filtering	Control Effort	Tracking Error Mean	Tracking Error Deviation
Mode 1 bandpass filtered	**0.62207**	**0.00046139**	**0.018780**
Mode 1 notch filtered	0.89711	0.00046872	0.029906
Mode 1 Bandpass and notch filtered	0.71577	0.00058161	0.020662

3.5. Comparing Mode 2 Filtering with Single-Sinusoidal Trajectories and No Feedforward

Similar to Section 3.4′s comparison of the compensation of the first mode, this section sequentially compensates for the second mode's anti-resonance, resonance, and then both modal components. The results are displayed in Figure 15 which reveal marginal qualitative differences that are validated by quantitative results in Table 10.

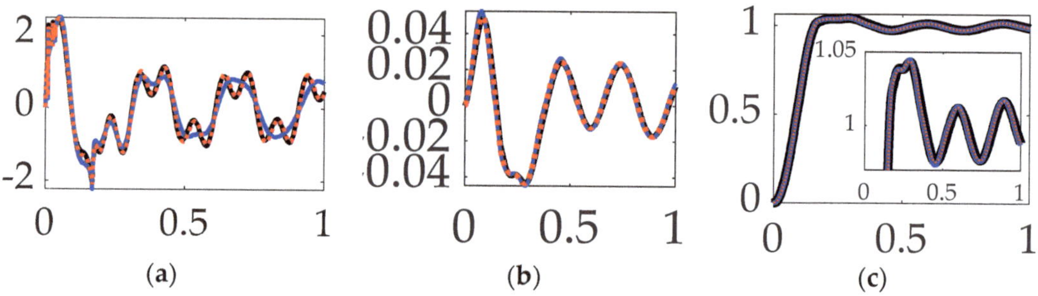

Figure 15. Selectable mode 2 feedback filtering comparative simulation experiments performed in Simulink® with normalized time on the abscissae. The solid black line indicates single-sine trajectory, no feedforward, mode 2 bandpass filtered; the solid blue line indicates single-sine trajectory, no feedforward, mode 2 notch filtered; the red dotted line indicates single-sine trajectory, no feedforward, mode 2 bandpass and notch filtered, where the ordinants display: (**a**) control in [Newton meters], (**b**) tracking error in [degrees], and (**c**) rotation angle in [degrees]. Qualitative results correspond to quantitative figures of merit in Table 10.

Table 10. Selectable mode 2 feedback filtering comparative simulation experiments performed in Simulink®. Quantitative figures of merit correspond to qualitative results in Figure 15.

Single-Sine Trajectory, No Feedforward, Iterated Feedback Filtering	Control Effort	Tracking Error Mean	Tracking Error Deviation
Mode 2 bandpass filtered	0.47683	0.00026452	**0.022876**
Mode 2 notch filtered	0.68856	**0.00013103**	0.023599
Mode 2 Bandpass and notch filtered	**0.41301**	0.00016559	0.023188

Interim summary. When comparing mode 2 filtering options bandpass and notch led to the least control effort, but surprisingly accurate tracking results were inconsistent.

3.6. Comparing Mode 3 Filtering with Single-Sinusoidal Trajectories and No Feedforward

Continuing the evaluation to the third structural mode, like Sections 3.4 and 3.5, this paragraph presents the experiment results of sequentially compensating for the third mode's anti-resonance, resonance, and then both modal components. Similar to the results achieved compensating for the first resonant mode, compensation of the third mode's bandpass alone proved superior. In this instance, the results are clearly marginal, evidenced in the qualitative displays of Figure 16, where the small differences are quantitatively displayed in Table 11.

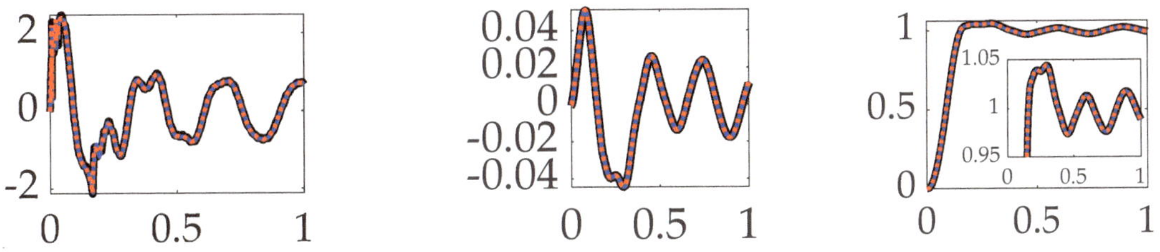

(a) (b) (c)

Figure 16. Selectable mode 3 feedback filtering comparative simulation experiments performed in Simulink® with normalized time on the abscissae. The solid black line indicates single-sine trajectory, no feedforward, mode 3 bandpass filtered; the solid blue line indicates single-sine trajectory, no feedforward, mode 3 notch filtered; the red dotted line indicates single-sine trajectory, no feedforward, mode 3 bandpass and notch filtered, where the ordinants display: (**a**) control in [Newton meters], (**b**) tracking error in [degrees], and (**c**) rotation angle in [degrees]. Qualitative results correspond to quantitative figures of merit in Table 10.

Table 11. Selectable mode 3 feedback filtering comparative simulation experiments performed in Simulink®. Quantitative figures of merit correspond to qualitative results in Figure 16.

Single-Sine Trajectory, No Feedforward, Iterated Feedback Filtering	Control Effort	Tracking Error Mean	Tracking Error Deviation
Mode 3 bandpass filtered	**0.67211**	**0.00015387**	**0.023116**
Mode 3 notch filtered	0.68804	0.00025072	0.023377
Mode 3 Bandpass and notch filtered	0.68986	0.00017818	0.023274

Interim summary. When comparing mode 3 filtering options bandpass led to the least control effort and also produced the most accurate tracking accuracy.

3.7. Comparing Mode 4 Filtering with Single-Sinusoidal Trajectories and No Feedforward

The three immediately preceding sections of this manuscript iteratively investigated the first three structural modes, while this paragraph iteratively compensates for the modal components of the fourth structural mode: the anti-resonance, resonance, and then both. Figure 17 (like Figure 16′s qualitative results) reveal marginal results, where inspection of the quantitative results in Table 12 indicate the repeated trend: compensation for the anti-resonance has the largest impact on target tracking errors.

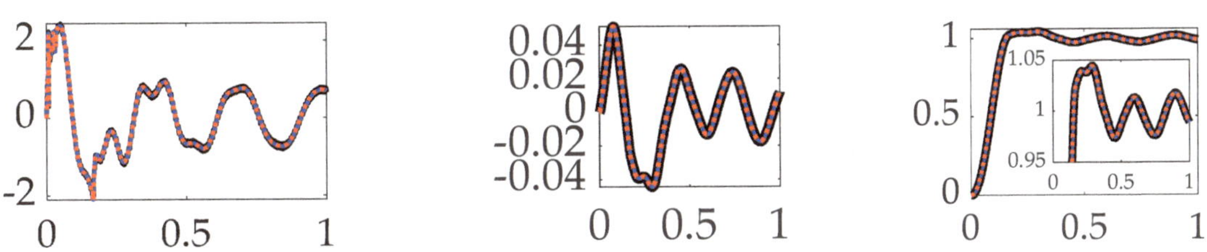

(a) (b) (c)

Figure 17. Selectable mode 4 feedback filtering comparative simulation experiments performed in Simulink® with normalized time on the abscissae. The solid black line indicates single-sine trajectory, no feedforward, mode 4 bandpass filtered; the solid blue line indicates single-sine trajectory, no feedforward, mode 4 notch filtered; the red dotted line indicates single-sine trajectory, no feedforward, mode 4 bandpass and notch filtered, where the ordinants display: (**a**) control in [Newton meters], (**b**) tracking error in [degrees], and (**c**) rotation angle in [degrees]. Qualitative results correspond to quantitative figures of merit in Table 12.

Table 12. Selectable mode 4 feedback filtering comparative simulation experiments performed in Simulink®. Quantitative figures of merit correspond to qualitative results in Figure 12.

Single-Sine Trajectory, No Feedforward, Iterated Feedback Filtering	Control Effort	Tracking Error Mean	Tracking Error Deviation
Mode 4 bandpass filtered	**0.65649**	**0.00075748**	**0.023116**
Mode 4 notch filtered	0.69137	0.00330010	0.023697
Mode 4 Bandpass and notch filtered	0.68994	0.00094136	0.023266

Interim summary. When comparing mode 4 filtering options bandpass led to the least control effort and also produced the most accurate tracking accuracy.

3.8. Comparing Modes 1–4 Filtering with Single-Sinusoidal Trajectories and No Feedforward

Sections 3.1–3.7 iteratively examined compensation of individual modal components, while this section simultaneously compensates for all four modes' anti-resonances followed by all round modes' resonances. Definite differences are immediately apparent in Figure 18, where the quantitative figures of merit in Table 13 re-validate the discoveries of Sections 3.1–3.7: compensation for the anti-resonance alone has the biggest impact on target tracking performance (with a reminder: the opposite result comes from compensating for stability rather than tracking performance).

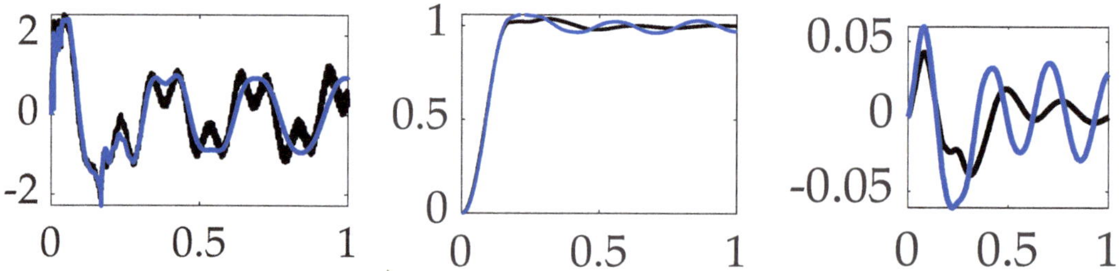

(**a**) (**b**) (**c**)

Figure 18. Selectable modes 1–4 feedback filtering comparative simulation experiments performed in Simulink® with normalized time on the abscissae. The solid black line indicates single-sine trajectory, no feedforward, modes 1–4 bandpass filtered; the solid blue line indicates single-sine trajectory, no feedforward, modes 1–4 notch filtered, where the ordinants display: (**a**) control in [Newton meters], (**b**) tracking error in [degrees], and (**c**) rotation angle in [degrees]. Qualitative results correspond to quantitative figures of merit in Table 13.

Table 13. Selectable mode feedback filtering comparative simulation experiments performed in Simulink®. Quantitative figures of merit correspond to qualitative results in Figure 13.

Single-Sine Trajectory, No Feedforward, Iterated Feedback Filtering	Control Effort	Tracking Error Mean	Tracking Error Deviation
Modes 1–4 bandpass filtered	**0.22672**	**0.0010466**	**0.017807**
Modes 1–4 notch filtered	**0.90941**	**0.0038787**	**0.030924**

Interim summary. When comparing modes 1–4 filtering options bandpass not only led to the least control effort but also produced the most accurate tracking accuracy.

3.9. Comparison of the Best Options Studies

This paragraph assembles and compares thirteen common options for controlling highly flexible space robotics providing advice to the readership: should filtered or unfiltered feedback be used? Should feedforward techniques be considered? Is command trajectory shaping effective? From the experiments presented, interesting mixtures of options illustrate efficacy. Table 14 contains a summary of figures of merit achieved by thirteen disparate combinations of available options: input shaping, feedforward and/or feedback control and optional structural filtering.

Table 14. Comparison of performance figures of merit: effort and tracking errors.

Control Methods [1]	Control Effort	Tracking Error Mean	Tracking Error Deviation
Bio-inspired whiplash trajectory, no feedforward, unfiltered feedback	1.0997	–0.026376	0.2936
Bio-inspired whiplash, no feedforward, filtered feedback	0.70804	–0.023355	0.27816
Rigid-body optimal trajectory, no feedforward, filtered feedback	3.0251	0.0020571	0.040348
Time-delayed input-shaped trajectory, no feedforward, unfiltered	1.0997	–0.026376	–0.27936
Step trajectory, no feedforward, unfiltered	0.27662	0.025967	0.29883
Step trajectory, no feedforward, filtered feedback	0.028957	0.026908	0.30033
Step trajectory, rigid-body optimal feedforward, unfiltered	**0.15916**	0.030193	0.29752
Single-sinusoid trajectory, no feedforward, unfiltered	0.69228	0.00052658	0.025702
Single-sinusoid trajectory, time-delay input-shaped feedforward, unfiltered	0.61639	–0.0014197	0.026812
Sinusoidal trajectories, no feedforward, mode 1 bandpass filtered	0.62207	0.00046139	0.018780
Sinusoidal trajectories, no feedforward, mode 2 bandpass filtered	0.47683	0.00026452	0.022876
Sinusoidal trajectories, no feedforward, mode 2 notch filtered	0.68856	**0.00013103**	0.023599
Sinusoidal trajectories, no feedforward, mode 2 bandpass and notch filtered	0.41301	0.00016559	0.023188
Sinusoidal trajectories, no feedforward, mode 3 bandpass filtered	0.67211	0.00015387	0.023116
Sinusoidal trajectories, no feedforward, mode 4 bandpass filtered	0.65649	0.00075748	0.023116
Sinusoidal trajectories, no feedforward, mode 1–4 bandpass filtered	0.22672	0.0010466	**0.017807**
Average	0.658023	0.007386	0.087848

[1] Rigid-body minimum-fuel input trajectory shaping performed so poorly as to not be presentable.

Interim summary. The least control effort was achieved with step trajectories, rigid-body optimal feedforward control and unfiltered feedback, and the effort was 75% less than the average. The best mean tracking error was achieved with sinusoidal trajectories, no feedforward, mode 2 notch filtered, and the tracking error mean was 98% better than the average, while the control effort was 330% higher than the minimum available option. The best tracking error deviation was achieved with sinusoidal trajectories, no feedforward, mode 1–4 bandpass filtered, and the tracking error deviation was 80% better than the average, while the control effort was 42% higher than the minimum available option.

4. Discussion

Pictures of the machine design are replicated in Figure 19 as a late reminder. Next, this section discusses the maneuver results and how they can be interpreted from the perspective of previous studies and of the working hypotheses. The findings and their implications are discussed in the broadest context possible. Future research directions are also highlighted.

The top seven revelations follow:

1. Bio-inspired trajectory shaping (modified from a time-minimization control per [43]) seems confounded in the presence of classical, unfiltered feedback. While the technique performed well, it was not the exemplary option when compared to the multitude of other available options examined.
2. When comparing commanded trajectories, step trajectories surprisingly led to the least control effort, while single-sinusoid trajectories produce the most accurate tracking, with 150% more control effort. The bio-inspired whiplash shaping was optimized in the cited literature for minimum time in a feedforward control sense, while this sequel reveals that the solution is not minimum effort (fuel), nor minimum time in the presence of feedback.
3. When comparing feedforward controls, rigid-body optimal feedforward with step trajectory command surprisingly led to the least control effort, while time-delay input-

shaped feedforward with single-sinusoid trajectories commanded produced the most accurate tracking, with 280% more control effort.

4. When comparing commanded trajectories with filtered feedback and no feedforward, step trajectory commands surprisingly led to the least control effort, while rigid-body optimal trajectories achieved an order of magnitude higher accuracy with 2345% more control effort.
5. When comparing mode 1 filtering options bandpass alone not surprisingly led to the least control effort, but surprisingly also produced the most accurate tracking.
6. When comparing mode 3 filtering options bandpass led to the least control effort and also produced the most accurate tracking accuracy.
7. When comparing mode 4 filtering options bandpass led to the least control effort and also produced the most accurate tracking accuracy.
8. The least control effort was achieved with step trajectories, rigid-body optimal feedforward control and unfiltered feedback, and the effort was 75% less than the average. The best mean tracking error was achieved with sinusoidal trajectories, no feedforward, mode 2 notch filtered, and the tracking error mean was 98% better than the average, while the control effort was 330% higher than the minimum available option. The best tracking error deviation was achieved with sinusoidal trajectories, no feedforward, mode 1–4 bandpass filtered, and the tracking error deviation was 80% better than the average, while the control effort was 42% higher than the minimum available option.

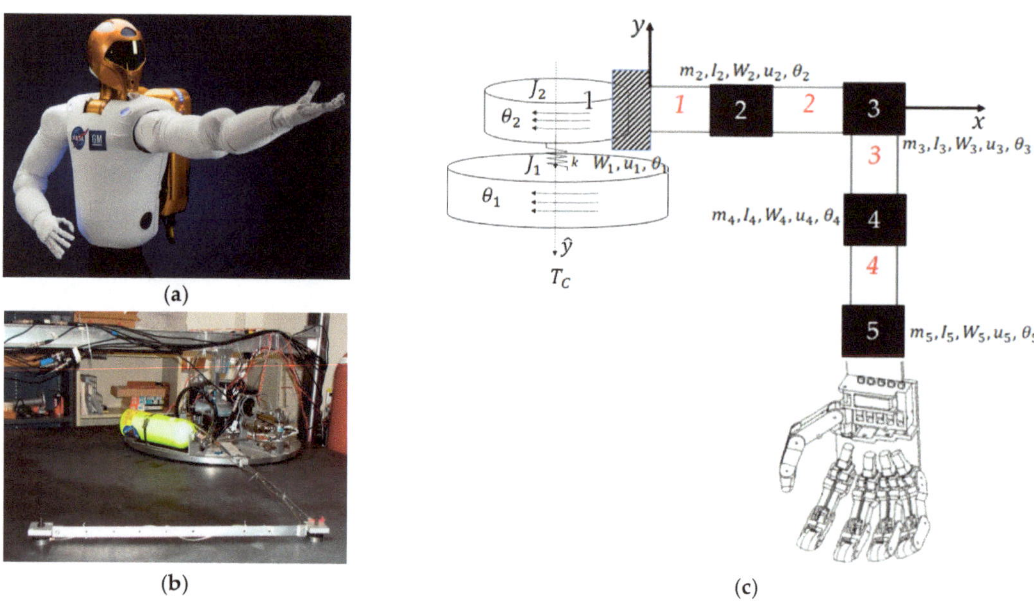

Figure 19. Space robots with cylindrical center rigid bodies and highly flexible appendages. (**a**) NASA's first humanoid space robot. Image credit: NASA [3]. (**b**) Laboratory flexible rotational spacecraft hub with a free-floating, planar air-bearing, very light robotic arm, the schematic of which is displayed in subfigure (**c**).

5. Conclusions

The research in [42] studied iterations of classical feedback control augmented with structural filters taken from the discipline of signal processing (e.g., notch filters and bandpass filters), where the options were iterated to maximize classical stability margins (i.e., gain margin and phase margin). Alternatively, the feedback-related sections of this investigation iterated the same options towards a goal of minimizing trajectory tracking error.

Meanwhile, open loop optimal feedforward in [43] was provided for minimizing the maneuver time without the presence of feedback or filtering. The results were bio-inspired, but the time-minimizing results did not rank in the top three performances in the present study seeking the best options for target tracking error.

5.1. Controversial or Unexpected Results

The study in [42] indicates the best selection of options to maximize classical stability margins (i.e., gain margin and phase margin) was attained by only compensating for the first complete flexible mode (resonance and anti-resonance) with notch and bandpass filters, respectively, with step commands shaped by novel single-sinusoidal trajectories.

Meanwhile, the results presented in the present manuscript indicate split results based on lowest control effort, least tracking error mean, or tracking error deviation.

5.1.1. Best Control Effort

For the best trajectory tracking measured by least control effort, unfiltered feedback (no notch or bandpass filters) is preferred when unshaped step inputs feed rigid-body optimal feedforward controls, as displayed in Table 15. Thus, unshaped rigid-body optimal feedforward with no filtering in the feedback channel is most appropriate for instances when fuel minimization is strictly required, particularly when traversing long trajectories from initial starting points towards targeted space items (as opposed to close-proximity operations).

Table 15. Top three performances in the present study.

Control Methods [1]	Control Effort	Tracking Error Mean	Tracking Error Deviation
Single-sinusoid trajectory, time-delay input-shaped feedforward, unfiltered	−6%	**−119%**	−69%
Step trajectory, rigid-body optimal feedforward, unfiltered	**−96%**	264%	242%
Sinusoidal trajectories, no feedforward, mode 1–4 bandpass filtered	−66%	−86%	**−80%**

[1] Rigid-body minimum-fuel input trajectory shaping performed so poorly as to not be presentable.

5.1.2. Best Tracking Error Mean

For the best trajectory tracking measured by the lowest mean, unfiltered feedback (no notch or bandpass filters) is the preferred step inputs and rigid-body optimal feedforward controls, as displayed in Table 15. Thus, step inputs with rigid-body optimal feedforward with unfiltered feedback are best for close-proximity operations, especially preparing for grasping operations.

5.1.3. Best Tracking Error Deviation

For the best trajectory tracking measured by lowest deviation, sinusoidally shaped commanded trajectories with no feedforward but bandpass- only filtering of modes should be used, as displayed in Table 15. During close-proximity operations, grasping necessitates precision pointing and minimal error deviation, and thus sinusoidal-shaped commanded trajectories with no feedforward and bandpass-only mode filtering should be used.

5.2. Recommended Future Reseach

A major motivation of the present study is to evaluate the various efficacies of disparate input shaping techniques available to the recently proposed deterministic artificial intelligence method that necessitates such.

Deterministic Artificial Intelligence

Following the introduction of the feedforward by Cooper/Heidlauf in 2017 [62] applied to chaotic circuits, the seminal introduction of the technique (including both normal optimal feedforward and feedback) was offered by Smeresky/Rizzo [63] applied to spacecraft, despite not yet bearing the name of the technique back in the year 2020. Following the publication of comparative prequels [42,43], the natural sequels are recommended here as future research. Utilize the trajectories evaluated here for spacecraft as inputs to the deterministic artificial intelligence method to formulate a seminal offering applied to highly flexible space robotics.

Funding: This research received no external funding.

Institutional Review Board Statement: Not applicable.

Data Availability Statement: Data supporting reported results can be obtained by contacting the corresponding author.

Conflicts of Interest: The author declares no conflict of interest.

Appendix A Elaboration of Modal System Identification on the Flexible Robot System

Substitute into Equation (19):

$$\ddot{\theta} + \frac{\sum_{i=1}^{n} D_i(-2\xi\omega_i I_{zz})}{I_{zz}(I_{zz} - D_i\sum_{i=1}^{n} D_i)}\dot{q}_i + \frac{\sum_{i=1}^{n} D_i(\omega_i^2 I_{zz})}{I_{zz}(I_{zz} - D_i\sum_{i=1}^{n} D_i)}q_i + \frac{\sum_{i=1}^{n} D_i(-D_i T)}{I_{zz}(I_{zz} - D_i\sum_{i=1}^{n} D_i)} = \frac{T}{I_{zz}} \quad \text{(A1)}$$

$$\ddot{\theta} = \frac{2\xi\omega_i\sum_{i=1}^{n} D_i}{I_{zz} - D_i\sum_{i=1}^{n} D_i}\dot{q}_i + \frac{\omega_i^2\sum_{i=1}^{n} D_i}{I_{zz} - D_i\sum_{i=1}^{n} D_i}q_i + \frac{T D_i\sum_{i=1}^{n} D_i}{I_{zz}(I_{zz} - D_i\sum_{i=1}^{n} D_i)} = \frac{T}{I_{zz}} \quad \text{(A2)}$$

$$\ddot{\theta} = \frac{2\xi\omega_i\sum_{i=1}^{n} D_i}{I_{zz} - D_i\sum_{i=1}^{n} D_i}\dot{q}_i + \frac{\omega_i^2\sum_{i=1}^{n} D_i}{I_{zz} - D_i\sum_{i=1}^{n} D_i}q_i + \frac{T(D_i\sum_{i=1}^{n} D_i + I_{zz} - D_i\sum_{i=1}^{n} D_i)}{I_{zz}(I_{zz} - D_i\sum_{i=1}^{n} D_i)} \quad \text{(A3)}$$

$$\ddot{\theta} = \frac{2\xi\omega_i\sum_{i=1}^{n} D_i}{I_{zz} - D_i\sum_{i=1}^{n} D_i}\dot{q}_i + \frac{\omega_i^2\sum_{i=1}^{n} D_i}{I_{zz} - D_i\sum_{i=1}^{n} D_i}q_i + \frac{T I_{zz}}{I_{zz}(I_{zz} - D_i\sum_{i=1}^{n} D_i)} \quad \text{(A4)}$$

$$\ddot{\theta} = \frac{2\xi\omega_i\sum_{i=1}^{n} D_i}{I_{zz} - D_i\sum_{i=1}^{n} D_i}\dot{q}_i + \frac{\omega_i^2\sum_{i=1}^{n} D_i}{I_{zz} - D_i\sum_{i=1}^{n} D_i}q_i + \frac{T}{(I_{zz} - D_i\sum_{i=1}^{n} D_i)} \quad \text{(A5)}$$

Recall the expressions for the rigid elastic coupling using modal coordinates: $D_i = \int_F \left(x_F\phi_i^y - y_F\phi_i^x\right)dm$, where ϕ's are mode shapes from finite element analysis using the eigenvalues of K/m (stiffness/mass). The system stiffness matrix is included in Table A1 and the mass matrix in Table A2, resulting in the natural frequencies and mode shapes for the flexible system in Tables A3 and A4.

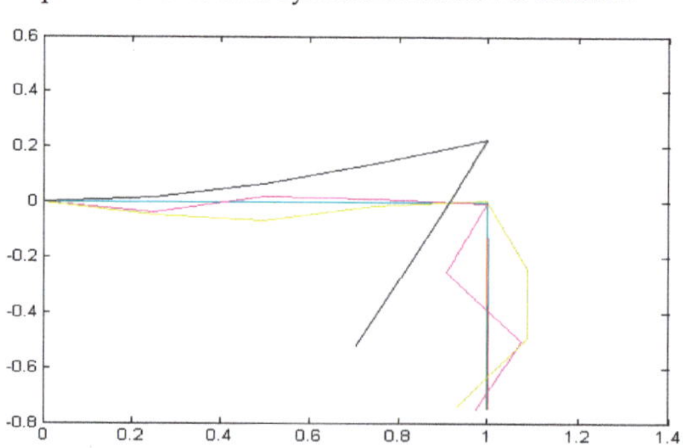

Figure A1. Normalized mode shapes (modal coordinates) displayed in physical coordinates with normalized length on the abscissa in meters and displacements in meters on the ordinant, where each mode shape is annotated by a different color.

Table A1. Stiffness matrix [K] [1].

	W2	θ2	W3	θ3	W4	θ4	W5	θ5	U6	θ6	u7	θ7	u8	θ8
W2	958.8179	0.0000	−479.409	59.9261	0	0	0	0	0	0	0	0	0	0
θ2	0.0000	19.9754	−59.9261	4.9938	0	0	0	0	0	0	0	0	0	0
W3	−479.409	−59.926	958.8179	0.0000	−479.409	59.9261	0	0	0	0	0	0	0	0
θ3	59.9261	4.9938	0.0000	19.9754	−59.9261	4.9938	0	0	0	0	0	0	0	0
W4	0	0	−479.409	−59.926	958.8179	0.0000	−479.409	59.9261	0	0	0	0	0	0
θ4	0	0	59.9261	4.9938	0.0000	19.9754	−59.9261	4.9938	0	0	0	0	0	0
W5	0	0	0	0	−479.409	−59.926	479.409	−59.926	0	0	0	0	0	0
θ5	0	0	0	0	59.9261	4.9938	−59.9261	19.9754	−59.9261	4.9938	0	0	0	0
U6	0	0	0	0	0	0	0	−59.926	958.8179	0.0000	−479.409	59.9261	0	0
θ6	0	0	0	0	0	0	0	4.9938	0.0000	19.9754	−59.9261	4.9938	0	0
U7	0	0	0	0	0	0	0	0	−479.409	−59.926	958.8179	0.0000	−479.409	59.9261
θ7	0	0	0	0	0	0	0	0	59.9261	4.9938	0.0000	19.9754	−59.9261	4.9938
U8	0	0	0	0	0	0	0	0	0	0	−479.409	−59.926	479.4089	−59.926
θ8	0	0	0	0	0	0	0	0	0	0	59.9261	4.9938	−59.9261	9.9877

[1] Notice state sequence alternates translation, then rotation at each node.

Table A2. Mass matrix [M] [1].

Mass	W2	θ2	W3	θ3	W4	θ4	W5	θ5	U6	θ6	u7	θ7	u8	θ8
W2	0.4761	0.0000	0.0037	−0.0002	0	0	0	0	0	0	0	0	0	0
θ2	0.0000	0.0000	0.0002	−0.0001	0	0	0	0	0	0	0	0	0	0
W3	0.0037	0.0002	4.76 × 10^{-1}	0.0000	0.0037	−0.0002	0	0	0	0	0	0	0	0
θ3	−0.0002	−0.0001	0.0000	0.0000	0.0002	−0.0001	0	0	0	0	0	0	0	0
W4	0	0	0.0037	0.0002	0.4761	0.0000	0.0037	−0.0002	0	0	0	0	0	0
θ4	0	0	−0.0002	−0.0001	0.0000	0.0000	0.0002	−0.0001	0	0	0	0	0	0
W5	0	0	0	0	0.0037	0.0002	2.63 × 10^{0}	−0.0004	0	0	0	0	0	0
θ5	0	0	0	0	−0.0002	−0.0001	−0.0004	0.0000	0.0002	−0.0001	0	0	0	0
U6	0	0	0	0	0	0	0	0.0002	4.76 × 10^{-1}	0.0000	0.0037	−0.0002	0	0
θ6	0	0	0	0	0	0	0	−0.0001	0.00 × 10^{0}	0.0000	0.0002	−0.0001	0	0
U7	0	0	0	0	0	0	0	0	0.0037	0.0002	0.4761	0.0000	0.0037	−0.0002
θ7	0	0	0	0	0	0	0	0	−0.0002	−0.0001	0.0000	0.0000	0.0002	−0.0001
U8	0	0	0	0	0	0	0	0	0	0	0.0037	0.0002	4.66 × 10^{-1}	−0.0004
θ8	0	0	0	0	0	0	0	0	0	0	−0.0002	−0.0001	−0.0004	0.0000

[1] Notice state sequence alternates translation, then rotation at each node. 4.76 × 10^{-1} and such indicates 4.76 × 10^{-1}, where the former notation is used to ameliorate table crowding.

Table A3. Natural frequencies, ω_n [rad/s] [1].

1809.5	1415.5	1042.2	774.3	596.8
478.8	410.9	54.9	43.7	30.9
15.8	10.2	0.7	2.1	

[1] Corresponding to mode shapes in Table 5.

Table A4. Normalized mode shapes [×10[4]].

−1	3	2	0	−3	−5	3	1501	383	1037	443	−692	181	240
−1097	3080	4481	4992	4505	3173	−1154	−4544	−136	2388	2221	4049	1395	1669
−1	1	−3	−7	−5	1	−2	−1569	−215	204	667	1425	−670	−712
−2158	4857	3958	28	−3814	−4883	2208	−943	−2040	−7064	−867	912	−2460	−1864
−1	−1	−5	0	7	3	2	1296	−125	−1076	125	995	−1385	−1057
−3111	4495	−1058	−4992	−1185	4481	−3142	5806	2118	368	−2572	4061	−3204	−683
0	0	0	1	1	−1	1	−99	30	113	−105	248	2247	954
−3902	2199	−4861	−47	4875	−2071	3937	−4288	−3426	4504	1878	4753	−3652	1689
0	−3	4	0	−6	6	1	536	−918	135	898	754	−946	736
−4493	−1062	−3138	4985	−3062	−1232	−4519	815	2292	−2423	3091	891	−3893	3998
0	2	5	−7	7	−4	0	−294	753	446	880	410	−1936	1905
−4872	−3903	2129	75	−2232	3920	4832	−2471	3385	−210	−3658	3392	−4013	5178
−1	−2	2	−3	4	−5	−6	90	−261	192	−699	−732	−2945	3254
−5031	−5052	5012	−5030	5062	−4984	−4965	3580	−7823	3941	−7649	−5153	−4046	5502

Equations of Motion in the Standard State Space Form

$$\begin{Bmatrix} \dot{x} \\ \ddot{x} \end{Bmatrix}_{[nx1]} = [A]_{nxn} \begin{Bmatrix} x \\ \dot{x} \end{Bmatrix}_{nx1} + [B]_{nx1} \{u\}_{1x1} \tag{A6}$$

Finite element analysis performed in MATLAB® (code is included in the appendix) generates the mode shapes used to calculate the rigid–elastic coupling terms. The program outputs the flexible system [A], [B], [C], and [D] matrices of the standard state space form. The results are:

$$\begin{Bmatrix} \dot{x} \\ \ddot{x} \end{Bmatrix}_{[nx1]} = \begin{bmatrix} 0 & 0 & 1 \; 0 & \\ 0 & 0 & 0 \; 1 & \\ \frac{-\omega_i^2 I_{zz}}{I_{zz} - D_i \sum_{i=1}^{n} D_i} & 0 & \frac{-2\xi\omega_i I_{zz}}{I_{zz} - D_i \sum_{i=1}^{n} D_i} & 0 \\ \frac{-\omega_i^2 \sum_{i=1}^{n} D_i}{I_{zz} - D_i \sum_{i=1}^{n} D_i} & 0 & \frac{-2\xi\omega_i I_{zz}}{I_{zz} - D_i \sum_{i=1}^{n} D_i} & 0 \end{bmatrix}_{nxn} \begin{Bmatrix} x \\ \dot{x} \end{Bmatrix}_{nx1} + \begin{bmatrix} 0 \\ 0 \\ \frac{-D_i T}{I_{zz} - D_i \sum_{i=1}^{n} D_i} \\ \frac{T}{I_{zz} - D_i \sum_{i=1}^{n} D_i} \end{bmatrix}_{nx1} \{u\}_{1x1} \tag{A7}$$

$$[C] = [1 \; 0 \; 0 \; 0 \; 0 \; 0 \; 0 \; 0 \; 0 \; 0 \; 0 \; 0] \qquad [D] = [0] \tag{A8}$$

Given these equations, the resultant state space matrices are:

Table A5. State space [A] matrix [1].

	1	2	3	4	5	6	7	8	9	10	11	12
1	0	0	0	0	0	0	1	0	0	0	0	0
2	0	0	0	0	0	0	0	1	0	0	0	0
3	0	0	0	0	0	0	0	0	1	0	0	0
4	0	0	0	0	0	0	0	0	0	1	0	0
5	0	0	0	0	0	0	0	0	0	0	1	0
6	0	0	0	0	0	0	0	0	0	0	0	1
7	0	0.099	−1.064	−3.382	−0.736	−26.635	0	1.392×10^{-4}	-5.066×10^{-4}	-3.298×10^{-4}	-4.662×10^{-5}	-8.609×10^{-4}
8	0	−0.659	1.642	5.218	1.136	41.100	0	-9.264×10^{-4}	7.817×10^{-4}	5.088×10^{-4}	7.194×10^{-5}	1.328×10^{-3}
9	0	0.188	−6.435	−6.433	−1.401	−50.670	0	2.649×10^{-4}	-3.064×10^{-3}	-6.273×10^{-4}	-8.869×10^{-5}	-1.638×10^{-3}
10	0	0.025	−0.270	−106.024	−0.187	−6.755	0	3.531×10^{-5}	-1.285×10^{-4}	-1.034×10^{-2}	-1.182×10^{-5}	-2.183×10^{-4}
11	0	0.002	−0.025	−0.079	−249.447	−0.620	0	3.241×10^{-6}	-1.179×10^{-5}	-7.677×10^{-6}	-1.579×10^{-2}	-2.004×10^{-5}
12	0	0.022	−0.233	−0.742	−0.162	−962.998	0	3.055×10^{-5}	-1.112×10^{-4}	-7.237×10^{-5}	-1.023×10^{-5}	-3.113×10^{-2}

[1] Flexible states where base (rigid-body) rotation is controlled. 4.76×10^{-1} and such indicates 4.76×10^{-1}, where the former notation is used to ameliorate table crowding.

$$[B] = \{0 \quad 0 \quad 0 \quad 0 \quad 0 \quad 0 \quad 0.126794 \quad -0.19565 \quad 0.126794 \quad 0.032156 \quad 0.002952 \quad 0.27828 \quad \}^T \tag{A9}$$

$$[C] = \{1 \quad 0 \quad 0 \quad 0 \quad 0 \quad 0 \quad 0 \quad 0 \quad 0 \quad 0 \quad 0 \quad 0 \quad \} \qquad [D] = [0] \tag{A10}$$

Appendix B Initialization Function Callbacks for Simulation

This appendix is an optional section containing details supplemental to the main text crucial to understanding and reproducing the research.

```
clear all; close all; clc;

%This block of code establishes the properties of each beam element
a=0.0254; b=0.0016; L=0.25;
E=72*10^9;   I=a*b^3/12;
Li=[12 6*L -12 6*L;6*L 4*L^2 -6*L 2*L^2;-12 -6*L 12 -6*L;6*L 2*L^2 -6*L 4*L^2];
k_beam=E*I/L^3*[Li];
rho_beam=2.8*10^3;        %Beam density kg/m^3
A_beam=a*b;               %Beam cross sectional area
mb=rho_beam*A_beam;       %Beam mass per unit length

%This block creates the empty stiffness matrix [k]
k=zeros(14,14);

% This block fills in the stiffness matrix components

% Row 1 components start at index=1
k(1,1)=k_beam(3,3)+k_beam(1,1);
k(1,2)=k_beam(3,4)+k_beam(1,2);
k(1,3)=k_beam(1,3);
k(1,4)=k_beam(1,4);

% Row 2 components start at index=15
k(2,1)=k_beam(4,3)+k_beam(2,1);
k(2,2)=k_beam(4,4)+k_beam(2,2);
k(2,3)=k_beam(2,3);
k(2,4)=k_beam(2,4);

% Row 3 components start at index=29
k(3,1)=k_beam(3,1);
k(3,2)=k_beam(3,2);
k(3,3)=k_beam(3,3)+k_beam(1,1);
k(3,4)=k_beam(3,4)+k_beam(1,2);
k(3,5)=k_beam(1,3);
k(3,6)=k_beam(1,4);

% Row 4 components start at index=43
k(4,1)=k_beam(4,1);
k(4,2)=k_beam(4,2);
k(4,3)=k_beam(4,3)+k_beam(2,1);
k(4,4)=k_beam(4,4)+k_beam(2,2);
k(4,5)=k_beam(2,3);
k(4,6)=k_beam(2,4);

% Row 5 components start at index=59
k(5,3)=k_beam(3,1);
k(5,4)=k_beam(3,2);
k(5,5)=k_beam(3,3)+k_beam(1,1);
k(5,6)=k_beam(3,4)+k_beam(1,2);
k(5,7)=k_beam(1,3);
k(5,8)=k_beam(1,4);

% Row 6 components start at index=73
k(6,3)=k_beam(4,1);
k(6,4)=k_beam(4,2);
k(6,5)=k_beam(4,3)+k_beam(2,1);
k(6,6)=k_beam(4,4)+k_beam(2,2);
k(6,7)=k_beam(2,3);
k(6,8)=k_beam(2,4);

% Row 7 components start at index=89
k(7,5)=k_beam(3,1);
k(7,6)=k_beam(3,2);
k(7,7)=k_beam(3,3);
k(7,8)=k_beam(3,4);
k(8,9)=k_beam(2,3);
k(8,10)=k_beam(2,4);

% Row 8 components start at index=103
k(8,5)=k_beam(4,1);
k(8,6)=k_beam(4,2);
k(8,7)=k_beam(4,3);
k(8,8)=k_beam(4,4)+k_beam(2,2);

% Row 9 components start at index=120
k(9,8)=k_beam(3,2);
k(9,9)=k_beam(3,3)+k_beam(1,1);
k(9,10)=k_beam(3,4)+k_beam(1,2);
k(9,11)=k_beam(1,3);
k(9,12)=k_beam(1,4);

% Row 10 components start at index=134
k(10,8)=k_beam(4,2);
k(10,9)=k_beam(4,3)+k_beam(2,1);
k(10,10)=k_beam(4,4)+k_beam(2,2);
k(10,11)=k_beam(2,3);
k(10,12)=k_beam(2,4);
```

```
% Row 11 components start at index=149
k(11,9)=k_beam(3,1);
k(11,10)=k_beam(3,2);
k(11,11)=k_beam(3,3)+k_beam(1,1);
k(11,12)=k_beam(3,4)+k_beam(1,2);
k(11,13)=k_beam(1,3);
k(11,14)=k_beam(1,4);

% Row 12 components start at index=163
k(12,9)=k_beam(4,1);
k(12,10)=k_beam(4,2);
k(12,11)=k_beam(4,3)+k_beam(2,1);
k(12,12)=k_beam(4,4)+k_beam(2,2);
k(12,13)=k_beam(2,3);
k(12,14)=k_beam(2,4);

% Row 13 components start at index=179
k(13,11)=k_beam(3,1);
k(13,12)=k_beam(3,2);
k(13,13)=k_beam(3,3);
k(13,14)=k_beam(3,4);

% Row 14 components start at index=193
k(14,11)=k_beam(4,1);
k(14,12)=k_beam(4,2);
k(14,13)=k_beam(4,3);
k(14,14)=k_beam(4,4);

%Display stiffness matrix to check
k=k;

%END STIFFNESS MATRIX.   START MASS MATRIX

%Assemble individual beam inertia matrix
I_beam=ones(1,8);          %Creates empty matrix of I's for eight node points
I_beam=[I_beam.*I];        %Fill in matrix values with beam inertia
I_beam(1)=0;               %First node point inertia = 0

%This block of code creates the individual beam mass matrix "m_beam"
mi=[156 22*L 54 -13*L;22*L 4*L^2 13*L -3*L^2;54 13*L 156 -22*L;-13*L -3*L^2 -22*L 4*L^2];
m_beam=mb*L/420*mi;

%This block of code establishes the value of each point mass (mp)
%and the system point mass matrix (M)
mp=0.455;                              %Point masses, M
M=[0 mp mp mp 2*mp mp mp mp];   %Matrix of 8 point masses (0 First point mass)

%Creates a 14x14 empty mass matrix [m]
m=zeros(14,14);

%Fill in the system mass matrix components

% Row 1 components start at index=1
m(1,1)=m_beam(3,3)+m_beam(1,1)+M(2);
m(1,2)=m_beam(3,4)+m_beam(1,2);
m(1,3)=m_beam(1,3);
m(1,4)=m_beam(1,4);

% Row 2 components start at index=15
m(2,1)=m_beam(4,3)+m_beam(2,1);
m(2,2)=m_beam(4,4)+m_beam(2,2);
m(2,3)=m_beam(2,3);
m(2,4)=m_beam(2,4);

% Row 3 components start at index=29
m(3,1)=m_beam(3,1);
m(3,2)=m_beam(3,2);
m(3,3)=m_beam(3,3)+m_beam(1,1)+M(3);
m(3,4)=m_beam(3,4)+m_beam(1,2);
m(3,5)=m_beam(1,3);
m(3,6)=m_beam(1,4);

% Row 4 components start at index=43
m(4,1)=m_beam(4,1);
m(4,2)=m_beam(4,2);
m(4,3)=m_beam(4,3)+m_beam(2,1);
m(4,4)=m_beam(4,4)+m_beam(2,2);
m(4,5)=m_beam(2,3);
m(4,6)=m_beam(2,4);

% Row 5 components start at index=59
m(5,3)=m_beam(3,1);
m(5,4)=m_beam(3,2);
m(5,5)=m_beam(3,3)+m_beam(1,1)+M(4);
m(5,6)=m_beam(3,4)+m_beam(1,2);
m(5,7)=m_beam(1,3);
m(5,8)=m_beam(1,4);

% Row 6 components start at index=73
m(6,3)=m_beam(4,1);
m(6,4)=m_beam(4,2);
m(6,5)=m_beam(4,3)+m_beam(2,1);
m(6,6)=m_beam(4,4)+m_beam(2,2);
m(6,7)=m_beam(2,3);
m(6,8)=m_beam(2,4);

% Row 7 components start at index=89
m(7,5)=m_beam(3,1);
m(7,6)=m_beam(3,2);
m(7,7)=m_beam(3,3)+3*mb+M(5)+M(6)+M(7)+M(8);
m(7,8)=m_beam(3,4);
```

```
% Row 8 components start at index=103
m(8,5)=m_beam(4,1);
m(8,6)=m_beam(4,2);
m(8,7)=m_beam(4,3);
m(8,8)=m_beam(4,4)+m_beam(2,2);
m(8,9)=m_beam(2,3);
m(8,10)=m_beam(2,4);

% Row 9 components start at index=120
m(9,8)=m_beam(3,2);
m(9,9)=m_beam(3,3)+m_beam(1,1)+M(6);
m(9,10)=m_beam(3,4)+m_beam(1,2);
m(9,11)=m_beam(1,3);
m(9,12)=m_beam(1,4);

% Row 10 components start at index=134
m(10,8)=m_beam(4,2);
m(10,9)=m_beam(4,3)+m_beam(2,1);
m(10,10)=m_beam(4,4)+m_beam(2,2);
m(10,11)=m_beam(2,3);
m(10,12)=m_beam(2,4);

% Row 11 components start at index=149
m(11,9)=m_beam(3,1);
m(11,10)=m_beam(3,2);
m(11,11)=m_beam(3,3)+m_beam(1,1)+M(7);
m(11,12)=m_beam(3,4)+m_beam(1,2);
m(11,13)=m_beam(1,3);
m(11,14)=m_beam(1,4);

% Row 12 components start at index=163
m(12,9)=m_beam(4,1);
m(12,10)=m_beam(4,2);
m(12,11)=m_beam(4,3)+m_beam(2,1);
m(12,12)=m_beam(4,4)+m_beam(2,2);
m(12,13)=m_beam(2,3);
m(12,14)=m_beam(2,4);

% Row 13 components start at index=179
m(13,11)=m_beam(3,1);
m(13,12)=m_beam(3,2);
m(13,13)=m_beam(3,3)+M(8);
m(13,14)=m_beam(3,4);

% Row 14 components start at index=193
m(14,11)=m_beam(4,1);
m(14,12)=m_beam(4,2);
m(14,13)=m_beam(4,3);
m(14,14)=m_beam(4,4);

%Display the system mass matrix to check
m=m;

%Calculate the natural frequencies and normal modes
[NormalModes,EigenValues]=eig(inv(m)*k);
NaturalFrequencies=diag(EigenValues^0.5);
ModeShapes=NormalModes;

%Check Orthogonality like Homework 1 confirm diagonal matrix of 1's
%to satisfy equation 24 on slide 17
OrthoMass=NormalModes'*m*NormalModes;
OrthoStiff=NormalModes'*k*NormalModes;
StiffCheck=OrthoStiff/EigenValues;
Equation24_OrthoCheck=diag(diag(StiffCheck/OrthoMass));

%Spacecraft Radius to be used designating rigid modal coordinate
R=0.381;
FeeE=NormalModes;             %Designate Elastic mode shapes array FeeE
Omega=NaturalFrequencies;     %Designate variable name 'Omega' as natural
frequencies
%Designate Rigid modal coordinate FeeR
FeeR=[R+L 1 R+L*2 1 R+L*3 1 R+L*4 1 -L 1 -L*2 1 -L*3 1];
Di=FeeE'*m*diag(FeeR);        %Calculate Rigid-Elastic Coupling Coefficient
DiCheck=det(Di);              %Confirm Di is singular...det(Di=0)

Z=0.0005;
Izz=14;
w=diag(NaturalFrequencies);    %Generate a diagonal matrix of natural frequency
Iw=0.0912;
Td=0;                          %Disturbance Torque
Tc=0.1;                        %Control Torque is Iw*qddot_wheel
T=Td+Tc;                       %Total Torque is sum of disturbance and control
torques
```

```
%Start State Space Development
NatFreq = diag(EigenValues).^0.5;

r = 0.381;                        %Radius of the wheel (large rigid body)
freqs = sqrt(EigenValues);   %
NatFreq = EigenValues(1:5,1:5);
freqs = freqs(1:5,1:5);
zeta = 0.0005;                    %Given damping ratio for all modes
Izz = 14;

phi_E = NormalModes(1:14,1:5);
phi_R = [r+L,1,r+2*L,1,r+3*L,1,r+4*L,1,-L,1,-2*L,1,-3*L,1]';

M_II = m;
Di = [phi_E'*M_II*phi_R];
M_state = [Izz Di';
Di   eye(5)];
C_damp = [zeros(6,6)];
C_damp(2:6,2:6) = 2*zeta*freqs;
K=[zeros(6,6)];
K(2:6,2:6) = NatFreq;

A = [zeros(6),eye(6,6);
-inv(M_state)*K, -inv(M_state)*C_damp];
Bprime = [1;0;0;0;0;0];
B = [0 0 0 0 0 0 (inv(M_state)*Bprime)']';
C = zeros(12,12); C(1,1)=1;
D = zeros(12,1);

[Gnum,Gden] = ss2tf(A,B,C,D);
G1 = tf(Gnum(1,:),Gden)

%Manually input Transfer Function to check
NUM=[1.998e-015 0.1268 0.007582 166.9 5.591 4.718e004 771 3.412e006 1.218e004
1.576e007 1.475e004 7.11e006];
DEN=[1 0.06125 1326 46.15 3.781e005 6683 2.808e007 1.388e005 1.813e008 2.065e005
9.954e007 0 0];
G=tf(NUM,DEN);

%Put PID controller Transfer function into workspace
It=14; Z=0.516931;
Bandwidth=4; wn=Bandwidth; T=10/Z/wn;
Kd=2*Z*wn*It+It/T;
Kp=wn^2+2*Z*wn/T;
Ki=wn^2/T;
PID=tf([Kd Kp Ki],[0 1 0]);

%DESIGN FILTERS TO SMOOTH OUT MODE 1
%Design Bandpass filter for w = 10^-0.1478 = 0.711541 Hz
wz=0.711541;Zz=0.1;wp=wz;Zp=0.0005;
BP1=tf([1/wz^2 2*Zz/wz 1],[1/wp^2 2*Zp/wp 1]);
PID_BP1=PID*BP1; %Design Notch filter for w = 10^-0.109 = 0.778037 Hz
wz=0.778037;Zz=0.0005;wp=wz;Zp=0.1;
Notch1=tf([1/wz^2 2*Zz/wz 1],[1/wp^2 2*Zp/wp 1]);
Mode_1=PID*BP1*Notch1;

%DESIGN FILTERS TO SMOOTH OUT MODE 2
%Design Bandpass filter for w = 10^0.3223
wz=10^0.3223;Zz=0.1;wp=wz;Zp=0.0005;
BP2=tf([1/wz^2 2*Zz/wz 1],[1/wp^2 2*Zp/wp 1]);

%Design Notch filter for w = 10^0.405
wz=10^0.405;Zz=0.0006;wp=wz;Zp=0.1;
Notch2=tf([1/wz^2 2*Zz/wz 1],[1/wp^2 2*Zp/wp 1]);
Mode_2=Mode_1*BP2*Notch2;
```

```
%Design Lead filter for wz~1, wp~3
%wz=1;Zz=1;wp=3;Zp=1;
%Lead=tf([1/wz^2 2*Zz/wz 1],[1/wp^2 2*Zp/wp 1]);
%Mode_2=Mode_2*Lead;

%DESIGN FILTERS TO SMOOTH OUT MODE 3
%Design Bandpass filter for w = 10^1.0110
wz=10^1.0110;Zz=0.1;wp=wz;Zp=0.0005;
BP3=tf([1/wz^2 2*Zz/wz 1],[1/wp^2 2*Zp/wp 1]);

%Design Notch filter for w = 10^1.0128
wz=10^1.0128;Zz=0.0005;wp=wz;Zp=0.1;
Notch3=tf([1/wz^2 2*Zz/wz 1],[1/wp^2 2*Zp/wp 1]);
Mode_3=Mode_2*BP3*Notch3;

%DESIGN FILTERS TO SMOOTH OUT MODE 4
%Design Bandpass filter for w = 10^1.49035
wz=10^1.49035;Zz=0.1;wp=wz;Zp=0.0005;
BP4=tf([1/wz^2 2*Zz/wz 1],[1/wp^2 2*Zp/wp 1]);

%Design Notch filter for w = 10^1.492
wz=10^1.492;Zz=0.0005;wp=wz;Zp=0.1;
Notch4=tf([1/wz^2 2*Zz/wz 1],[1/wp^2 2*Zp/wp 1]);
Mode_4=Mode_3*BP4*Notch4;

%CALCULATE SYSTEM NATURAL FREQUENCIES
[NaturalFrequencies,Damping,EigenValue]=damp(G);
NaturalFrequencies=NaturalFrequencies;
```

Appendix C Stop Function Callbacks for Simulation

This appendix is an optional section containing details supplemental to the main text crucial to understanding and reproducing the research.

```
[mag1,phase1,wout1] = bode(G);   Mag1=20*log10(mag1(:)); Phase1=phase1(:);

[mag2,phase2,wout2] = bode(G*PID);   Mag2=20*log10(mag2(:)); Phase2=phase2(:);

[mag3,phase3,wout3] = bode(G*PID*BP1);   Mag3=20*log10(mag3(:)); Phase3=phase3(:);
[mag4,phase4,wout4] = bode(G*PID*BP1*Notch1);   Mag4=20*log10(mag4(:));
Phase4=phase4(:);

[mag5,phase5,wout5] = bode(G*PID*BP1*Notch1*BP2);   Mag5=20*log10(mag5(:));
Phase5=phase5(:);
[mag6,phase6,wout6] = bode(G*PID*BP1*Notch1*BP2*Notch2);
Mag6=20*log10(mag6(:)); Phase6=phase6(:);

[mag7,phase7,wout7] = bode(G*PID*BP1*Notch1*BP2*Notch2*BP3);
Mag7=20*log10(mag7(:)); Phase7=phase7(:);
[mag8,phase8,wout8] = bode(G*PID*BP1*Notch1*BP2*Notch2*BP3*Notch3);
Mag8=20*log10(mag8(:)); Phase8=phase8(:);

[mag9,phase9,wout9] = bode(G*PID*BP1*Notch1*BP2*Notch2*BP3*Notch3*BP4);
Mag9=20*log10(mag9(:)); Phase9=phase9(:);
[mag10,phase10,wout10] =
bode(G*PID*BP1*Notch1*BP2*Notch2*BP3*Notch3*BP4*Notch4);
Mag10=20*log10(mag10(:)); Phase10=phase10(:);

figure(1);   hold on;
semilogx(wout1,Mag1,'--','LineWidth',1);
semilogx(wout2,Mag2,'LineWidth',1);
semilogx(wout3,Mag3,'--','LineWidth',3);
semilogx(wout4,Mag4,':','LineWidth',2);
hold off; grid on; axis([0.5,40, -100, 150 ]); set(gca, 'FontSize',28, 'FontName','Palatino
Linotype');
legend('Flexible space robot','PID','PID + Bandpass','PID + Notch + Bandpass')
```

```
figure(2); hold on; set(gca, 'FontSize',28, 'FontName','Palatino Linotype');
semilogx(wout1,Phase1,'--','LineWidth',1);
semilogx(wout2,Phase2,'LineWidth',1);
semilogx(wout3,Phase3,'--','LineWidth',3);
semilogx(wout4,Phase4,':','LineWidth',2);
hold off; grid on;

figure(3);    hold on;
semilogx(wout2,Mag2,'--','LineWidth',1);
semilogx(wout4,Mag4,'LineWidth',1);
semilogx(wout5,Mag5,'--','LineWidth',3);
semilogx(wout6,Mag6,':','LineWidth',2);
hold off; grid on; axis([0.5,40, -100, 150 ]); set(gca, 'FontSize',28, 'FontName','Palatino
Linotype');
legend('PID controlled Flexible space robot','PID+Mode 1','PID + Mode 1 +
Bandpass','PID + Mode 1 + Notch + Bandpass')

figure(4); hold on; set(gca, 'FontSize',28, 'FontName','Palatino Linotype');
semilogx(wout2,Phase2,'--','LineWidth',1);
semilogx(wout4,Phase4,'LineWidth',1);
semilogx(wout5,Phase5,'--','LineWidth',3);
semilogx(wout6,Phase6,':','LineWidth',2);
hold off; grid on;

figure(5);    hold on;
semilogx(wout2,Mag2,'--','LineWidth',1);
semilogx(wout4,Mag4,'LineWidth',1);
semilogx(wout6,Mag6,'--','LineWidth',3);
semilogx(wout7,Mag7,':','LineWidth',2);
semilogx(wout8,Mag8,':','LineWidth',2);
hold off; grid on; axis([0.5,40, -100, 150 ]); set(gca, 'FontSize',28, 'FontName','Palatino
Linotype');
legend('PID controlled Flexible space robot','PID + Mode 1','PID + Mode 2','PID +
Mode 1 + Mode 2 + Bandpass','PID + Mode 1 + Mode 2 + Bandpass + Notch')

figure(6); hold on; set(gca, 'FontSize',28, 'FontName','Palatino Linotype');
semilogx(wout2,Phase2,'--','LineWidth',1);
semilogx(wout4,Phase4,'LineWidth',1);
semilogx(wout6,Phase6,'--','LineWidth',3);
semilogx(wout7,Phase7,':','LineWidth',2);
semilogx(wout8,Phase8,':','LineWidth',2);
hold off; grid on;

figure(7);    hold on;
semilogx(wout2,Mag2,'--','LineWidth',1);
semilogx(wout4,Mag4,'LineWidth',1);
semilogx(wout6,Mag6,'--','LineWidth',3);
semilogx(wout8,Mag8,':','LineWidth',2);
semilogx(wout9,Mag9,':','LineWidth',2);
semilogx(wout10,Mag10,':','LineWidth',2);
hold off; grid on; axis([0.5,40, -100, 150 ]); set(gca, 'FontSize',28, 'FontName','Palatino
Linotype');
legend('PID controlled Flexible space robot','PID + Mode 1','PID + Mode 2','PID +
Mode 1 + Mode 2 + Mode 3','PID + Mode 1 + Mode 2 + Mode 3 + Bandpass + Notch')

figure(8); hold on; set(gca, 'FontSize',28, 'FontName','Palatino Linotype');
semilogx(wout2,Phase2,'--','LineWidth',1);
semilogx(wout4,Phase4,'LineWidth',1);
semilogx(wout6,Phase6,'--','LineWidth',3);
semilogx(wout8,Phase8,':','LineWidth',2);
semilogx(wout9,Phase9,':','LineWidth',2);
semilogx(wout10,Phase10,':','LineWidth',2);
hold off; grid on;
```

```
sys1=G*PID/(1+G*PID);
sys2=(G*PID*BP1/(1+G*PID*BP1));
sys3=(G*PID*BP1*Notch1/(1+G*PID*BP1*Notch1));
sys4=(G*PID*BP1*Notch1*BP2/(1+G*PID*BP1*Notch1*BP2));
sys5=(G*PID*BP1*Notch1*BP2*Notch2/(1+G*PID*BP1*Notch1*BP2*Notch2));
sys6=(G*PID*BP1*Notch1*BP2*Notch2*BP3/(1+G*PID*BP1*Notch1*BP2*Notch2*BP3));

sys7=(G*PID*BP1*Notch1*BP2*Notch2*BP3*Notch3/(1+G*PID*BP1*Notch1*BP2*Notch2
*BP3*Notch3));
sys8=(G*PID*BP1*Notch1*BP2*Notch2*BP3*Notch3*BP4/(1+G*PID*BP1*Notch1*BP2
*Notch2*BP3*Notch3*BP4));
sys9=(G*PID*BP1*Notch1
*BP2*Notch2*BP3*Notch3*BP4*Notch4/(1+G*PID*BP1*Notch1*BP2*Notch2*BP3*Notch3
*BP4*Notch4));

figure(9); step(sys1,sys2);   legend('PID','PID + BP1');set(gca, 'FontSize',28,
'FontName','Palatino Linotype');
figure(10);   step(sys1,sys3); legend('PID','PID + BP1+Notch1'); set(gca, 'FontSize',28,
'FontName','Palatino Linotype');
figure(11); step(sys1,sys4); legend('PID','PID + Mode 1 + BP2'); set(gca, 'FontSize',28,
'FontName','Palatino Linotype');
figure(12);   step(sys1,sys5); legend('PID','PID + Mode 1 + BP2 + Notch 2'); set(gca,
'FontSize',28, 'FontName','Palatino Linotype');
figure(13); step(sys1,sys6); legend('PID','PID + Mode 1 + Mode 2 + BP3'); set(gca,
'FontSize',28, 'FontName','Palatino Linotype');
figure(14);   step(sys1,sys7); legend('PID','PID + Mode 1 + Mode 2 + BP3 + Notch3');
set(gca, 'FontSize',28, 'FontName','Palatino Linotype');
figure(15);   step(sys1,sys8); legend('PID','PID + Mode 1 + Mode 2 + Mode 3 + BP4');
set(gca, 'FontSize',28, 'FontName','Palatino Linotype');
figure(16);   step(sys1,sys9); legend('PID','PID + Mode 1 + Mode 2 + Mode 3 + BP4 +
Notch4'); set(gca, 'FontSize',28, 'FontName','Palatino Linotype');
```

References

1. AF, Navy Baseball Teams Square off for 2018 Freedom Classic. Available online: https://www.nellis.af.mil/News/Article/1452330/af-navy-baseball-teams-square-off-for-2018-freedom-classic/ (accessed on 12 December 2023).
2. Department of Defense Policy. U.S. Department of Defense Photographs and Imagery, Unless Otherwise Noted, Are in the Public Domain. Available online: https://www.defense.gov/Contact/Help-Center/Article/article/2762906/use-of-department-of-defense-imagery (accessed on 12 December 2023).
3. Nile, R. First Humanoid Robot in Space Receives NASA Government Invention of the Year. 17 June 2015. Available online: https://www.nasa.gov/mission_pages/station/research/news/invention_of_the_year (accessed on 17 February 2023).
4. NASA Image Use Policy. NASA Content (Images, Videos, Audio, etc.) Are Generally Not Copyrighted and May Be Used for Educational or Informational Purposes without Needing Explicit Permissions. Available online: https://gpm.nasa.gov/image-use-policy (accessed on 8 October 2023).
5. Nesnas, I.; Fesq, L.; Volpe, R. Autonomy for Space Robots: Past, Present, and Future. *Curr. Robot. Rep.* **2021**, 2, 251–263. [CrossRef]
6. Sagan, C.; Reddy, R. Machine Intelligence and Robotics: Report of the NASA Study Group Final Report. 715-32. Carnegie Mellon University. 1980. Available online: http://www.rr.cs.cmu.edu/NASA%20Sagan%20Report.pdf (accessed on 7 September 2011).
7. NASA Autonomous Systems—Systems Capability Leadership Team. Autonomous Systems Taxonomy. NASA Technical Reports Server. 14 May 2018; p. 20180003082. Available online: https://ntrs.nasa.gov/citations/20180003082 (accessed on 8 February 2024).
8. Marshall Spaceflight Center, Advanced Space Transportation Program: Paving the Highway to Space. Available online: https://www.nasa.gov/centers/marshall/news/background/facts/astp.html (accessed on 15 January 2022).
9. Curiosity Bot, Advantages and Disadvantages of Using Robots Instead of Astronauts. Available online: https://mycuriositybot.wordpress.com/2019/02/18/advantages%E2%80%93and%E2%80%93disadvantages%E2%80%93of%E2%80%93using%E2%80%93robots%E2%80%93instead%E2%80%93of%E2%80%93astronauts/ (accessed on 18 February 2019).
10. NASA Is Building a Mission That Will Refuel and Repair Satellites in Orbit. Available online: https://www.universetoday.com/155863/nasa-is-building-a-mission-that-will-refuel-and-repair-satellites-in-orbit/ (accessed on 8 October 2023).
11. DARPA Teams with Northrop Grumman to Build Robotic Service Satellites. Available online: https://newatlas.com/space/robotic-service-satellite-darpa-northrop-grumman/ (accessed on 8 October 2023).
12. Usage Policy. Images Posted on the DARPA Website May Be Used for Educational or Informational Purposes. Available online: https://www.darpa.mil/policy/usage-policy (accessed on 8 October 2023).

13. Holzinger, M.; Chow, C.; Garretson, P. *A Primer on Cislunar Space*; Technical Report; Air Force Research Labs: Wright-Patterson Air Force Base, OH, USA, 7 May 2021. Available online: https://www.afrl.af.mil/Portals/90/Documents/RV/A%20Primer%20on%20Cislunar%20Space_Dist%20A_PA2021-1271.pdf?ver=vs6e0sE4PuJ51QC-15DEfg== (accessed on 9 October 2023).
14. Available online: https://www.ssc.spaceforce.mil/Portals/3/SDA%20Briefings/13.%20Bates_Tyler_SDA%20Industry%20Day_Cislunar%20Space%20SDA_V4.pdf?ver=fhTzhW7RFhI1MDcP90KAuQ== (accessed on 14 December 2023).
15. Image Created by NASA. Available online: https://en.wikipedia.org/wiki/Halo_orbit#/media/File:Lagrange_points2.svg (accessed on 31 December 2023).
16. Ellery, A. Tutorial Review on Space Manipulators for Space Debris Mitigation. *Robotics* **2019**, *8*, 34. [CrossRef]
17. Zhang, W.; Li, F.; Li, J.; Cheng, Q. Review of On-Orbit Robotic Arm Active Debris Capture Removal Methods. *Aerospace* **2023**, *10*, 13. [CrossRef]
18. Jiang, D.; Cai, Z.; Peng, H.; Wu, Z. Coordinated Control Based on Reinforcement Learning for Dual-Arm Continuum Manipulators in Space Capture Missions. *J. Aero. Eng.* **2021**, *34*, 04021087. [CrossRef]
19. Yang, J.; Peng, H.; Zhang, J.; Wu, Z. Dynamic modeling and beating phenomenon analysis of space robots with continuum manipulators. *Chin. J. Aero.* **2022**, *35*, 226–241. [CrossRef]
20. Jiang, D.; Cai, Z.; Liu, Z.; Peng, H. An Integrated Tracking Control Approach Based on Reinforcement Learning for a Continuum Robot in Space Capture Missions. *J. Aero. Eng.* **2022**, *35*, 04022065-1. [CrossRef]
21. Yang, J.; Peng, H.; Zhou, W.; Wu, Z. Integrated Control of Continuum-Manipulator Space Robots with Actuator Saturation and Disturbances. *J. Guid. Con. Dyn.* **2022**, *45*, 2379. [CrossRef]
22. Ai, H.; Zhu, A.; Wang, J.; Yu, X.; Chen, L. Buffer Compliance Control of Space Robots Capturing a Non-Cooperative Spacecraft Based on Reinforcement Learning. *Appl. Sci.* **2021**, *11*, 5783. [CrossRef]
23. Armanini, C.; Boyer, F.; Mathew, A.; Duriez, C.; Renda, F. Soft robots modeling: A structured overview. *IEEE Trans. Robot.* **2022**, *39*, 1–21. [CrossRef]
24. Encyclopaedia Britannica, Britannica, Euclidean Space. Available online: https://www.britannica.com/science/Euclidean%E2%80%93space (accessed on 10 October 2022).
25. Newton, I. *Principia, Jussu Societatis Regiæ ac Typis Joseph Streater*; Cambridge University Library: London, UK, 1687.
26. Mazalov, V.; Parilina, E. The Euler–equation approach in average–oriented opinion dynamics. *Mathematics* **2020**, *8*, 355. [CrossRef]
27. Sands, T. Comparison and Interpretation Methods for Predictive Control of Mechanics. *Algorithms* **2019**, *12*, 232. [CrossRef]
28. A.C.M. Operations, What–Are–the–6–Degrees–of–Freedom, Industrial Inspection & Analysis (IIA). Available online: https://industrial%E2%80%93ia.com/what%E2%80%93are%E2%80%93the%E2%80%936%E2%80%93degrees%E2%80%93of%E2%80%93freedom%E2%80%936dof%E2%80%93explained/ (accessed on 21 March 2023).
29. Doyle, J.; Francis, A.; Tannenbaum, A. *Feedback Control Theory*; Dover: Mineola, NY, USA, 2009.
30. Johansson, R.; Robertsson, A.; Nilsson, K.; Verhaegen, M. State–Space System Identification of Robot Manipulator Dynamics. *Mechatronics* **2000**, *10*, 403–418. [CrossRef]
31. Corke, P. High–Performance Visual Closed–Loop Robot Control. Ph.D. Thesis, Mechanical and Manufacturing Engineering, University of Melbourne, Melbourne, Australia, 1994.
32. Li, Y.; Ang, K.; Chong, G.C.Y. PID Control System Analysis and design. *IEEE Control Syst.* **2006**, *26*, 32–41.
33. Bemporad, A.; Morari, M.; Dua, V.; Pistikopoulos, E. The explicit linear quadratic regulator for constrained systems. *Automatica* **2002**, *38*, 3–20. [CrossRef]
34. Tao, K.; Kosut, R.; Aral, G. Learning feedforward control. In Proceedings of the 1994 American Control Conference—ACC '94, Baltimore, MD, USA, 29 June–1 July 1994.
35. Chang, T.; Seufert, C.; Eminaga, O.; Shkolyar, E.; Hu, J.; Liao, J. Current trends in artificial intelligence application for endourology and robotic surgery. *Urol. Clin. N. Am.* **2021**, *48*, 151–160. [CrossRef]
36. Sands, T. Inducing Performance of Commercial Surgical Robots in Space. *Sensors* **2023**, *23*, 1510. [CrossRef]
37. Fu, S.; Bhavsar, P. Robotic Arm Control based on internet of things. In Proceedings of the 2019 IEEE Long Island Systems, Applications and Technology Conference (LISAT), Long Island, NY, USA, 3 May 2019.
38. Liu, X.-F.; Zhang, X.-Y.; Cai, G.-P.; Chen, W.-J. Capturing a Space Target Using a Flexible Space Robot. *Appl. Sci.* **2022**, *12*, 984. [CrossRef]
39. Li, J.; Ma, K.; Wu, Z. Tracking control via switching and learning for a class of uncertain flexible joint robots with variable stiffness actuators. *Neurocomputing* **2022**, *469*, 130–137. [CrossRef]
40. Pao, L. Multi-input shaping design for vibration reduction. *Automatica* **1999**, *35*, 81–89. [CrossRef]
41. Yavuz, H.; Mıstıkoğlu, S.; Kapucu, S. Hybrid input shaping to suppress residual vibration of flexible systems. *J. Vib. Control* **2012**, *18*, 132–140. [CrossRef]
42. Sands, T. Flattening the Curve of Flexible Space Robotics. *Appl. Sci.* **2022**, *12*, 2992. [CrossRef]
43. Sands, T. Optimization Provenance of Whiplash Compensation for Flexible Space Robotics. *Aerospace* **2019**, *6*, 93. [CrossRef]
44. Carabis, D.; Wen, J. Trajectory Generation for Flexible-Joint Space Manipulators. *Front. Robot. AI* **2022**, *9*, 687595. [CrossRef]
45. Menon, C.; Broschart, M.; Lan, N. Biomimetics and robotics for space applications: Challenges and emerging technologies. In Proceedings of the 2007 IEEE International Conference on Robotics and Automation—Workshop on Biomimetic Robotics, Roma, Italy, 10–14 April 2007.

46. Bar–Cohen, Y. Planetary Exploration Using bio–Inspired Technologies. Presentation of the Jet Propulsion Laboratory, California Institute of Technology. 2016. Available online: https://www1.grc.nasa.gov/wp-content/uploads/Bar-Cohen_JPL_2016_biomimetics_presentation.pdf (accessed on 2 February 2024).
47. Ellery, A. Tutorial Review of Bio-Inspired Approaches to Robotic Manipulation for Space Debris Salvage. *Biomimetics* **2020**, *5*, 19. [CrossRef]
48. Ellery, A. How to Build a Biological Machine Using Engineering Materials and Methods. *Biomimetics* **2020**, *5*, 35. [CrossRef]
49. Li, Q.; Pang, Y.; Wang, Y.; Han, X.; Li, Q.; Zhao, M. CBMC: A Biomimetic Approach for Control of a 7-Degree of Freedom Robotic Arm. *Biomimetics* **2023**, *8*, 389. [CrossRef]
50. Menezes, J.; Sands, T. Discerning Discretization for Unmanned Underwater Vehicles DC Motor Control. *J. Mar. Sci. Eng.* **2023**, *11*, 436. [CrossRef]
51. Banken, E.; Oeffner, J. Biomimetics for innovative and future-oriented space applications—A review. *Front. Space Technol.* **2022**, *3*, 1000788. [CrossRef]
52. Yang, T.; Xu, F.; Zeng, S.; Zhao, S.; Liu, Y.; Wang, Y. A Novel Constant Damping and High Stiffness Control Method for Flexible Space Manipulators Using Luenberger State Observer. *Appl. Sci.* **2023**, *13*, 7954. [CrossRef]
53. Hamad, R. Modelling and Feed-Forward Control of Robot Arms with Flexible Joints and Flexible Links. Master's Thesis, Chalmers University of Technology, Gothenburg, Sweden, June 2016.
54. Li, X.; Yang, D.; Liu, H. China's space robotics for on-orbit servicing: The state of the art. *Natl. Sci. Rev.* **2023**, *10*, 129. [CrossRef]
55. Papadopoulos, E.; Aghili, F.; Ma, O.; Lampariello, R. Robotic Manipulation and Capture in Space: A Survey. *Front. Robot. AI* **2021**, *228*, 686723. [CrossRef] [PubMed]
56. Wie, B. *Space Vehicle Dynamics and Control*, 2nd ed.; American Institute of Aeronautics and Astronautics: Reston, VA, USA, 2008.
57. Singhose, W.; Seering, W.; Singer, M. Input Shaping for Vibration Reduction with Specified Insensitivity to Modeling Errors. In Proceedings of the Japan-USA Symposium on Flexible Automation, Boston, MA, USA, 7–10 July 1996.
58. Gorinevsky, D.; Vukovich, G. Nonlinear Input Shaping Control of Flexible Spacecraft Reorientation Maneuver. *J. Guid. Con. Dyn* **1998**, *21*, 264–270. [CrossRef]
59. Kong, X.; Yang, Z. Combined feedback control and input shaping for vibration suppression of flexible spacecraft. In Proceedings of the International Conference on Mechatronics and Automation, Changchun, China, 18 September 2009; pp. 3257–3262. [CrossRef]
60. Pontryagin, L.; Boltyanskii, V.; Gamkrelidze, R.; Mischenko, E. *The Mathematical Theory of Optimal Processes*; Neustadt, L.W., Ed.; Wiley: New York, NY, USA, 1962.
61. Banginwar, P.; Sands, T. Autonomous Vehicle Control Comparison. *Vehicles* **2022**, *4*, 1109–1121. [CrossRef]
62. Cooper, M.; Heidlauf, P.; Sands, T. Controlling Chaos—Forced van der Pol Equation. *Mathematics* **2017**, *5*, 70. [CrossRef]
63. Smeresky, B.; Rizzo, A.; Sands, T. Optimal Learning and Self-Awareness Versus PDI. *Algorithms* **2020**, *13*, 23. [CrossRef]
64. Osburn, P.; Whitaker, H.; Kezer, A. *New Developments in the Design of Model Reference Adaptive Control Systems*; Institute of the Aerospace Sciences: Reston, VA, USA, 1961; Volume 61.
65. MDPI Image Use Policy. No Special Permission Is Required to Reuse All or Part of Article Published by MDPI, Including Figures and Tables. For Articles Published under an Open Access Creative Common CC BY License, any Part of the Article May Be Reused without Permission Provided that the Original Article Is Clearly Cited. Available online: https://www.mdpi.com/openaccess#:~:text=Permissions-,No%20special%20permission%20is%20required%20to%20reuse%20all%20or%20part,original%20article%20is%20clearly%20cited (accessed on 14 February 2023).

biomimetics

MDPI

Article

Discrete-Time Impedance Control for Dynamic Response Regulation of Parallel Soft Robots

Ameer Hamza Khan [1,2] and Shuai Li [3,*]

1 Smart City Research Institute (SCRI), Hong Kong Polytechnic University, Kowloon, Hong Kong; ameer.h.khan@connect.polyu.hk or ahakhan@polyu.edu.hk
2 Department of Land Surveying and Geo-Informatics (LSGI), Hong Kong Polytechnic University, Kowloon, Hong Kong
3 Faculty of Information Technology and Electrical Engineering (ITEE), University of Oulu, 90570 Oulu, Finland
* Correspondence: shuaili@ieee.org

Abstract: Accurately controlling the dynamic response and suppression of undesirable dynamics such as overshoots and vibrations is a vital requirement for soft robots operating in industrial environments. Pneumatically actuated soft robots usually undergo large overshoots and significant vibrations when deactuated because of their highly flexible bodies. These large vibrations not only decrease the reliability and accuracy of the soft robot but also introduce undesirable characteristics in the system. For example, it increases the settling time and damages the body of the soft robot, compromising its structural integrity. The dynamic behavior of the soft robots on deactuation needs to be accurately controlled to increase their utility in real-world applications. The literature on pneumatic soft robots still does not sufficiently address the issue of suppressing undesirable vibrations. To address this issue, we propose the use of impedance control to regulate the dynamic response of pneumatic soft robots since the superiority of impedance control is already established for rigid robots. The soft robots are highly nonlinear systems; therefore, we formulated a nonlinear discrete sliding mode impedance controller to control the pneumatic soft robots. The formulation of the controller in discrete-time allows efficient implementation for a high-order system model without the need for state-observers. The simplification and efficiency of the proposed controller enable fast implementation of an embedded system. Unlike other works on pneumatic soft robots, the proposed controller does not require manual tuning of the controller parameters and automatically calculates the parameters based on the impedance value. To demonstrate the efficacy of the proposed controller, we used a 6-chambered parallel soft robot as an experimental platform. We presented the comparative results with an existing state-of-the-art controller in SMC control of pneumatic soft robots. The experiment results indicate that the proposed controller can effectively limit the amplitude of the undesirable vibrations.

Keywords: soft robots; sliding mode; impedance control

1. Introduction

Soft robots, i.e., robots made from soft materials, have attained tremendous research attention in recent years [1–5]. They have shown great potential as a research platform as well as practical applications in industry [6,7] and healthcare [8–11]. Inspired by worms, snakes, and aquatic life [12,13], the main advantage offered by the soft robots are their flexible bodies, which impart them the ability to adapt to environmental variations automatically. They can easily navigate and interact [14,15] with previously unknown and irregular environments without any posing any danger to the interacting objects. Pneumatically driven soft robots, i.e., soft robots which use air pressure for actuation [16–20], have found most applications as compared to other soft robots because of their high flexibility, agility, and ease of fabrication [21]. This quality makes the pneumatic soft robots ideal for industrial applications [17,22,23] where safe handling of delicate objects is needed, as compared to rigid robots [24].

However, despite their advantages, the flexible structure of soft robots also contributes to some of its inherent weaknesses. Foremost among those weaknesses is the lack of stiffness, which results in undesirable dynamic, i.e., vigorous vibrations and overshooting when actuated or deactuated rapidly as pointed by [25,26]. These undesirable dynamics increases the overall settling time of the pneumatic soft robots, which not only decrease their accuracy and productivity but also reduce their useful lifespan by creating structural defects in their body. Figure 1 illustrates the problem of excessive overshooting and vibrations in soft robots. In the figure, the soft robot undergoes a vibration amplitude of about 50 degrees and a settling time of about 0.8 s. These dynamics can create the following issues in industrial applications

- Large settling time reduces productivity.
- Large amplitude oscillations reduce accuracy and can potentially cause damage to a delicate product.
- Oscillations accelerate the process of wear and tear in soft robots, thus increasing maintenance costs.

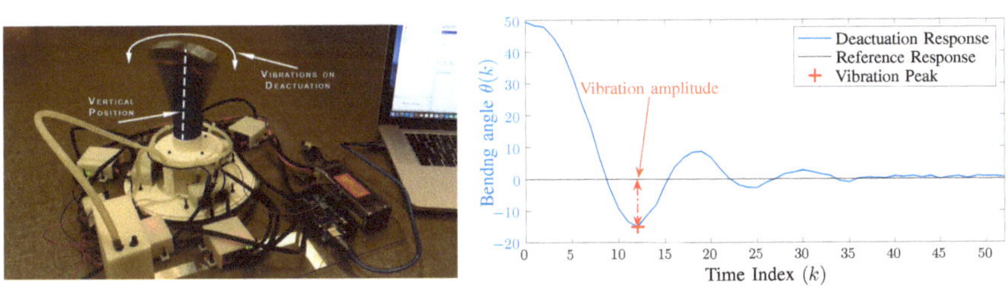

(a) (b)

Figure 1. The vibration problem in soft robots. (**a**) shows the image of vibration amplitude, and (**b**) shows the response in the form of a graph. (Sampling Time: $T_s = 0.02$).

These issues are inhibiting the large-scale adaption of soft robots in industrial applications. Most of the industrial applications require the soft robot to operate at an increased speed with a reasonable degree of accuracy and robustness [27]. In order to increase the utility of the soft robots in industrial applications, the issues posed by vibrations, overshooting, and large settling time need to be addressed.

Several approaches have been proposed in soft robotic literature to control the motion of soft robots. Finite Element Method (FEM) [28] is a popular technique for the kinematic control of soft robots. However, such methods require high computation cost and cannot be realistically used for controlling the dynamic response in real-time. The use of a model-free controller such as PI [29] has also been investigated for the control of pneumatic soft robots. However, in the presence of model nonlinearities, the tuning of PI parameters becomes a time-consuming and labor-intensive task. Passive approaches have been proposed in the literature to regulate the dynamic behavior of the pneumatic soft robots by using the natural damping properties of the materials [30]. For example, Ni et al. [25] propose attaching an external damper to the soft robot, while Li et al. [26] propose particle damping approach by attaching an additional grain-filled chamber to the soft robot. These external components help in regulating the dynamic behavior by dissipating excessive kinetic energy. Although these methods effectively suppress the oscillations, but at the cost of attaching the additional mechanical component. This mechanical overhead makes the system heavy, bulky, and less portable and increases the overall fabrication cost of soft robots. Luo et al. [31] make further progress in this direction by proposing a Sliding-Mode Controller (SMC) to regulate the dynamic response; however, their control law is essentially a PD controller with a deadband and does not account for external disturbances. Similar to the PI controller proposed by [29], their proposed SMC controller relies heavily on manual parameter tuning. Additionally, once tuned, the controller parameters remain constant and do not adapt to variation in the soft robot. Soft robots undergo inevitable wear and tear [32] after several usage cycles, which primarily impact the model of soft robots. The controllers mentioned above are non-adaptive to the model variations, which results in gradual degrading of dynamic response.

In this paper, we propose a novel algorithm to regulate the dynamic characteristics of the soft robot by using impedance control [33–37] approach. The impedance control is used to simultaneously regulate both the position and force in a mechanical system. By controlling the impedance of a mechanical system, the required dynamic response can be achieved. In order to enforce the required impedance on the pneumatic soft robot, a nonlinear sliding mode controller (SMC) [1,31,36] is proposed in this paper. The sliding mode surface for the controller is defined according to the desired dynamic response, i.e., on the sliding surface system, the impedance of the system closely matches the desired impedance. Due to model irregularities and nonlinearities, the system can quickly diverge from the sliding surface. Therefore, a nonlinear switching control action term [38–40] is added to the controller, which always drives the system toward the sliding surface and guarantees its convergence. Additionally, the proposed SMC is discretized to avoid state observers [41], which significantly reduces the computational load. The discrete controller can easily be extended to higher-order systems and efficiently implemented on embedded processors.

The novelty of our proposed controller is the ability to achieve the desired dynamic response by directly specifying the impedance parameters instead of relying on manual parameter tuning. Our proposed controller uses the identified model of the robot to calculate the controller parameter automatically. To the best of our knowledge, no literature work related to soft pneumatic robot focuses on the impedance control to regulate their dynamic response. The main contributions of this paper can be summarized as follows

- Formulation of a discrete sliding mode controller based on impedance control for regulating the dynamic response of soft robots, suppressing undesirable dynamics and reducing settling time.
- Automatic calculation of the controller parameters based on the required impedance of the soft robot. It removes the reliance on a human operator to manually tune the controller parameters as required by the traditional controller.
- Formulation of the proposed sliding mode controller in discrete time to avoid the state observers. It reduces the computational complexity and makes the controller easily extendable to higher order systems models.

The remainder of this paper is distributed as follow. The impedance control problem is formally stated in Section 2, and then we propose the formulation of the SMC controller in Section 3. Later in Section 4, we present the design of our experimental platform used to test the proposed SMC controller on the soft parallel robot. Section 5 presents the experimental results and Section 6 concludes the paper.

2. Problem Formulation

In this section, we will present the formal mathematical definition of the impedance control problem.

2.1. Dynamic Model

The Sliding Mode Controller (SMC) is based on the dynamic model of the system. Consequently, it requires the construction of state observers for the estimation of higher-order terms in the system model. The computational burden and complexity of state observers can be avoided by formulating the model and controller in discrete-time. In this paper, we will formulate a generalized discrete SMC, which can easily be extended to high order systems models. The generalized discrete-time transfer function of a practical system, whose next state depends only on the present and past values of state and input can be written as follows,

$$\mathbf{H}(z) = \frac{\boldsymbol{\theta}(z)}{\mathbf{U}(z)} = \frac{b_1 z^{-1} + b_2 z^{-2} + ... + b_m z^{-m}}{a_0 + a_1 z^{-1} + ... + a_n z^{-n}}, \tag{1}$$

where θ is the bending angle of our soft robot, and u is the input air pressure. The above discrete transfer function is equivalent to the following state-space model with added component $d(k)$ for modeling errors and disturbances,

$$\theta(k+1) = \sum_{i=0}^{n-1} A_i \theta(k-i) + \sum_{j=0}^{m-1} B_j u(k-j) + d(k) \tag{2}$$

where $A_i = -a_{i+1}/a_0$ and $B_j = b_{j+1} \cdot a_i$ and b_i are the model parameters. The above equation represents an n-th order system model. Based on the perturbation method, we can generate an approximation of $d(k)$ during runtime, following the approach of [34,41–44], using one-step delayed estimation as follow

$$\begin{aligned} \hat{d}(k) &= d(k-1) \\ &= \theta(k) - \sum_{i=0}^{n-1} A_i \theta(k-i-1) - \sum_{j=0}^{m-1} B_j u(k-j-1). \end{aligned} \tag{3}$$

where $\hat{d}(k)$ denotes an estimation of the real disturbance value $d(k)$. Using the above estimation, the dynamic model in (2) can be written as,

$$\theta(k+1) = \sum_{i=0}^{n-1} A_i \theta(k-i) + \sum_{j=0}^{m-1} B_j u(k-j) + \hat{d}(k) + \tilde{d}(k), \tag{4}$$

where $\tilde{d}(k) = d(k) - \hat{d}(k)$ is the disturbance estimation error.

Assumption 1. *The rate of change in disturbance $d(k)$ is bounded i.e., $|\dot{d}(t)| < \rho$ for some constant value of ρ. This can be discretized using Euler backward approximation method as follow,*

$$|\dot{d}(t)| \approx \left| \frac{d(k+1) - d(k)}{T_s} \right| \leq \rho,$$

where T_s is the sampling time. The above equation can be further written as,

$$|d(k+1) - d(k)| \leq \rho T_s. \tag{5}$$

This relation mathematically models the assumption that the maximum difference between current and the future value of real disturbance $d(k)$ depends on sampling time T_s and bounded by a maximum value of ρT_s. This assumption will be used later in the stability analysis of the controller.

2.2. Impedance Control

The relation between system motion and the applied force is defined as the impedance of the system. Impedance control deals with the simultaneous control of motion as well as the force to regulate the dynamic impedance of the system [35]. By appropriately regulating the force-motion relations of a system, the dynamic response of the system, i.e., overshoot, settling time, and vibrations, can be accurately controlled. Now we will formally define the impedance control problem.

Let f be the force being applied to the soft robot. The force will generate some bending motion θ in the soft robot. By following the strategy of [34,45,46] the generalized impedance model for the trajectory tracking of the soft robot can be derived as [33,35,47–49]

$$m\ddot{e}_\theta(t) + b\dot{e}_\theta(t) + ke_\theta(t) = k_f f(t), \tag{6}$$

where $e_\theta(t) = \theta(t) - \theta_r(t)$ is the position error. $\theta_r(t)$ denotes the desired bending angle, also called reference signal. The m, b, k and k_f are the design parameters and chosen according to the required impedance. These parameters control the final dynamic response of the controlled system. The above equation can be discretized by using Euler backward difference as [34]

$$e_\theta(t) = e_\theta(k+1), \quad \dot{e}_\theta(t) = \frac{e_\theta(k+1) - e_\theta(k)}{T},$$
$$\ddot{e}_\theta(t) = \frac{e_\theta(k+1) - 2e_\theta(k) + e_\theta(k-1)}{T^2}.$$

Using the above approximations with (6), the continuous impedance model can be discretized as follow,

$$Me_\theta(k-1) + Be_\theta(k) + Ke_\theta(k+1) = K_f f(k+1), \tag{7}$$

where the parameters M, B, K and K_f are defined as,

$$M = \frac{m}{T^2}, B = -\frac{2m}{T^2} - \frac{b}{T}, K = \frac{m}{T^2} + \frac{b}{T} + k, K_f = k_f.$$

If the system (2) is forced to follow the dynamics of the impedance model of (7), then the desired transient response will be achieved. However, in practical scenarios, because of the significant disturbance $d(k)$, it is quite challenging to make the system follow the desired impedance response using a linear controller. The nonlinear discrete SMC solves this issue by using a nonlinear term in control law $u(k)$, which guarantees that system closely follow the required dynamic response of (7). The convergence of system response to desired dynamic behavior will be proven in the next section.

3. Discrete SMC Formulation

In this section, we will first present the formulation of discrete SMC controller and then prove its stability under the influence of disturbances $d(k)$.

3.1. Controller Design

The SMC is formulated by considering that the control objective is to enforce the dynamics of (7) on system (2). Therefore the sliding function [44,50,51] is defined as follows

$$s(k) = c_1 e_\theta(k-1) + c_2 e_\theta(k) + c_3 e_\theta(k+1) + c_4 f(k+1), \tag{8}$$

where we take $c_1 = M$, $c_2 = B$, $c_3 = K$ and $c_4 = -K_f$.

It can be seen that if the sliding function $s(k)$ converges to 0, then the system will produce the desired impedance behavior of (7). The SMC can help in achieving this objective by maintaining the system on the sliding surface $s(k) = 0$, thus giving the desired impedance behavior. In the following, we will drive the SMC to reach the sliding surface in finite time and then prove its stability.

According to (8), the sliding surface $s(k) = 0$ can be defined as,

$$c_1 e_\theta(k-1) + c_2 e_\theta(k) + c_3 e_\theta(k+1) + c_4 f(k+1) = 0. \tag{9}$$

Different definitions of discrete sliding mode are used in literature [52]. According to our definition of the sliding surface $s(k)$, we will use $s(k+1) = s(k) = 0$ in this paper. The resultant controller will try to direct the system toward the sliding surface in a single iteration. It results in faster convergence and robust performance.

Equation (9) can be rewritten as follows

$$c_1 e_\theta(k-1) + c_2 e_\theta(k) + c_3\{\theta(k+1) - \theta_r(k+1)\} + c_4 f(k+1) = 0. \tag{10}$$

Since the control objective is to enforce the dynamics of above equation to the soft robot, we will replace the value of $\theta(k+1)$ from robot model (4) to the above equation. After algebraic manipulation and ignoring the unknown disturbance estimation error $\tilde{d}(k)$, the final expression for equivalent control action $u_{eq}(k)$ can be obtained as follow [34,42,48]

$$\begin{aligned} u_{eq}(k) = &- B_0^{-1}\Big\{ \sum_{i=0}^{n-1} A_i\theta(k-i) + \sum_{j=1}^{m-1} B_j u(k-j) \\ &+ \hat{d}(k) - \theta_r(k+1)\Big\} \\ &- (c_3B_0)^{-1}\Big\{c_1e_\theta(k-1) + c_2e_\theta(k) + c_4f(k+1)\Big\}. \end{aligned} \tag{11}$$

The above control action will give the required dynamic performance if the disturbance estimation $\hat{d}(k)$ is ideal i.e., $\hat{d}(k) = d(k)$. This control action will keep the system on the sliding surface if it is initialized on the sliding surface, and there are no unknown external disturbances. If the initial starting point starts away from the sliding surface or if there a disturbance estimation error, i.e., $\hat{d}(k) \neq d(k)$, the equivalent control action cannot drive the system back to the sliding surface. A nonlinear switching control action is defined, to overcome this issue, as follows

$$u_{sw}(k) = -G(c_3B_0)^{-1}\text{sign}(s(k-1)), \tag{12}$$

where sign(.) is the signum function and G is a positive gain. Factor of $(c_3B_0)^{-1}$ is added to make notation simpler in the stability analysis, later in this section. The actual control action $u(k)$ is defined as the combination of the equivalent control action $u_{eq}(k)$ and the nonlinear switching control $u_{sw}(k)$,

$$u(k) = u_{eq}(k) + u_{sw}(k).$$

By putting the values from (11) and (12), the final form of the control law becomes,

$$\begin{aligned} u(k) = &- B_0^{-1}\Big\{ \sum_{i=0}^{n-1} A_i\theta(k-i) + \sum_{j=1}^{m-1} B_j u(k-j) \\ &+ \hat{d}(k) - \theta_r(k+1)\Big\} \\ &- (c_3B_0)^{-1}\Big\{c_1e_\theta(k-1) + c_2e_\theta(k) + c_4f(k+1)\Big\} \\ &- G(c_3B_0)^{-1}\text{sign}(s(k-1)). \end{aligned} \tag{13}$$

3.2. Stability Analysis

Now we will show that the the controller (13) will drive the system (4) toward the sliding surface (9) in finite number of steps.

Replacing the value of $u(k)$ from (13) into robot model (4) and after simplification, we get,

$$\begin{aligned} \theta(k+1) = &\,\theta_r(k+1) + \tilde{d}(k) - Gc_3^{-1}\text{sign}(s(k-1)) \\ &- c_3^{-1}\{c_1e_\theta(k-1) + c_2e_\theta(k) + c_4e_\theta(k-1)\}. \end{aligned}$$

If we replace this value of $\theta(k+1)$ in (8) by using the $\theta(k+1) - \theta_r(k+1) = e_\theta(k+1)$, we get the following expression for the sliding function $s(k)$ after simplification,

$$s(k) = -G\text{sign}(s(k-1)) + c_3\tilde{d}(k). \tag{14}$$

We now need to show that $s(k)$ converges to the sliding surface i.e., $s(k) = 0$ in finite time-steps. This can be shown by following the scheme of [34]. Lets assume that $G = K_3|\tilde{d}(k)| + \sigma$, where σ is an arbitrary positive constant. Based on such value of G and (14), the following conclusion can be drawn,

$$\sigma < |s(k)| < 2c_3|\tilde{d}(k)| + \sigma.$$

According to (5) as stated in assumption 1, the disturbance is bounded

$$|\tilde{d}(k)| \leq \rho T,$$

therefore, we have the following bound on the value of $s(k)$,

$$\sigma < |s(k)| < 2c_3\rho T + \sigma.$$

This shows that the value of $s(k)$ will reach in $2c_3\rho T + \sigma$ neighborhood of sliding surface $s(k) = 0$ within finite number of steps.

Remark 1. *The $sign(s)$ function is a discontinuous function at $s = 0$. The discontinuity introduces a chattering in the control input $u(k)$. The chattering can be avoided by gradually changing the value of the switching function. There are several ways to smooth the function as shown by [38–40,53]. For our controller, we used the saturation function [34] for its simplicity and computational efficiency*

$$sat(s) = \begin{cases} +1, & \text{if} \quad s > +\delta \\ s/\delta, & \text{if} \quad -\delta < s < +\delta \\ -1, & \text{if} \quad s < -\delta \end{cases} \tag{15}$$

where δ corresponds to the width of the ramp in saturation function.

4. Experimental Platform

In this section, we will first describe the fabrication, actuation, sensing, and control mechanism for our soft robot and then present the system identification method for estimating the parameters A_i's and B_i's in (4) for our soft robot.

4.1. Soft Parallel Robot

4.1.1. Design and Fabrication

For experiments, we used a pneumatically actuated soft parallel robot, as shown in Figure 2. The soft robot consists of a cylindrical body, and 6-parallel inflatable chambers radially distributed inside its body. The soft robot is capable of bending motion when actuated using pressurized air. Since the soft robot consists of 6 inflatable chambers, it is capable of performing a motion in 3D-space.

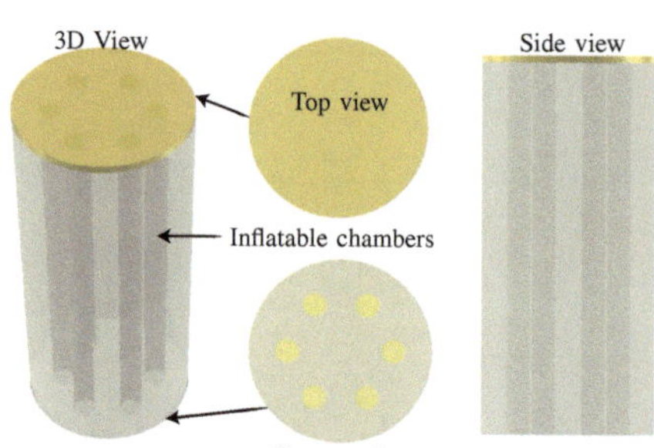

Figure 2. 3D model of the soft robot. Left: 3d view, middle-up: top view, middle-down: bottom view, and right: side view of the soft robot.

We designed the molds for the fabrication of our soft parallel robot. The molds were prepared using a 3D printer. We chose Ecoflex-30 (Smooth-On Inc., Macungie, PA, USA) silicone material as the fabrication material of our soft robot because of its high flexibility and excellent strength. Ecoflex-30 consists of two separate liquid components. The liquid silicone mixtures were prepared by mixing both components in equal proportion.

The mixture was then poured in the 3d-printed molds and left for curing at atmospheric temperature and pressure for about 8 h. After the curing of silicone is complete, the soft robot was removed from the molds.

4.1.2. Actuation and Sensing

Our soft parallel robot is based on pneumatics and uses air pressure for the actuation of inflatable chambers. To provide the necessary air pressure, we used a 12V-DC vacuum-pump. Since the input $u(k)$ to the soft robot is the air pressure, therefore we also used pneumatic-valves to regulate the air pressure from the pump to the inflatable chambers of the soft robot. The schematic diagram of the actuation mechanism is shown in Figure 3. A 3-port 3-position valves connected between each chamber and the vacuum pump. A 3-port 3-state valve can exist in 3 states: state one corresponds to airflow from pump to chambers, state 2 corresponds to holding the air inside the chamber, and position three is used to direct airflow out from the chambers to the atmosphere. By controlling the valve states, the air pressure inside the chambers can be controlled as required by the control law $u(k)$.

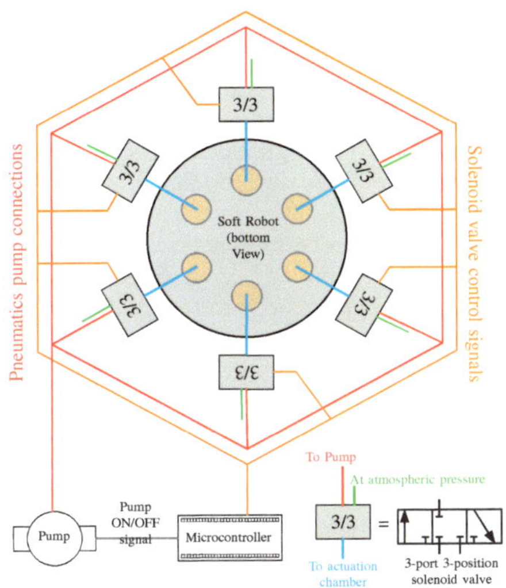

Figure 3. Schematic diagram of the actuation mechanism and control components used in the experimental platform for our soft parallel robot.

The valves require high electric power for switching between state, which cannot be directly provided by the embedded microcontrollers. To overcome this issue, we used MOSFET switches between the valves and the microcontrollers. The MOSFET switches have a high power rating and fast switching speed, which makes them ideal for controlling the valve states.

Our soft robot is capable of performing bending motion in the 3D-space. Therefore, we used an orientation sensor to measure its bending motion. The orientation sensor is mounted on the top surface of the soft robot, as shown in Figure 4. The orientation sensor gives the magnitude of the bending angle relative to the horizontal top surface of the soft robot. In the normal state, when the soft robot is the vertical position, the orientation gives zero angles of bending. When a bending motion is created in soft robot and the top surface orient at an angle relative to the horizontal plane. The mounted sensor provides this magnitude of this orientation angle. The sensor is connected to the microcontroller wirelessly using a Bluetooth module. The wireless sensing can even allow a separate server for data acquisition and processing.

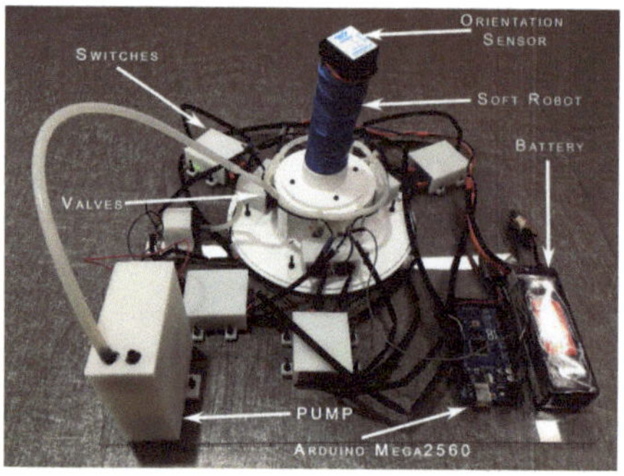

Figure 4. Experimental platform to test the effectiveness of the proposed controller for our soft parallel robot. All the major components are labeled in the image.

4.1.3. Controller Implementation

For the digital implementation of the discrete SMC controller, we used an Arduino Mega2560 microcontroller (RS Components, Hong Kong). We used an additional HC-05 Bluetooth module for wireless connection with the orientation sensor using the microcontroller's serial port. The digital pins of the microcontroller are connected with the MOSFET switches to drive the pneumatic valves. The high clock speed provided by the Arduino mega2560 microcontroller provides a reliable solution for real-time data acquisition and processing for the SMC control algorithm.

The commonly used approaches for implementing a controller on an embedded processor requires writing the C code. This approach is prone to bugs and increases the implementation time. To increase the implementation speed and improve the reliability of the implemented control algorithm, we used Simulink to design and test our control algorithm. Once the controller design is tested and finalized, we used Simulink Coder to generate C code for the Arduino mega2560 target automatically. This approach offers several advantages over manually writing the C code. Simulink is widely accepted as an excellent platform for control design, simulation, and testing [54]. Combining it with the Simulink coder allows for rapid prototyping and testing of the control algorithm while avoiding the complexity and bugs of the underlying C code. This increases the speed and reliability of the controller implementation stage. The sampling time (T_s) of the control loop was set at 0.02 s. The schematic diagram of the controller implementation is shown in Figure 5.

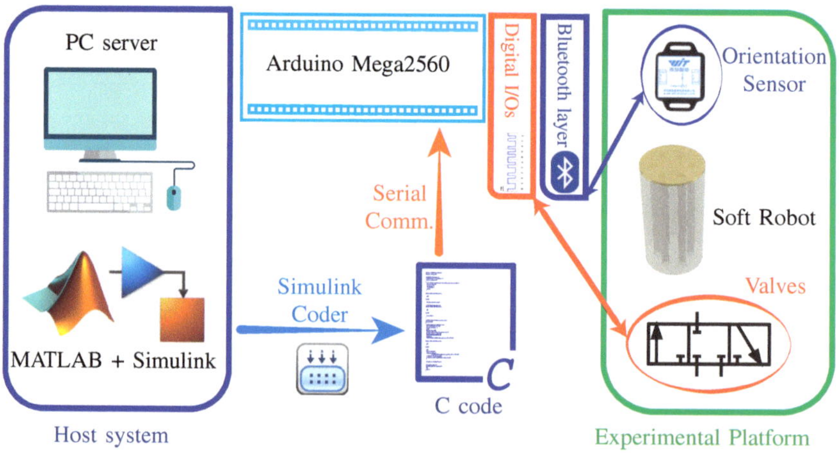

Figure 5. Schematic diagram of the control architecture used in experiments. It shows the interaction of software and hardware components of the platform.

4.2. Model Identification

After the construction of the experimental platform for our soft parallel robot, our next task was to estimate the model of the soft robot according to system model (4). The objective of this task was to estimate the model parameters A_i's and B_i's. Since all the chambers are identical, therefore we only need to run the estimation experiment for a single chamber. The disturbance estimation $\hat{d}(k)$ will account for minor differences between chambers. To estimate the parameters, we repeated the inflate-hold-deflate experiment with a time-period of 2 s and recorded the bending angles of the soft robot. During the experiment, a valve was switched to state 1 to apply high-pressure air to the corresponding chamber of the soft robot, then the valve was switched to state 2 to maintain the air pressure, and at the end, the valve was switched to state 3 to release the air pressure. The measured system response is shown in Figure 6. Once the input-output data was collected, we used MATLAB (R2022a) system identification toolbox (R2022a) [55] to estimate the model parameters. We estimated a 3rd order system model because its calculated response closely matched the measured response.

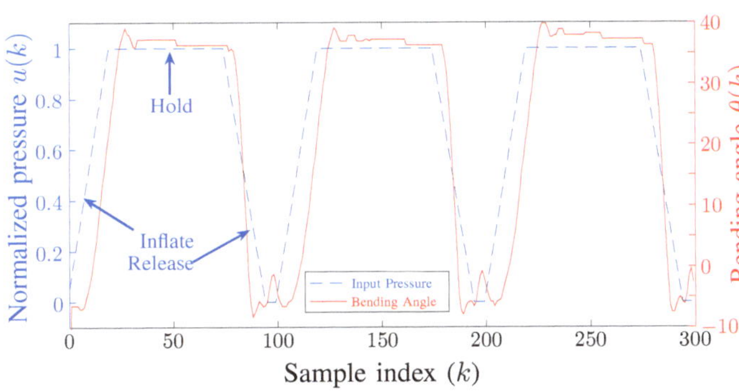

Figure 6. System response for inflate-hold-release experiment used to estimate the robot's model parameters A_i's and B_i's according to (4). (Sampling Time: $T_s = 0.02$).

5. Simulations and Experiments

In this section, we will present the experimental methodology and the obtained results. The experimental results show the efficacy of the presented SMC control algorithm in regulating the dynamic response and suppress the vibration amplitude. We will first present the simulation results and then present the results of experiments on our experimental platform.

5.1. Simulations

We first conducted a simulation-based evaluation of the discrete SMC to verify the effectiveness of the proposed algorithms. The simulation results help in assessing the stability and safety of the controller before trying it on the experimental platform. We used Simulink [54] as the simulation environment since later, we can directly use the simulation files for hardware implementation using the Simulink Coder (R2022a). We conducted a series of simulations using different system models, impedance parameters, and initial conditions.

One of the important choices for the impedance control was the selection of vibration damping force $f(t)$ profile along the robot trajectory, according to (6). For the purpose of simulation and experimentation, we chose the damping force to be proportional to the position error i.e., $f(t) = ce_\theta(t)$. It allows a large vibration-damping force when the soft robot is far from the vertical position, which gradually decreases as the robot return to the vertical position. The equivalent impedance equation can be written as,

$$m\ddot{e}_\theta(t) + b\dot{e}_\theta(t) + \bar{k}e_\theta(t) = 0, \tag{16}$$

where $\bar{k} = k - ck_f$. The results of varying the parameters m, b, and $\bar{k}$ are shown in Figure 7. The impact of different parameter values on the settling time and the vibration amplitude can be seen. Figure 7a shows that increasing the value of parameter m increases both the settling time and the vibration amplitude; thus, it is desirable to decrease the value of m. However, its lower value is bounded by the actuator's capacity, i.e., pump and valves speed, and cannot be made infinitesimally small. Similarly, it can be observed in Figure 7c that increasing b decreases the vibration amplitude. Figure 7d shows that increasing $\bar{k}$ will correspondingly increases the vibration amplitude. These correlations between values of m, b, and $\bar{k}$ also matches with the theoretical analysis of the second-order impedance Equation (7).

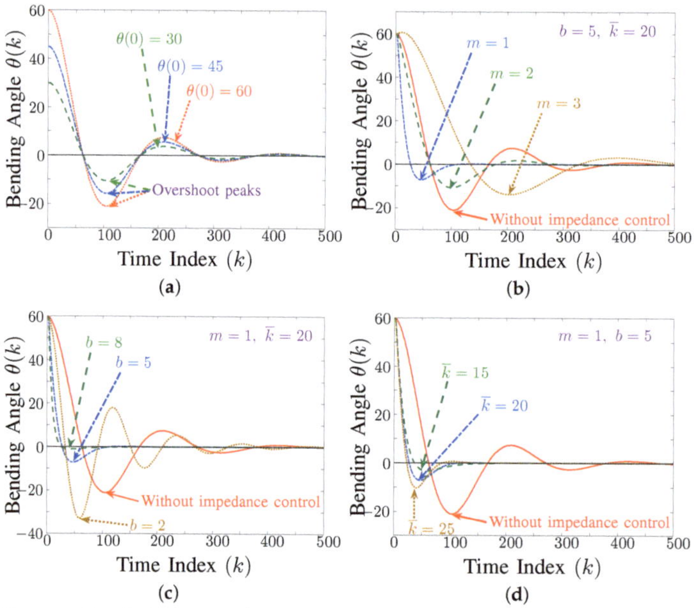

Figure 7. Simulation results showing impact of parameter variation on the settling time and vibration amplitude. (**a**) shows the deactuation response of the system for different initial conditions without any impedance control. (**b–d**) shows the impact of variation in parameter m, b, and $\bar{k}$ on system response. (Sampling Time: $T_s = 0.02$).

5.2. Experimental Results

We conducted experiments on our soft robot to demonstrate the effectiveness and performance of the proposed discrete SMC. These experiments check the performance of the controller on the deactuation of the soft robots since most vibrations are created on deactuation. The first set of experiments consists of a single axis motion of the soft robot in which one of the chambers is initially actuated with air and then deactuated using impedance control. The comparison for three experimental trials is shown in Figure 8. Figure 8a shows the performance of the soft robot without using impedance control. It can be seen that there is a large overshoot and vibrations as the soft robot approaches reference angle $\theta = 0$. We then implemented the impedance control with values of impedance parameters be $m = 1$, $b = 20$ and $\bar{k} = 500$. We chose these values based on the experimental trials and the estimated model of the soft robot. Figure 8b shows the results of using the impedance control on our soft robot. It can be seen that the vibration amplitude is smaller as compared to Figure 8a, which illustrates the effectiveness of the proposed SMC scheme in suppressing the vibration amplitude. In the next experiment, we modified the impedance parameters of the controller by changing $b = 25$. As already shown in the simulation results, increasing the value of b increases the overall damping force in the system. The experimental result in Figure 8c also shows agreement with the simulation results. It can be seen that there is no overshooting in this case and minimal vibrations.

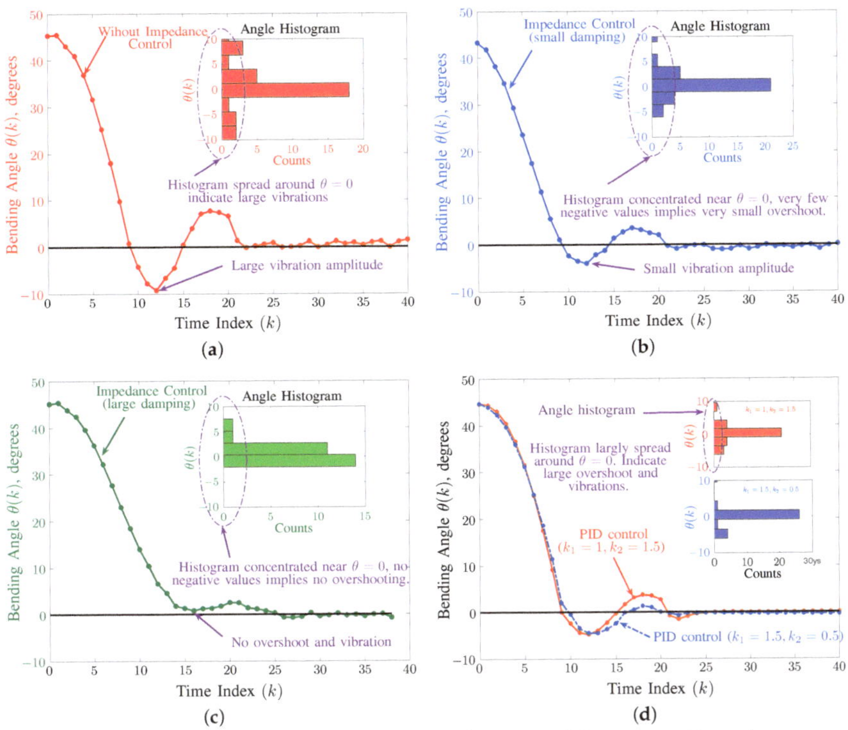

Figure 8. The experimental results, showing the performance of the proposed impedance control scheme. (**a**) the performance without the impedance control, shows large overshoot and vibrations. (**b**) the performance with low impedance ($m = 1, b = 20, \bar{k} = 500$), shows small overshoot. (**c**) the performance for high impedance ($m = 1, b = 25, \bar{k} = 500$) with almost no overshoot and minimal vibration. (**d**) The performance of PID controller in (17) for two different values of parameters k_1 and k_2, overshooting similar to (**b**). (Sampling Time: $T_s = 0.02$).

To further illustrate the difference between the three experimental cases, a histogram of angle values around $\theta = 0$ ($[-10, 10]$) is shown inside each plot. In the case of no impedance control, as shown in Figure 8a, the histogram is spread in the entire range from -10 to 10. The distribution of angle values indicates a large deviation from $\theta = 0$. In the second case, when the damping force is small, as shown in Figure 8b, the angle values are concentrated around $\theta = 0$. Whereas in the third case, when damping force is large, as shown in Figure 8c, the angle values are again more concentrated around $\theta = 0$ with no data points below $\theta = 0$.

The above-presented simulation and experimental results sufficiently demonstrate the efficacy of the proposed discrete SMC based impedance control scheme. We then ran a trajectory tracking experiment for our soft robot. In this experiment, three of the chambers of the soft robots were actuated and then deactuated sequentially using impedance control. Figure 9 shows the result of these experimental trails in the form of a 3D graph. Figure 9a shows the result without impedance control; it can be seen that the vibration (shown as red) have a large magnitude. In comparison, Figure 9b shows the results with the impedance control. It can be seen that the vibration cloud, in this case, is tiny as compared to Figure 9a.

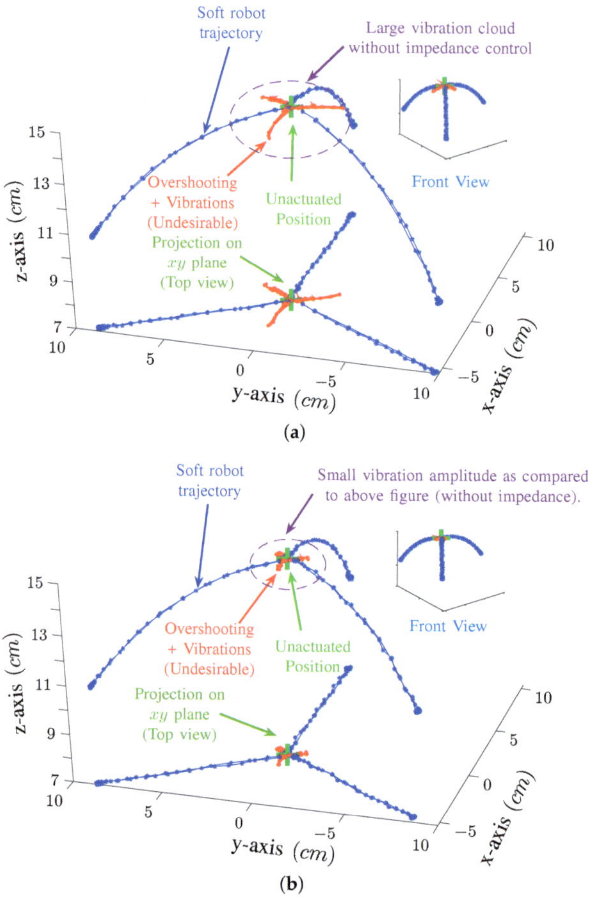

Figure 9. Illustration of 3D motion of the soft robot. Three chambers of the soft robot were actuated and then deactuated in consecutive turns. This figure shows a comparison between natural deactuation and impedance control. (**a**) shows result without the impedance control and (**b**) show response with impedance control. (Sampling Time: $T_s = 0.02$).

5.3. Comparative Results

We conducted experiments to determine the comparative performance of the proposed controller with the state-of-art SMC dynamic response controller [31]. As mentioned in Section 1, their proposed continuous-time controller is essentially a PD controller with a deadband, which can be formulated as follow,

$$\sigma(t) = \dot{e}_\theta(t) + De_\theta(t)$$

where D is the controller parameter need to be manually tuned. The input $u(t) = \sigma(t)$ is only applied if $|\sigma(t)| > \epsilon$, where ϵ corresponds to width of deadband. If $\sigma(t) < 0$ then $u(t) = 0$. Based on their strategy, we used the following general structure of PID controller in our experiments,

$$\sigma(t) = \dot{e}_\theta(t) + k_1 e_\theta(t) + k_2 \int_0^t e_\theta(\tau) d\tau$$

k_1 and k_2 are the parameter of the PID controller. The above controller can be converted to following discrete time controller for digital implementation,

$$\sigma(k) = \left[\frac{e_\theta(k) - e_\theta(k-1)}{T_s}\right] + k_1 e(k) + k_2 \sum_{i=0}^{k} e_\theta(k) T_s$$

and the deadband is added as follow,

$$u(k) = \begin{cases} \sigma(t) & \text{if} \quad |\sigma(t)| > \epsilon \\ 0 & \text{if} \quad |\sigma(t)| < \epsilon, \end{cases} \tag{17}$$

The experimental results are shown in Figure 8d. The figure shows the response of the system for two different values of parameters, k_1 and k_2. To present a fair comparison, we tuned the values of the PID parameters using the Ziegler–Nichols algorithm [56]. It can be seen that the response of the system in both cases shows overshooting. The first response (shown in red) corresponds to the parameter values: $k_1 = 1, k_2 = 1.5$, shows large overshoot as compared to the second response (shown in blue), which corresponds to $k_1 = 1.5, k_2 = 0.5$. In both cases, the bending angle overshoot to a maximum angle of about -4 degrees. The PID response is comparative to the response of Figure 8b, which shows approximately the same overshoot. However, the response of the PID controller is worse, then the SMC impedance controller in Figure 8c. The histograms of bending angles inside Figure 8d. The spread of histogram also affirms that the response is similar to Figure 8b. The presented experimental establishes the superiority of the proposed controller against the current state-of-the-art in SMC control for pneumatic soft robots.

6. Conclusions

In this paper, we presented an impedance control scheme for pneumatic soft robots. A nonlinear discrete-time sliding-mode controller based on impedance control was used to enforce the desired impedance dynamics on a soft pneumatic robot. The proposed controller automatically calculates the controller parameter based on the required impedance of the soft robot. It is in contrast with the traditional work on soft robotics, which uses manual parameter tuning to tune the parameters. The stability analysis of the proposed controller is also presented and proved that the controller is asymptotically stable. The 3D-design of our 6-chambered soft parallel robot was also presented along with the design and construction of the experimental platform used to conduct the experiments. It was proven through theoretical analysis and extensive comparative experimentation with the state-of-the-art PID controller that the proposed control scheme shows an outstanding performance in regulating the dynamic behavior of the soft pneumatic robot and suppressing the undesirable dynamic behavior, e.g., overshooting and vibrations. It is worth mentioning that this is the first work to apply the impedance control scheme to the recently developing field of soft robots, the inherent uncertainty of which makes the extension from rigid-body systems to soft-body ones a challenging topic.

Author Contributions: Conceptualization, A.H.K. and S.L.; methodology, A.H.K.; software, A.H.K.; validation, A.H.K. and S.L.; formal analysis, A.H.K. and S.L.; investigation, A.H.K.; resources, A.H.K.; data curation, A.H.K.; writing—original draft preparation, A.H.K.; writing—review and editing, A.H.K.; visualization, A.H.K.; supervision, S.L.; project administration, S.L.; funding acquisition, S.L. All authors have read and agreed to the published version of the manuscript.

Funding: This research received no external funding.

Institutional Review Board Statement: Not applicable.

Data Availability Statement: The raw data supporting the conclusions of this article will be made available by the authors on request.

Conflicts of Interest: The authors declare no conflict of interest.

References

1. Yu, L.; Huang, J.; Luo, W.; Chang, S.; Sun, H.; Tian, H. Sliding-mode control for PMLSM position control—A review. *Actuators* **2023**, *12*, 31. [CrossRef]
2. Armanini, C.; Boyer, F.; Mathew, A.T.; Duriez, C.; Renda, F. Soft robots modeling: A structured overview. *IEEE Trans. Robot.* **2023**, *39*, 1728–1748. [CrossRef]
3. Della Santina, C.; Duriez, C.; Rus, D. Model-based control of soft robots: A survey of the state of the art and open challenges. *IEEE Control Syst. Mag.* **2023**, *43*, 30–65. [CrossRef]
4. Haggerty, D.A.; Banks, M.J.; Kamenar, E.; Cao, A.B.; Curtis, P.C.; Mezić, I.; Hawkes, E.W. Control of soft robots with inertial dynamics. *Sci. Robot.* **2023**, *8*, eadd6864. [CrossRef] [PubMed]
5. Cheng, L.; Liu, W.; Yang, C.; Huang, T.; Hou, Z.G.; Tan, M. A neural-network-based controller for piezoelectric-actuated stick–slip devices. *IEEE Trans. Ind. Electron.* **2017**, *65*, 2598–2607. [CrossRef]
6. Lin, H.; Guo, F.; Wang, F.; Jia, Y.B. Picking up a soft 3D object by "feeling" the grip. *Int. J. Robot. Res.* **2015**, *34*, 1361–1384. [CrossRef]
7. Overvelde, J.; Kloek, T.; D'haen, J.; Bertoldi, K. Amplifying the response of soft actuators by harnessing snap-through instabilities. *Proc. Natl. Acad. Sci. USA* **2015**, *112*, 10863–10868. [CrossRef] [PubMed]
8. Cianchetti, M.; Laschi, C.; Menciassi, A.; Dario, P. Biomedical applications of soft robotics. *Nat. Rev. Mater.* **2018**, *3*, 143–153. [CrossRef]
9. Polygerinos, P.; Wang, Z.; Galloway, K.; Wood, R.; Walsh, C. Soft robotic glove for combined assistance and at-home rehabilitation. *Robot. Auton. Syst.* **2015**, *73*, 135–143. [CrossRef]
10. Sridar, S.; Qiao, Z.; Muthukrishnan, N.; Zhang, W.; Polygerinos, P.P. A soft-inflatable exosuit for knee rehabilitation: Assisting swing phase during walking. *Front. Robot. AI* **2018**, *5*, 44. [CrossRef]
11. Ling, S.; Wang, H.; Liu, P.X. Adaptive fuzzy dynamic surface control of flexible-joint robot systems with input saturation. *IEEE/CAA J. Autom. Sin.* **2019**, *6*, 97–107. [CrossRef]
12. Marchese, A.; Onal, C.; Rus, D. Autonomous soft robotic fish capable of escape maneuvers using fluidic elastomer actuators. *Soft Robot.* **2014**, *1*, 75–87. [CrossRef]
13. Laschi, C.; Cianchetti, M.; Mazzolai, B.; Margheri, L.; Follador, M.; Dario, P. Soft robot arm inspired by the octopus. *Adv. Robot.* **2012**, *26*, 709–727. [CrossRef]
14. Hawkes, E.W.; Blumenschein, L.H.; Greer, J.D.; Okamura, A.M. A soft robot that navigates its environment through growth. *Sci. Robot.* **2017**, *2*, eaan3028. [CrossRef] [PubMed]
15. Kim, J.; Alspach, A.; Yamane, K. 3D printed soft skin for safe human-robot interaction. In Proceedings of the IEEE/RSJ International Conference on Intelligent Robots and Systems (IROS), Hamburg, Germany, 28 September–2 October 2015; IEEE: Piscataway, NJ, USA, 2015; pp. 2419–2425.
16. Yang, Y.; Chen, Y.; Li, Y.; Chen, M.Z.; Wei, Y. Bioinspired robotic fingers based on pneumatic actuator and 3D printing of smart material. *Soft Robot.* **2017**, *4*, 147–162. [CrossRef] [PubMed]
17. Hao, Y.; Gong, Z.; Xie, Z.; Guan, S.; Yang, X.; Ren, Z.; Wang, T.; Wen, L. Universal soft pneumatic robotic gripper with variable effective length. In Proceedings of the 2016 35th Chinese Control Conference (CCC), Chengdu, China, 27–29 July 2016; IEEE: Piscataway, NJ, USA, 2016; pp. 6109–6114.
18. Doumit, M.; Fahim, A.; Munro, M. Analytical modeling and experimental validation of the braided pneumatic muscle. *IEEE Trans. Robot.* **2009**, *25*, 1282–1291. [CrossRef]
19. Polygerinos, P.; Wang, Z.; Overvelde, J.T.; Galloway, K.C.; Wood, R.J.; Bertoldi, K.; Walsh, C.J. Modeling of soft fiber-reinforced bending actuators. *IEEE Trans. Robot.* **2015**, *31*, 778–789. [CrossRef]
20. Yang, C.; Luo, J.; Liu, C.; Li, M.; Dai, S.L. Haptics electromyogrphy perception and learning enhanced intelligence for teleoperated robot. *IEEE Trans. Autom. Sci. Eng.* **2018**, *16*, 1512–1521. [CrossRef]
21. Bilodeau, R.A.; White, E.L.; Kramer, R.K. Monolithic fabrication of sensors and actuators in a soft robotic gripper. In Proceedings of the 2015 IEEE/RSJ International Conference on Intelligent Robots and Systems (IROS), Hamburg, Germany, 28 September–2 October 2015; IEEE: Piscataway, NJ, USA, 2015; pp. 2324–2329.
22. Wang, H.; Liu, P.X.; Zhao, X.; Liu, X. Adaptive fuzzy finite-time control of nonlinear systems with actuator faults. *IEEE Trans. Cybern.* **2019**, *50*, 1786–1797. [CrossRef]
23. Wang, H.; Liu, P.X.; Xie, X.; Liu, X.; Hayat, T.; Alsaadi, F.E. Adaptive fuzzy asymptotical tracking control of nonlinear systems with unmodeled dynamics and quantized actuator. *Inf. Sci.* **2018**, *575*, 779–792. [CrossRef]
24. Hochberg, L.R.; Bacher, D.; Jarosiewicz, B.; Masse, N.Y.; Simeral, J.D.; Vogel, J.; Haddadin, S.; Liu, J.; Cash, S.S.; van der Smagt, P.; et al. Reach and grasp by people with tetraplegia using a neurally controlled robotic arm. *Nature* **2012**, *485*, 372. [CrossRef]
25. Ni, F.; Henning, A.; Tang, K.; Cai, L. Soft damper for quick stabilization of soft robotic actuator. In Proceedings of the IEEE International Conference on the Real-Time Computing and Robotics (RCAR), Angkor Wat, Cambodia, 6–10 June 2016; IEEE: Piscataway, NJ, USA, 2016; pp. 466–471.
26. Li, Y.; Chen, Y.; Ren, T.; Hu, Y. Passive and Active Particle Damping in Soft Robotic Actuators. In Proceedings of the 2018 IEEE International Conference on Robotics and Automation (ICRA), Brisbane, QLD, Australia, 21–25 May 2018; IEEE: Piscataway, NJ, USA, 2018; pp. 1547–1552.

27. Jeon, J.W. An efficient acceleration for fast motion of industrial robots. In Proceedings of the 1995 IEEE IECON 21st International Conference on Industrial Electronics, Control, and Instrumentation, Orlando, FL, USA, 6–10 November 1995; IEEE: Piscataway, NJ, USA, 1995; Volume 2, pp. 1336–1341.
28. Largilliere, F.; Verona, V.; Coevoet, E. Real-time control of soft-robots using asynchronous finite element modeling. In Proceedings of the 2015 IEEE International Conference on Robotics and Automation (ICRA), Seattle, WA, USA, 26–30 May 2015; IEEE: Piscataway, NJ, USA, 2015; pp. 2550–2555.
29. Gerboni, G.; Diodato, A.; Ciuti, G.; Cianchetti, M.; Menciassi, A. Feedback control of soft robot actuators via commercial flex bend sensors. *IEEE/ASME Trans. Mechatron.* **2017**, *22*, 1881–1888. [CrossRef]
30. De Silva, C.W. *Vibration Damping, Control, and Design*; CRC Press: Boca Raton, FL, USA, 2007.
31. Luo, M.; Skorina, E.H.; Tao, W.; Chen, F.; Ozel, S.; Sun, Y.; Onal, C.D. Toward Modular Soft Robotics: Proprioceptive Curvature Sensing and Sliding-Mode Control of Soft Bidirectional Bending Modules. *Soft Robot.* **2017**, *4*, 117–125. [CrossRef]
32. Terryn, S.; Brancart, J.; Lefeber, D.; Van Assche, G.; Vanderborght, B. Self-healing soft pneumatic robots. *Sci. Robot.* **2017**, *2*, eaan4268. [CrossRef] [PubMed]
33. He, W.; Dong, Y.; Sun, C. Adaptive neural impedance control of a robotic manipulator with input saturation. *IEEE Trans. Syst. Man Cybern. Syst.* **2016**, *46*, 334–344. [CrossRef]
34. Xu, Q. Robust Impedance Control of a Compliant Microgripper for High-Speed Position/Force Regulation. *IEEE Trans. Ind. Electron.* **2015**, *62*, 1201–1209. [CrossRef]
35. Li, Z.; Huang, Z.; He, W.; Su, C.Y. Adaptive impedance control for an upper limb robotic exoskeleton using biological signals. *IEEE Trans. Ind. Electron.* **2017**, *64*, 1664–1674. [CrossRef]
36. Xu, Q. Adaptive discrete-time sliding mode impedance control of a piezoelectric microgripper. *IEEE Trans. Robot.* **2013**, *29*, 663–673. [CrossRef]
37. Yang, C.; Peng, G.; Li, Y.; Cui, R.; Cheng, L.; Li, Z. Neural Networks Enhanced Adaptive Admittance Control of Optimized Robot–Environment Interaction. *IEEE Trans. Cybern.* **2018**, *49*, 2568–2579. [CrossRef]
38. Chen, M.S.; Hwang, Y.R.; Tomizuka, M. A state-dependent boundary layer design for sliding mode control. *IEEE Trans. Autom. Control* **2002**, *47*, 1677–1681. [CrossRef]
39. Wang, Y.; Gu, L.; Xu, Y.; Cao, X. Practical tracking control of robot manipulators with continuous fractional-order nonsingular terminal sliding mode. *IEEE Trans. Ind. Electron.* **2016**, *63*, 6194–6204. [CrossRef]
40. Yu, S.; Yu, X.; Shirinzadeh, B.; Man, Z. Continuous finite-time control for robotic manipulators with terminal sliding mode. *Automatica* **2005**, *41*, 1957–1964. [CrossRef]
41. Abidi, K.; Xu, J.X.; Xinghuo, Y. On the discrete-time integral sliding-mode control. *IEEE Trans. Autom. Control* **2007**, *52*, 709–715. [CrossRef]
42. Wang, G.; Xu, Q. Sliding mode control with disturbance rejection for piezoelectric nanopositioning control. In Proceedings of the 2018 Annual American Control Conference (ACC), Milwaukee, WI, USA, 27–29 June 2018; IEEE: Piscataway, NJ, USA, 2018; pp. 6144–6149.
43. Li, Y.; Xu, Q. Adaptive sliding mode control with perturbation estimation and PID sliding surface for motion tracking of a piezo-driven micromanipulator. *IEEE Trans. Control Syst. Technol.* **2009**, *18*, 798–810. [CrossRef]
44. Du, H.; Yu, X.; Chen, M.Z.; Li, S. Chattering-free discrete-time sliding mode control. *Automatica* **2016**, *68*, 87–91. [CrossRef]
45. Liaw, H.C.; Shirinzadeh, B. Robust generalised impedance control of piezo-actuated flexure-based four-bar mechanisms for micro/nano manipulation. *Sens. Actuators A Phys.* **2008**, *148*, 443–453. [CrossRef]
46. Khan, A.T.; Cao, X.; Li, S.; Hu, B.; Katsikis, V.N. Quantum beetle antennae search: A novel technique for the constrained portfolio optimization problem. *Sci. China Inf. Sci.* **2021**, *64*, 152204. [CrossRef]
47. Lee, T.H.; Harris, C.J. *Adaptive Neural Network Control of Robotic Manipulators*; World Scientific: Singapore, 1998; Volume 19.
48. Garofalo, G.; Werner, A.; Loeffl, F.; Ott, C. Joint-Space Impedance Control Using Intrinsic Parameters of Compliant Actuators and Inner Sliding Mode Torque Loop. *IEEE Control Syst. Lett.* **2018**, *3*, 1–6. [CrossRef]
49. Katsikis, V.N.; Mourtas, S.D.; Stanimirović, P.S.; Zhang, Y. Solving complex-valued time-varying linear matrix equations via QR decomposition with applications to robotic motion tracking and on angle-of-arrival localization. *IEEE Trans. Neural Netw. Learn. Syst.* **2021**, *33*, 3415–3424. [CrossRef]
50. Su, X.; Liu, X.; Shi, P.; Yang, R. Sliding mode control of discrete-time switched systems with repeated scalar nonlinearities. *IEEE Trans. Autom. Control* **2016**, *62*, 4604–4610. [CrossRef]
51. Xu, Q.; Li, Y. Micro-/nanopositioning using model predictive output integral discrete sliding mode control. *IEEE Trans. Ind. Electron.* **2011**, *59*, 1161–1170. [CrossRef]
52. Yu, X.; Wang, B.; Li, X. Computer-Controlled Variable Structure Systems: The State-of-the-Art. *IEEE Trans. Ind. Inform.* **2012**, *8*, 197–205. [CrossRef]
53. Katsikis, V.N.; Pappas, D.; Petralias, A. An improved method for the computation of the Moore–Penrose inverse matrix. *Appl. Math. Comput.* **2011**, *217*, 9828–9834. [CrossRef]
54. Xue, D.; Chen, Y. *Modeling, Analysis and Design of Control Systems in MATLAB and Simulink*; World Scientific Publishing: Singapore, 2015.

55. Ljung, L. *MATLAB: System Identification Toolbox: User's Guide Version 4*; The Mathworks: Natick, MA, USA, 1995.
56. Åström, K.J.; Hägglund, T. Revisiting the Ziegler–Nichols step response method for PID control. *J. Process Control* **2004**, *14*, 635–650. [CrossRef]

biomimetics

MDPI

Article

Coordinated Transport by Dual Humanoid Robots Using Distributed Model Predictive Control

Shengjun Wen [1,*], Zhaoyuan Shi [2,*] and Hongjun Li [1]

[1] School of Zhongyuan-Petersburg Aviation, Zhongyuan University of Technology, Zhengzhou 450007, China; li.hongjun@zut.edu.cn
[2] School of Electronic Information, Zhongyuan University of Technology, Zhengzhou 450007, China
* Correspondence: wsj@zut.edu.cn (S.W.); 2021106208@zut.edu.cn (Z.S.)

Abstract: Dual humanoid robot collaborative control systems possess better flexibility and adaptability in complex environments due to their similar structures to humans. This paper adopts a distributed model predictive controller based on the leader–follower approach to address the collaborative transportation control issue of dual humanoid robots. In the dual-robot collaborative control system, network latency issues may arise due to unstable network conditions, affecting the consistency of dual-robot collaboration. To solve this issue, a communication protocol was constructed through socket communication for dual-robot collaborative consistency, thereby resolving the problem of consistency in dual humanoid robot collaboration. Additionally, due to the complex structure of humanoid robots, there are deficiencies in position tracking accuracy during movement. To address the poor accuracy in position tracking, this paper proposes a distributed model predictive control that considers historical cumulative error, thus enhancing the position tracking accuracy of dual-robot collaborative control.

Keywords: dual-robot system; distributed model predictive control; trajectory tracking; collaborative transportation

1. Introduction

Humanoid robots represent a complex dynamical system characterized by multiple variables, nonlinearity, strong coupling, and variable structures. They are an important branch in the field of robotics research. Compared to single-robot technology, multi-robot systems have numerous advantages. They operate within a broader workspace, offer greater maneuverability, and possess higher degrees of freedom, which can compensate for the limitations of individual robots and solve a wider range of tasks. In recent years, this area has become a significant focus within the realms of intelligent control and autonomous unmanned systems. With advancing research, the application scenarios and functions of multi-robot cooperative control continue to expand. Currently, multi-robot collaborative control is widely applied to complex tasks such as drone swarm control [1], formation control [2], and multi-robot target tracking [3]. They have become key technologies for collaborative transportation, surveillance search, and autonomous driving, holding substantial academic importance and practical value. The current multi-robot collaborative control algorithms encompass a variety of methods, including robust control [4], consensus control [5], linear quadratic regulator control [6], and reinforcement learning [7]. Comparatively, model predictive control (MPC) [8] can easily handle control constraints and perform rolling optimization. Through an internal model, MPC predicts the future performance of the system and solves for a suboptimal solution within a shorter time horizon, continuously updating it. Therefore, MPC can progressively adjust for model errors in real time, reducing the discrepancy between model outputs and references, ensuring the effective implementation of the control law. In multi-input multi-output (MIMO) systems like multi-robot collaboration, MPC can achieve better application.

Model predictive control (MPC) is an advanced control method that offers several advantages: (1) it is applicable to multi-input multi-output systems; (2) it exhibits strong robustness and stability; and (3) it can handle a variety of constraint conditions, among others. Currently, numerous researchers have employed MPC in cooperative control issues. SHEN et al. [9] designed a trajectory tracking control strategy based on the analysis of dynamic characteristics of autonomous underwater vehicles. By employing a distributed MPC approach, they decomposed a large optimization problem into several smaller subproblems and processed them separately, significantly reducing the number of floating-point operations required. Mohseni et al. [10] proposed a cooperative control strategy for the decentralized optimization control problem of autonomous vehicle systems, aimed at performing various complex traffic maneuvers. They then used a nonlinear MPC algorithm to solve it, incorporating collision avoidance constraints to ensure safety among vehicles.

Distributed model predictive control (DMPC), as a variant within MPC, has garnered widespread attention due to its flexible architecture, lower computational costs, and reduced communication requirements, among other advantages. Cao et al. [11] designed a robust DMPC considering external disturbances, introducing robust constraints to handle external perturbations, and analyzed the stability of the strategy as well as the overall multi-agent system. Wei et al. [12] proposed a robust DMPC method for a class of heterogeneous autonomous ground vehicles with input constraints and bounded external disturbances. By solving the DMPC problem, the optimal nominal control inputs were obtained. Dai et al. [13], considering the challenges of disturbances and coupling constraints to distributed control algorithms, proposed a distributed robust MPC algorithm for multi-agent systems with external disturbances and obstacle avoidance constraints. Pan et al. [14] proposed a DMPC scheme for maintaining the desired formation while a virtual leader tracks the reference trajectory. A virtual lead was designed to follow the reference path, with other agents maintaining the required distances and angles relative to the virtual lead to achieve trajectory tracking, overcoming the disadvantage where the failure of the traditional leader in a leader–follower model could destabilize the entire system.

The problem of robotic collaborative transport is one of the main application scenarios for robot coordination control. It plays a significant role across various industrial sectors. Robotic cooperative transport strategies are broadly divided into two categories: leader–follower [15] and encirclement cooperation. The leader–follower approach is primarily used in the coordination between two robots, where the leader plans the movement path and the follower maintains a specific relative positional relationship with the leader through certain constraint conditions. Munir et al. [16] addressed the consensus problem in leader–follower multi-robot cooperative systems by proposing an innovative arbitrary order finite-time sliding mode control design, which ensured the enforcement of sliding mode in finite time and maintained the stability of error dynamics. The effectiveness of the proposed approach was corroborated through rigorous closed-loop stability analysis and simulation examples. Ullah et al. [17] adopted a leader–follower strategy and proposed an innovative design for a distributed fixed-time synchronization controller based on neuro-adaptive non-singular terminal sliding mode control for higher-order multi-agent nonlinear systems. The effectiveness of this approach was demonstrated through simulation examples. Zhang et al. [18] adopted the leader–follower strategy, transforming the formation control problem into a follower's trajectory tracking issue with respect to a virtual leader. They achieved position control among robots and verified the system's stability using a Lyapunov function, ultimately enabling the transport of large items. Liu et al. [19], in the realm of control strategies, put forth a collaborative model that embodies the concept of a virtual leader and actual follower, segmenting the overall system's control architecture into distinct hierarchies for the leader and follower robots. Additionally, they implemented a dual closed-loop kinematic control framework, transmuting the motion management of both robotic categories into trajectory tracking control problems. The efficacy of the proposed structure and control mechanism was corroborated through empirical testing. Wu et al. [20] investigated the guidance control method for multi-robot collaborative transportation of large components.

They proposed various transportation configurations, including monocular–multidrive, binocular–bidrive, and binocular–multimotion–pair omnidirectional systems, along with motion control models. Additionally, they designed a path-tracking control method based on a leader–follower strategy, which incorporates homogeneous and heterogeneous architectures involving fuzzy logic and model predictive control. A prototype system was developed to conduct experiments, confirming the efficacy of the proposed methods.

In the problem of collaborative control of robots, most scholars have used wheeled robots as the research object, while less research has been carried out on the collaborative handling control of humanoid robots. Therefore, this paper establishes a dual humanoid robot collaborative handling control system. Due to the communication delay caused by network fluctuations in the process of dual-robot communication, which affects the consistency of dual-robot collaboration, this paper constructs a consensus communication protocol to solve the consistency problem of robot collaboration. Further investigation has revealed that model predictive control (MPC) methods, as utilized in the referenced literature, commonly exhibit deficiencies in control precision and lack the requisite level of accuracy. To address the issue of insufficient control precision within the dual-humanoid robot collaborative control system, this paper proposes a DMPC approach that takes into account historical cumulative error. This method is devised to enhance the steady-state performance in the collaborative control of the dual robots, thereby significantly improving control precision and enhancing accuracy.

2. Robot Motion Model

2.1. Humanoid Robot Kinematics Analysis

This paper employs the NAO robot as the research platform, as illustrated in Figure 1. The NAO robot [21], produced by Aldebaran Robotics, is a bipedal intelligent robot measuring 58 cm in height and weighing 5.4 kg. It features a total of 25 degrees of freedom, and its hardware components include a CPU, ultrasonic sensors, gyroscopes, and infrared devices, among others, as detailed in Table 1, which presents an overview of some of the NAO robot's hardware specifications.

The NAO robot features a hand with three fingers capable of performing various grasping and transporting tasks. The motions of the robot are actuated by brushless DC motors. The motor specifications for the finger joints include a no-load speed of 8400 RPM (revolutions per minute) and a rated torque of 4.9 mNm (millinewton meters), while the motor specifications for the arm joints include a no-load speed of 10,700 RPM and a rated torque of 6.2 mNm.

Table 1. NAO robot hardware parameters.

Hardware Components	Parameters
CPU	2 × Intel Atom Z530 Processors
Memory	1 GB RAM, 2 GB Flash Memory
Network Connection	Ethernet, Wi-Fi
Battery	Lithium Battery

Figure 1. NAO robot.

The right arm of the NAO robot, for instance, possesses five degrees of freedom and is a type of serial kinematic structure. According to the Denavit–Hartenberg (D-H) convention [22], the data presented in Table 2 can be derived. On this basis, a linkage coordinate system for the NAO robot's right arm is established. As shown in Figure 2, s, e, and w represent the shoulder joint, elbow joint, and wrist joint of the NAO robot, respectively. The dimensions of the links are obtained through the parameters of the NAO robot's right arm, where d3 = 90 mm and d5 = 108.55 mm.

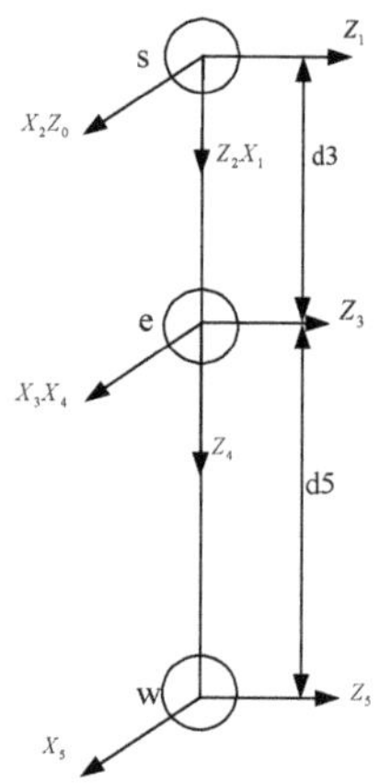

Figure 2. NAO robot right arm D-H model.

Table 2. D-H parameters of the right arm of NAO robot.

Link	$\theta_i/°$	d_i/mm	a_i/mm	$\alpha_i/°$
1	θ_1	0	0	90
2	$\theta_2(-90°)$	0	0	−90
3	θ_3	d3	0	90
4	θ_4	0	0	−90
5	$\theta_5(-90°)$	d5	0	90

The NAO robot has five degrees of freedom in each leg, with four force sensors installed on the sole of each foot. The forward, lateral, and vertical directions of the robot are defined as the X-axis, Y-axis, and Z-axis directions, respectively. When the robot is standing, the origin of the world coordinate system is located at the center position between the robot's two feet. To facilitate subsequent analysis, a D-H model is established for the robot's legs, as shown in Figure 3.

When constructing the model for the supporting leg, taking the left leg as an example, let O_0 be the base coordinate system and O_4 be the end-effector coordinate system. Using this as a reference, the position matrix is obtained by incorporating the translational and rotational transformation matrices along the X-axis and Z-axis into the homogeneous transformation matrix defined by the D-H method, as shown in Equation (1).

$$A_i = \begin{bmatrix} \cos\theta_i & -\sin\theta_i \sin\alpha_i & \sin\theta_i \sin\alpha_i & a_i \cos\theta_i \\ \sin\theta_i & \cos\theta_i \cos\alpha_i & -\cos\theta_i \sin\alpha_i & a_i \sin\theta_i \\ 0 & \sin\alpha_i & \cos\alpha_i & d_i \\ 0 & 0 & 0 & 1 \end{bmatrix} \quad (1)$$

In Equation (1), a_i represents the link length, α_i represents the link twist angle, d_i represents the distance between joints, and θ_i represents the joint twist angle. Based on the chain rule of homogeneous transformation, the homogeneous transformation matrix of O_4

relative to O_0 is derived, as shown in Equation (2). ${}_4^0R_{3\times3}$ is the rotation matrix, and ${}_4^0P_{3\times1}$ is the position vector.

$$ {}_4^0T = A_1A_2A_3A_4 = \begin{bmatrix} {}_4^0R_{3\times3} & {}_4^0P_{3\times1} \\ 0_{3\times3} & 1_{1\times3} \end{bmatrix} \quad (2) $$

When constructing the swing leg model, using the left leg as an example, let O_0 be the end-effector coordinate system and O_4 be the base coordinate system. It is easy to derive the homogeneous transformation matrix. Due to the symmetry between the left and right legs, when the right leg is the swing leg, the homogeneous transformation matrix of O_9 relative to O_5 can be determined as matrix ${}_5^9T$, which is similar in form to matrix ${}_4^0T$. Similarly, when the right leg is used as the supporting leg, the homogeneous transformation matrix of O_5 relative to O_9 can be determined as matrix ${}_5^9T$.

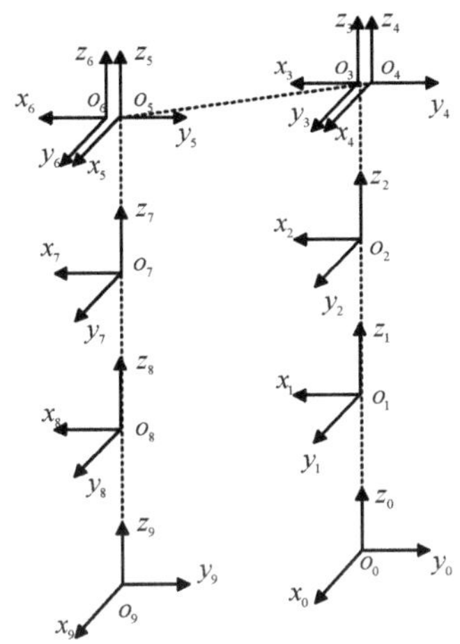

Figure 3. NAO robot leg D-H model.

2.2. NAO Robot Kinematics and Dynamics Modeling

The robot's walking is simplified to a linear inverted pendulum model for dynamic analysis [23], as shown in Figure 4. In this model, the robot's trunk is simplified to a point mass COM that concentrates all the mass, and the robot's leg is simplified to a massless, telescopic link of length r connecting the trunk and the foot. τ represents the torque at the robot's ankle, and f represents the reactive force received by the robot when pushing off the ground. Subsequently, the motion of the center of mass is driven through the bending and stretching of the legs.

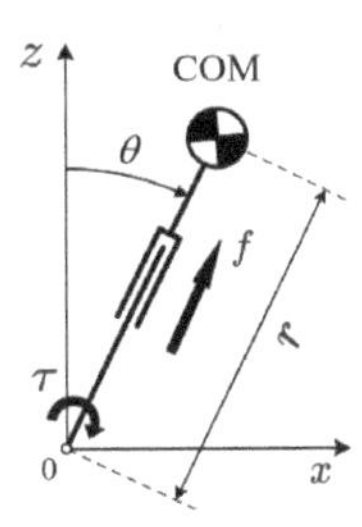

Figure 4. Linear inverted pendulum model.

After the initial state is determined, the motion state of the inverted pendulum can be described as shown in Equation (3), where T_c is a constant ratio of gravitational acceleration to the height of the center of mass; x_0 and $\dot{x}_0$ are, respectively, the displacement and velocity of the center of mass in the X direction at zero time, that is, the initial conditions.

$$\begin{cases} x_t = x_0 \cosh(t/T_c) + T_c\dot{x}_0 \sinh(t/T_c) \\ \dot{x}_t = x_0/T_c \sinh(t/T_c) + \dot{x}_0 \cosh(t/T_c) \\ T_c = \sqrt{z/g} \end{cases} \quad (3)$$

The humanoid robot can be simplified to an inverted pendulum model, with the robot's movement illustrated in Figure 5. During locomotion, the robot's center of mass moves horizontally, and ground contact is managed by alternating leg support, ensuring that there is always one leg in contact with the ground.

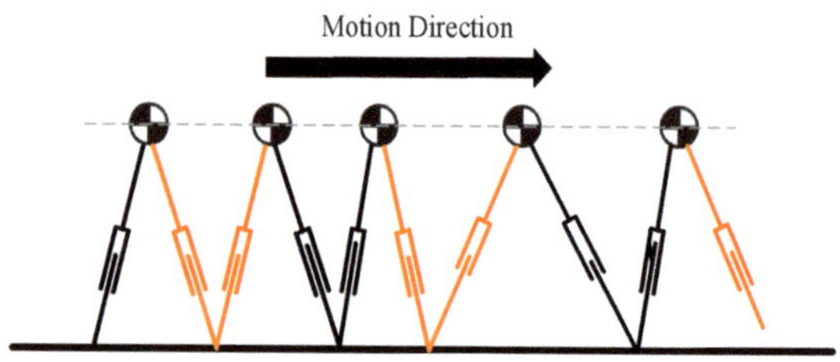

Figure 5. Linear inverted pendulum gait diagram.

2.3. *State Space Equation Based on ZMP*

For bipedal robots, maintaining balanced and stable walking is very important. The ZMP (zero moment point) [24] is the criterion for static or dynamic stable walking of bipedal robots. Its essence is to calculate the combined force point of the ground on the robot under the condition that the robot will not topple over, and when the robot has acceleration, the ground's acting force balances with gravity and inertial force at the ZMP. Since gravity remains constant, when the height of the center of mass does not change, the position of the ZMP corresponds to the motion acceleration of the center of mass. The support point of the linear inverted pendulum is its ZMP. Therefore, the mathematical expression of ZMP can be analyzed as shown in Equation (4).

$$\begin{cases} \ddot{x} = \frac{g}{z_c}x - \frac{1}{mz_c}\tau_x \\ \ddot{y} = \frac{g}{z_c}y - \frac{1}{mz_c}\tau_y \end{cases} \quad (4)$$

In the equation, m represents the mass of the center of gravity, g represents the constant of gravitational acceleration, and τ_x and τ_y, respectively, represent the torques of the inverted pendulum model rotating around the X-axis and Y-axis. The position equation of ZMP for a robot under the constraints of horizontal plane motion is shown in Equation (5).

$$\begin{cases} p_x = -\frac{\tau_y}{mg} \\ p_y = \frac{\tau_x}{mg} \end{cases} \quad (5)$$

Taking the derivative of acceleration $\dddot{u}_x$ as the input to the ZMP equation, and considering horizontal displacement x, velocity $\dot{x}$, and acceleration $\ddot{x}$ as state variables, combined with Equations (1)–(3), we can obtain the state space representation of the ZMP equation, as shown in Equations (6) and (7).

$$\frac{d}{dx}\begin{bmatrix} x \\ \dot{x} \\ \ddot{x} \end{bmatrix} = \begin{bmatrix} 0 & 1 & 0 \\ 0 & 0 & 1 \\ 0 & 0 & 0 \end{bmatrix}\begin{bmatrix} x \\ \dot{x} \\ \ddot{x} \end{bmatrix} + \begin{bmatrix} 0 \\ 0 \\ 1 \end{bmatrix} u_x \quad (6)$$

$$p_x = \begin{bmatrix} 1 & 0 & -z_c/g \end{bmatrix} \begin{bmatrix} x \\ \dot{x} \\ \ddot{x} \end{bmatrix} \tag{7}$$

After obtaining the state space equations in the continuous time domain, when using model predictive control methods, it is necessary to perform discretization. Setting the sampling time as T_s, the velocity of the center of mass in the horizontal direction at the kth sampling instance is shown in Equation (8). The acceleration in the horizontal direction is shown in Equation (9).

$$\dot{x}(k) = \frac{x(k+1) - x(k)}{T_s} \tag{8}$$

$$\ddot{x}(k) = \frac{\dot{x}(k+1) - \dot{x}(k)}{T_s} \tag{9}$$

In these equations, $x(k)$ is the horizontal position of the centroid at the kth sampling instant; $x(k+1)$ is the horizontal position of the centroid at the $(k+1)^{st}$ sampling instant. $\dot{x}(k)$ and $\dot{x}(k+1)$ represent the velocity of the centroid's horizontal movement at the kth and $(k+1)^{st}$ sampling instances, respectively; $\ddot{x}(k)$ is the acceleration in the horizontal direction of the centroid at the kth sampling instance.

Let the state variables be defined as displacement, velocity and acceleration, represented as $X(k) = [x(k), \dot{x}(k), \ddot{x}(k)]^T$; with the second derivative of velocity as the control input $u(k) = \dddot{x}(k)$, and the output as the ZMP trajectory $p(k) = x_{zmp}(k)$ of the NAO robot, the state-space equation of the system can be represented by Equation (10).

$$\begin{cases} X(k+1) = AX(k) + BU(k) \\ P(k) = CX(k) \end{cases} \tag{10}$$

In Equation (10):

$$A = \begin{bmatrix} 1 & T_s & T_s^2/2 \\ 0 & 1 & T_s \\ 0 & 0 & 1 \end{bmatrix}; B = \begin{bmatrix} T_s^3/6 & T_s^2/2 & T_s \end{bmatrix}^T; C = \begin{bmatrix} 1 & 0 & -z_c/g \end{bmatrix}.$$

3. Dual Robot Cooperative Control

3.1. Dual-Robot Communication System

The research objective of this paper is to design a DMPC controller to address the cooperative control issue of bipedal humanoid robot systems, thereby achieving consensus on trajectory tracking and position control. Before establishing the cooperative control system model for the dual robots, it is first necessary to establish a communication system between the robots.

This research employs the TCP/IP [25] communication protocol to achieve inter-robot communication. To establish the communication diagram shown in Figure 6, where robot1 acts as the leader robot and robot2 as the follower robot, robot1 receives control signals from the higher-level controller and sends commands to robot2, which receives and transmits information back to robot1.

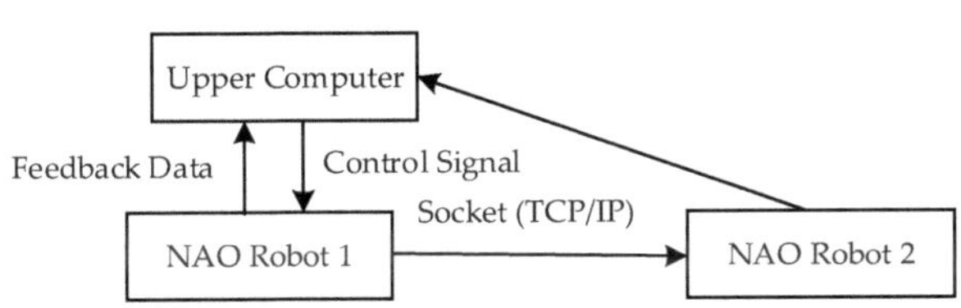

Figure 6. Dual-robot system communication diagram.

Therefore, we can obtain the adjacency matrix of the dual-robot system, as shown in Equation (11).

$$A = \begin{bmatrix} 0 & 0 & 0 \\ 1 & 1 & 0 \\ 0 & 1 & 0 \end{bmatrix} \tag{11}$$

In the context of dual-robot communication, to address network communication delays, which may mitigate the impact of communication delays on coordination consistency, a consensual communication protocol is established based on the collaborative actions of the dual-robot system, as shown in Table 3. Initially, action commands for the robots are pre-planned and set according to task requirements, including actions such as object grasping and coordinated movement. Subsequently, the actions to be executed by the robots, along with their respective start execution times, are compiled into structured data, with multiple structures forming an array that represents a series of consistent actions. Following confirmation of the accuracy of the structured array, the leader robot transmits a sequence of action commands to the follower. Ultimately, upon reaching the designated time points, both the leader and follower initiate the corresponding action commands to execute the planned actions. This approach ensures the consistency of collaboration between the two robots.

Table 3. Consensus protocol.

Protocol Field	Data Type
Command Content	string
Execution Time	datetime

3.2. Leader DMPC Controller Design

MPC predicts system behavior and performs optimization through a predictive model, thus obtaining an optimal control sequence at each time step [26]. Therefore, the construction of the prediction model is very important for the entire control allocation process, and its accuracy and applicability largely determine the overall performance of the system.

The obtained discrete state-space equation is shown in Equation (10). Given $p^{ref}(k)$ as the reference position for the ZMP, in order to make the system's output $p(k)$ track the target ZMP position $p^{ref}(k)$ as accurately as possible, we define a cost function to serve as the loss function, thereby constructing an optimization problem. The cost function is shown in Equation (12).

$$J = \sum_{i=k}^{\infty} [Q_e e(i)^2 + \Delta X^T(i) Q_x \Delta X(i) + R\Delta u^2(i)] \tag{12}$$

where $e(i) \equiv p(k) - p^{ref}(k)$ represents the trajectory error, $\Delta X(k) \equiv X(k) - X(k-1)$ is the increment of the state vector, and $\Delta u(k) \equiv u(k) - u(k-1)$ is the increment of the input. The weights Q_e, R and Q_x stand for the ZMP trajectory error, minimum control effort, and state variables, respectively, and $Q_e > 0$ and $R > 0$. At any time k, the solution $u(k)$ that minimizes J is the optimal solution required.

At each sampling time, if the reference values of the ZMP for the next N_L steps can be known in advance, then the optimal controller that minimizes the evaluation index is shown in Equation (13).

$$u(k) = -G_i \sum_{i=0}^{k} e(k) - G_x X(k) - \sum_{j=1}^{N_L} G_p(j) P^{ref}(k+j) \tag{13}$$

In the equations, G_i, G_x, and G_p represent the corresponding gains for the error, state variables, and output. As shown in Equations (14)–(16), the parameters can be solved using the dare() function in MATLAB, and the model predictive control is realized through continuous iteration.

$$G_i = A^T G_i A - A^T G_i B (B^T G_i B + R)^{-1} B^T G_i A + Q_x \tag{14}$$

$$G_x = eig(A - B \times G_p) \tag{15}$$

$$G_p = (B^T G_i + R)^{-1} B^T G_i A \tag{16}$$

3.3. Follower DMPC Controller Design

To establish the distributed control block diagram based on MPC control as shown in Figure 7, first, a ZMP reference trajectory is input to the leader. After going through the MPC controller, an optimal control sequence for the kth moment is obtained. Then, by solving the robot's inverse kinematics, motion control of the leader is achieved. The ZMP values output by the leader are then transformed into secondary coordinates and used as the follower's reference trajectory. Subsequently, through the MPC controller, the follower tracks the leader's trajectory, thus realizing the coordinated position control of the dual robots.

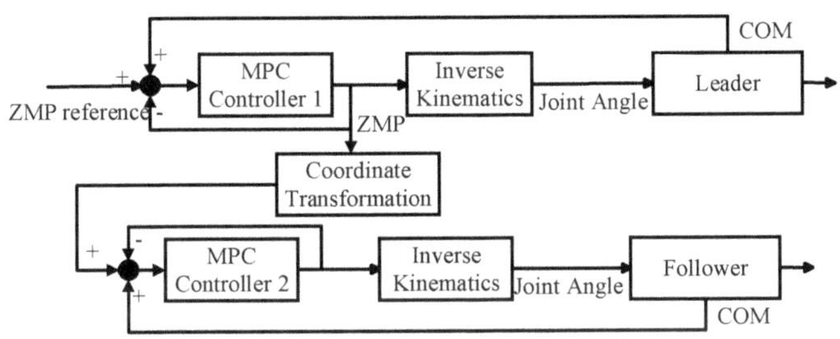

Figure 7. Leader—follower robots cooperative control block diagram.

Since the two robots stand face-to-face while performing tasks, their paths are not exactly the same, and the follower's reference path is the actual trajectory of the leader robot. Therefore, it is necessary to perform a secondary coordinate transformation on the leader's movement trajectory to use it as the reference path for the follower robot.

Relative to the follower robot's own coordinate system, the reference trajectory is shown in Equation (17). Here, P_1^{ref} represents the reference trajectory for the follower, P_1 represents the actual movement trajectory of the leader robot, and R_z denotes the rotation matrix around the Z-axis.

$$P_1^{ref} = -P_1 R_z = -P_1 \begin{bmatrix} \cos\theta & \sin\theta & 0 \\ -\sin\theta & \cos\theta & 0 \\ 0 & 0 & 1 \end{bmatrix} \tag{17}$$

In the dual-robot collaborative task, the control objective of the follower robot is to autonomously control its movement path to maintain relative direction and position with the leader. In the system, the state space equations for the follower and leader robots are the same and can be represented by Equation (18):

$$\begin{cases} X_1(k+1) = AX_1(k) + BU_1(k) \\ P_1(k) = CX_1(k) \end{cases} \tag{18}$$

The dual-robot formation matrix is shown in Equation (19).

$$D = \begin{bmatrix} 0 & \Delta x \\ 0 & \Delta y \\ 0 & \Delta v \\ 0 & \Delta a \end{bmatrix} \tag{19}$$

The first column of the matrix represents the motion state information of the leader robot relative to itself, with all elements being zero. The second column of the matrix represents the motion state of the follower with respect to the leader robot in the system, containing four rows of data: relative longitudinal coordinate difference, relative lateral coordinate difference, relative velocity error, and relative acceleration error.

The weighted sum of the formation error output between the follower and the leader robots at time k is shown in Equation (20).

$$error(k) = \sum_{j}^{3} a_{3j}[P(k) - P_1(k) + \Delta] \tag{20}$$

In Equation (20), a_{3j} represents the specific elements of the adjacency matrix A, which reflects the communication relationships between robots in the dual-robot system, as established by Equation (11); Δ represents the position, velocity, and acceleration errors between the leader robot and the follower robot, which can be obtained by subtracting the second column from the first column in the formation matrix D.

When designing the DMPC controller for coordinated motion, the follower must maintain consistent displacement, velocity, and acceleration with the leader. Thus, the cost function for the follower at time k can be designed as shown in Equation (21).

$$J_1 = \sum_{i=k}^{\infty} [Q_e error(i)^2 + \Delta X_1{}^T(i) Q_x \Delta X_1(i) + R\Delta u_1{}^2(i)] \tag{21}$$

In Equation (21), Q_e, R and Q_x represent the state error weight, minimum control weight, and state error weight, respectively. The first and second terms on the right side of the equation represent the error between the follower and the leader in the coordinated control and the state error output of the follower, respectively, while the third term is the weighted sum of changes in control inputs.

By constructing the follower's cost function, the problem can be transformed into a Riccati equation to solve. Similarly, the MATLAB function dare() can be used to calculate the weights. Thus, the mathematical model of the follower's DMPC (dynamic movement profile criterion) controller is obtained as shown in Equation (22).

$$u_1(k) = -G_i \sum_{i=0}^{k} error_1(k) - G_x X_1(k) - \sum_{j=1}^{N_L} G_p(j) P_1{}^{ref}(k+j) \tag{22}$$

4. Experimental Results and Analysis

4.1. Collaborative Control Experimental Design

The dual-humanoid robot system exhibits higher flexibility and adaptability in complex environments. This paper is based on the premise that dual-humanoid robots can conduct rescue operations for victims in complex situations such as natural disasters and accident scenes with casualties. An experiment was designed for the dual-humanoid robots to cooperatively carry a stretcher, aiming to validate the feasibility of the DMPC controller.

Figure 8 illustrates the overall experimental design framework. Initially, the two robots identify the stretcher using their onboard visual sensors. Subsequently, they measure the distance between themselves and the stretcher using a monocular ranging algorithm and move close to the stretcher. During locomotion, errors may occur due to factors such as slippery ground surfaces and motor aging. Therefore, upon reaching the designated spot, the robots adjust their postures. After posture adjustment, the follower robot sends feedback to the leader, employing the NAO robot's built-in setAngles() function to control the arm movements and simultaneously perform the grasping task. Finally, the two robots lift the stretcher together, with the follower tracking the leader's trajectory to reach the target location.

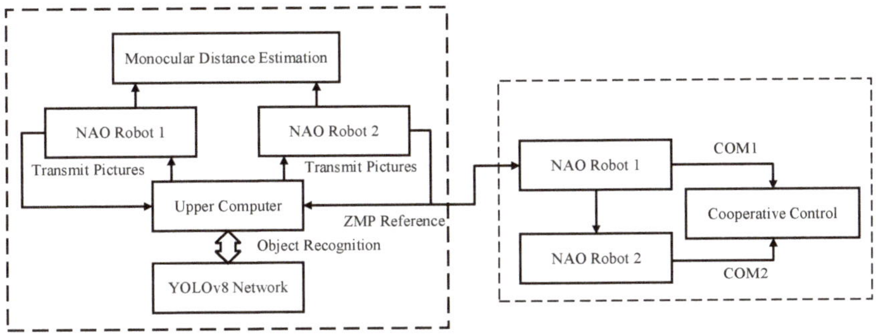

Figure 8. Overall experimental design.

4.2. Object Positioning and Grasping

This experiment involves collecting 300 images of the target stretcher from different angles using the NAO robot's lower camera and processing them with rotations, mirroring, etc. Then, the Yolov8 network is trained for a total of 800 rounds.

Firstly, using the lower cameras of the NAO robots, the two robots capture raw images as shown in Figure 9a,c and send them back to the control computer. The Yolov8 network in the Python3 environment is then used to identify the target pole pieces. The binary images of the target objects are transmitted to Python2 via Socket communication using the TCP/IP protocol, as shown in Figure 9b,d.

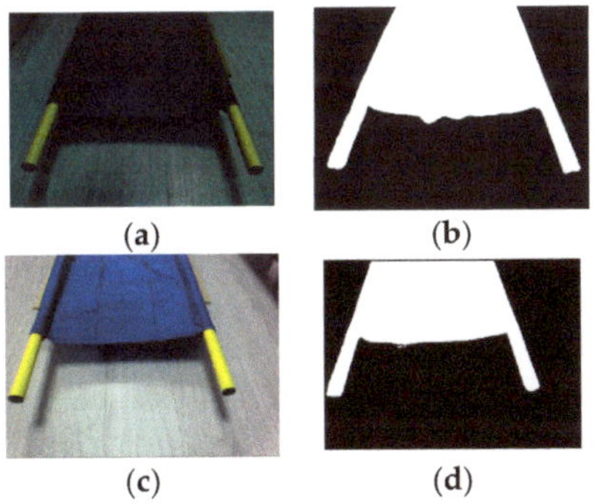

Figure 9. Target recognition: (**a**) original image captured by leader; (**b**) leader target recognition results; (**c**) original image captured by follower; (**d**) follower target recognition results.

Based on the results shown in Figure 9b,d, monocular ranging is performed using the center point of the line connecting the two handles at each side of the stretcher as the target point. After obtaining the distances in the X and Y directions relative to the robot itself, it moves close to the target stretcher.

Upon the robots reaching the designated positions and after postural adjustments, the handle positions are measured again, and the arms are controlled to perform the grasping action. During the grasping process, the end-effectors of the arms employ a motion strategy of acceleration-constant speed-deceleration through the setAngles() function, ensuring that the NAO robot's hands can smoothly reach the target position.

In the robot target positioning and grasping experiment, as shown in Figure 10a,b, the blue NAO robot acts as the leader, while the grey NAO robot is the follower. Both robots use monocular vision to identify and locate the stretcher simultaneously and move to the designated positions. After adjusting their postures, they determine the grasping locations and control the end-effectors of their arms to reach the grasping points.

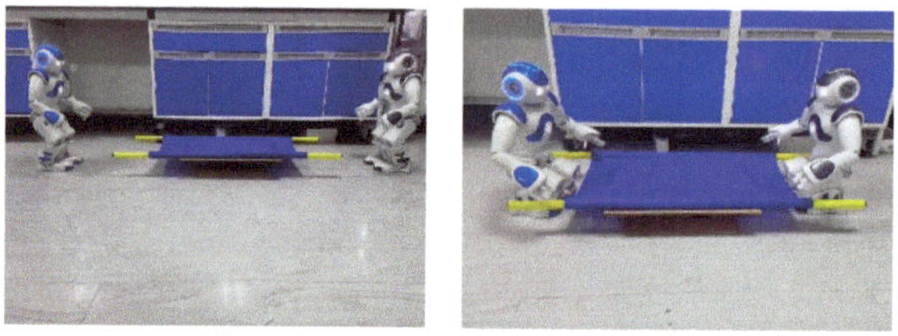

Figure 10. Object positioning and grasping process: (**a**) two robots positioning towards the target stretcher; (**b**) two robots grasping the target stretcher.

4.3. Dual-Robot Collaborative Transportation Experiment

Reference [24] proposes. a model predictive control method for bipedal robots based on a five-centroid model. However, reference [24] does not consider the error effect caused by historical cumulative error in MPC control, which leads to the lack of trajectory tracking control accuracy. Therefore, this paper proposes a DMPC control method based on error accumulation to compare and analyze with the method adopted in reference [24]. A coordinate system for the dual-robot system is established with the leader's position as the origin. In the xoy plane, the leader's coordinates are (0,0), and the follower's initial position is (1 m, 0). The reference trajectory for ZMP is set as [(0,0), (0.08,0.03), (0.16,−0.03), (0.24,0.03), (0.32,−0.03), (0.40,0.03), (0.48,−0.03)]. At point (0,0), the robot maintains a bipedal standing position, followed by alternating left and right leg movements, with the coordinates in the reference trajectory representing the landing points of the left and right feet, respectively. The leader's ZMP trajectory tracking was compared using the control method employed in reference [24] and the error accumulation-based DMPC control method proposed in this paper, with the experimental results presented in Figures 11–14, which include the ZMP trajectory of the robot in the X and Y directions as well as the trajectory of the robot's center of mass during locomotion.

Figures 11 and 12 illustrate the results of tracking the ZMP trajectory using the MPC control method proposed in reference [24]. Figure 11 shows the movement of the robot's center of mass in the X and Y directions in the time domain. The red line represents the trajectory of the center of mass, the black line indicates the reference trajectory, and the green line denotes the leader's tracking trajectory. Figure 12 shows the result of the robot's trajectory tracking on the XY plane for the robot in the leader's coordinate system.

Figures 13 and 14 demonstrate the experimental results obtained by using the error accumulation-based DMPC control method proposed in this paper. Figure 13 displays the tracking of the trajectory in the X and Y directions in the time domain, considering the cumulative error. Figure 14 presents the outcome of the robot's trajectory tracking under the leader's coordinate system based on accumulated error.

Comparing Figures 11 and 13, the method used in reference [24] fails to perfectly track the ZMP reference trajectory in both the X and Y axis directions during the steady-state phase. In contrast, the method proposed in this paper achieves better tracking of the reference trajectory once the system reaches the steady state, enhancing the steady-state performance of the system.

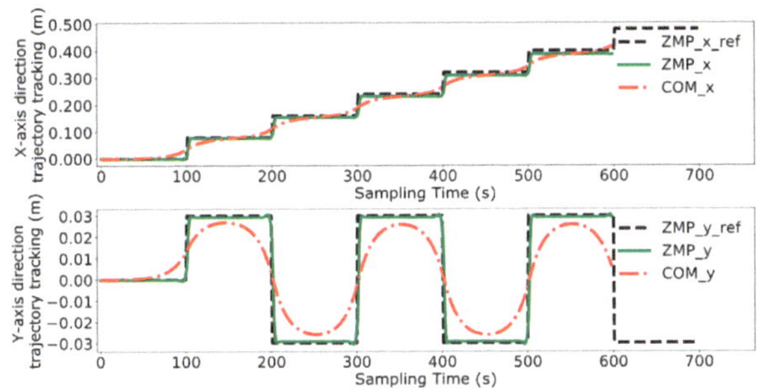

Figure 11. Time-domain ZMP trajectory tracking under current error.

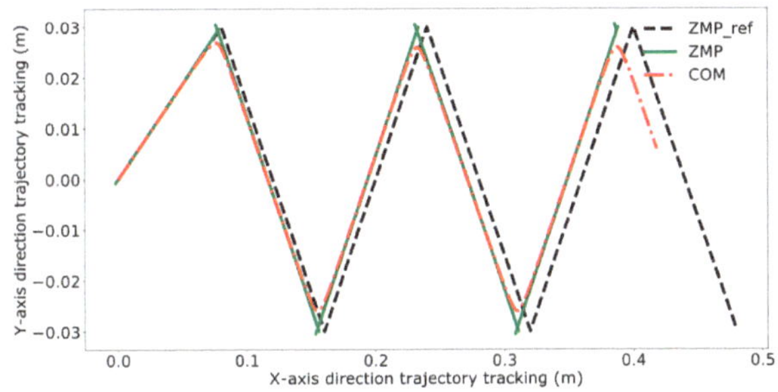

Figure 12. ZMP trajectory tracking under current error.

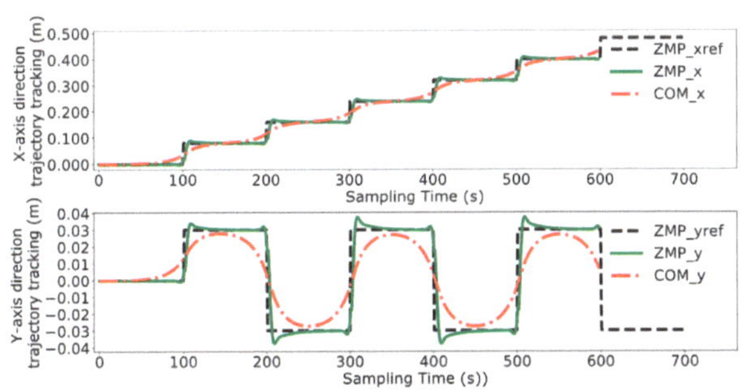

Figure 13. Time-domain ZMP trajectory tracking under accumulated error.

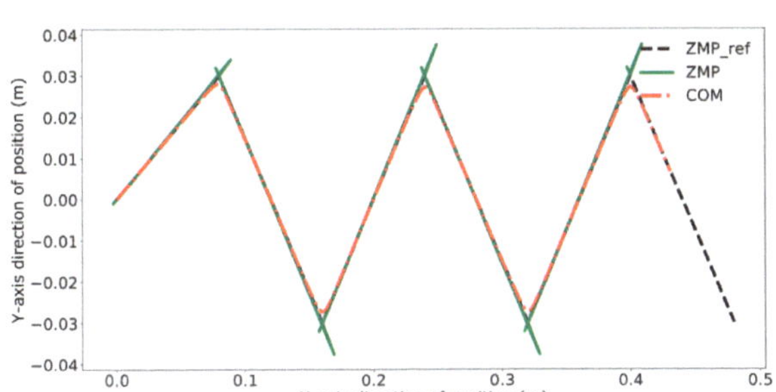

Figure 14. ZMP trajectory tracking under accumulated error.

A comparative analysis between Figures 12 and 14 reveals that the method utilized in reference [24] exhibits increasing tracking error with the increment of displacement. In contrast, the results depicted in Figure 14 indicate that the DMPC control method, which considers cumulative errors, consistently maintains effective tracking of the reference trajectory. Figure 15 more vividly illustrates the error variation between the two methods. The method used in reference [24] fails to accurately track the ZMP trajectory in the X-axis direction, whereas the improved DMPC control method proposed in this paper can accurately and stably track the trajectory, demonstrating superior control precision. The abrupt changes observed in the figure are due to the leg exchange during the robot's walking process.

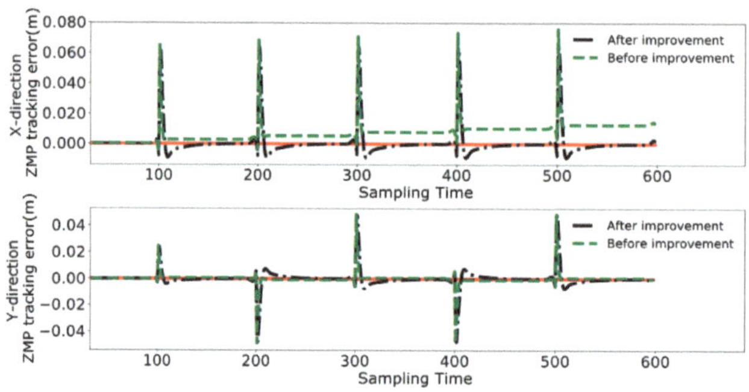

Figure 15. Leader's trajectory tracking error.

To achieve position coordination control between two robots, the follower robot employs the trajectory of the leader robot transformed through Equation (10) as its reference trajectory. Since the robots are facing each other during the collaborative task, in the follower's coordinate system, the follower is moving in the negative direction of the X-axis.

Figures 16–19 present the tracking performance of a follower robot using the leader's ZMP trajectory as a reference, comparing two different methods. Figures 16 and 17 utilize the method from reference [24], while Figures 18 and 19 display the results from the improved DMPC controller. Comparing the outcomes for both control methods on the follower reveals a pattern similar to that observed with the leader. The improved DMPC controller shows significantly improved tracking accuracy, whereas the unmodified method continues to exhibit an increase in tracking error with displacement, ultimately reaching an error margin of approximately 2 cm. This further corroborates the efficacy of the improved DMPC controller in accurately following the leader's ZMP trajectory.

The two methods are integrated into the leader–follower collaborative control framework designed in this paper, with the forward direction of the leader robot designated as the positive X-axis and from right to left as the positive Y-axis. Since the two robots face each other during transportation, when the leader robot steps forward with its left leg, the follower robot steps forward with its right leg, ensuring alternating progress that maintains consistency of the center of mass along the Y-axis, thereby enhancing stability during transport. Given that the stretcher being carried is 1 m in length, it is necessary for the two robots to maintain a distance of approximately 1 m in the X direction throughout the collaborative carrying process.

Initially employing the method used in reference [24], Figures 20–22, respectively, represent the collaborative control of the displacement, velocity, and acceleration of the center of mass for the leader and follower. The blue lines indicate the results of the follower's center of mass movements, while the green lines denote the movements of the leader.

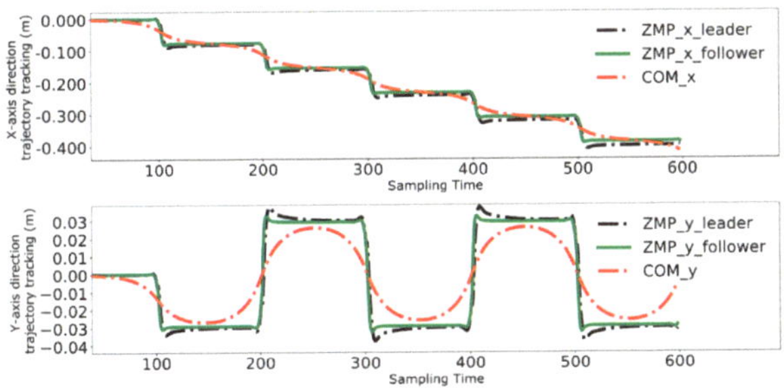

Figure 16. Pre-improvement ZMP trajectory tracking of the follower in the time domain.

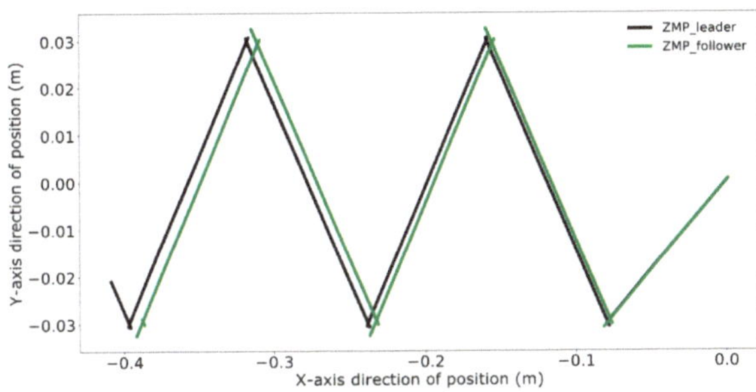

Figure 17. Pre-improvement follower's ZMP trajectory tracking.

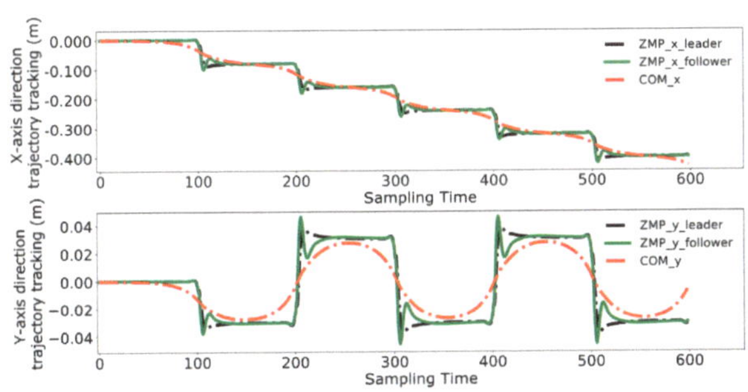

Figure 18. Post-improvement ZMP trajectory tracking of the follower in the time domain.

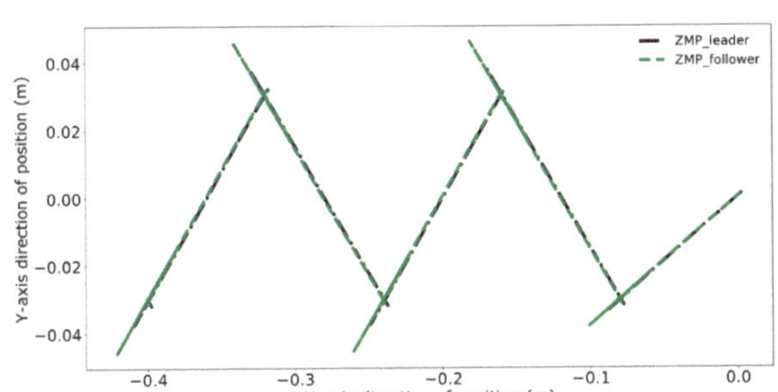

Figure 19. Post-improvement follower's ZMP trajectory tracking.

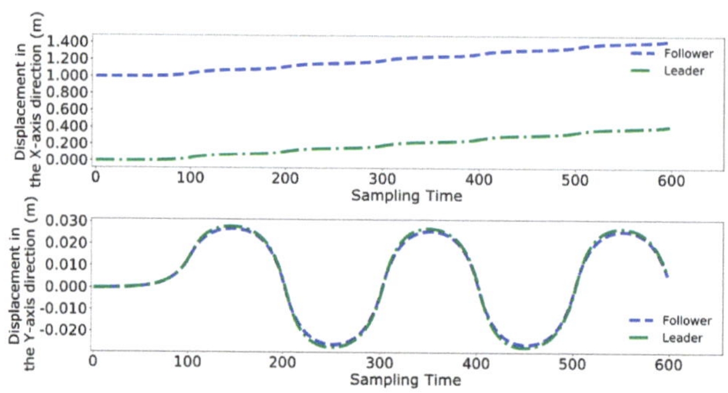

Figure 20. Pre-improvement dual-robot centroid displacement.

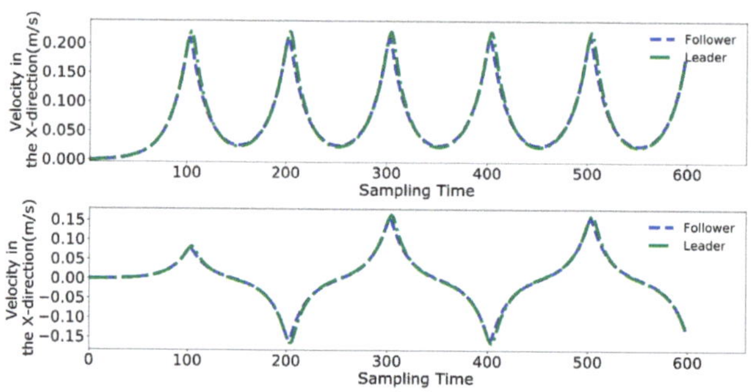

Figure 21. Pre-improvement dual-robot centroid velocity.

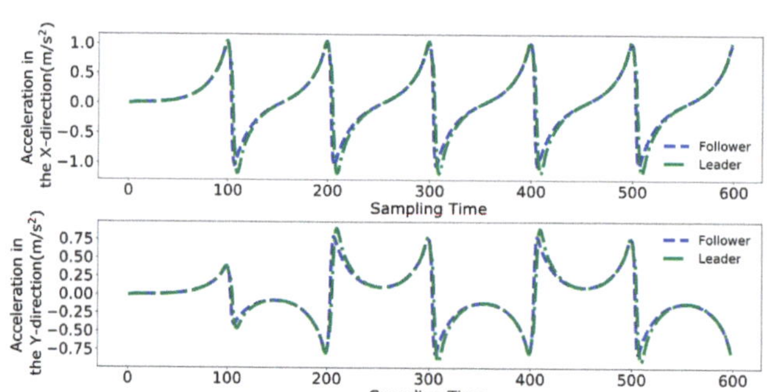

Figure 22. Pre-improvement dual-robot centroid acceleration.

To better analyze the coordination between the two robots, Figures 23–25 display the displacement, velocity, and acceleration errors during the collaborative control process. Figure 23 illustrates the displacement errors in the X and Y directions for both robots, with the error in the X direction gradually increasing over time. The error in the Y direction oscillates around zero. Figure 24 shows the maximum errors in the centroid velocities of the two robots in the X and Y directions, with a maximum error of 0.05 m/s in the X direction and 0.03 m/s in the Y direction. Figure 25 presents the acceleration errors for both robots, with the maximum errors in the X and Y directions being 1 m/s^2 and 0.7 m/s^2, respectively.

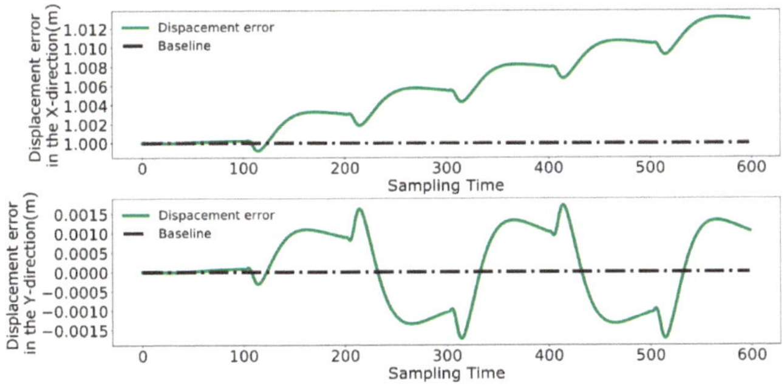

Figure 23. Pre-improvement dual-robot centroid displacement error.

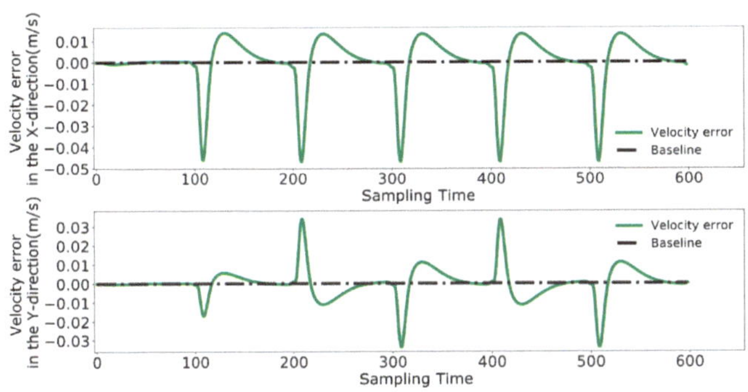

Figure 24. Pre-improvement dual-robot centroid velocity error.

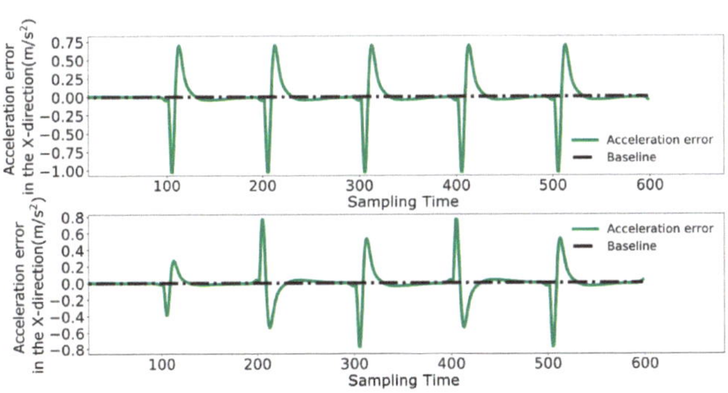

Figure 25. Pre-improvement dual-robot centroid acceleration error.

Figures 26–28 demonstrate the collaborative control motions of the leader and follower based on the enhanced DMPC control method. Figure 26 illustrates the displacements in the X and Y directions for both robots, indicating that a consistent distance of 1 m in the X direction is maintained between them. Figures 27 and 28, respectively, show the uniformity of the centroid velocity and acceleration in the X and Y directions for both robots during system control.

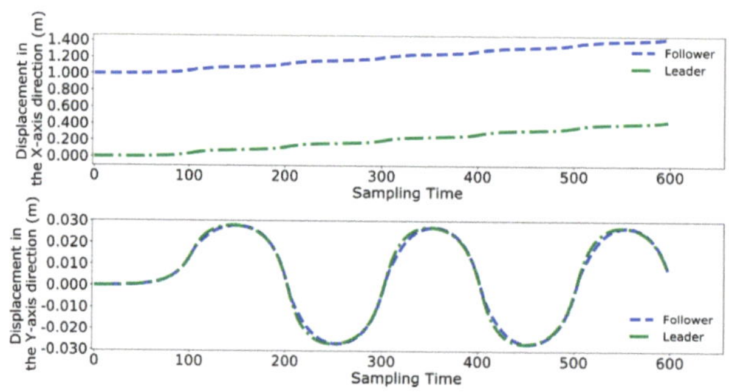

Figure 26. Post-improvement dual-robot centroid displacement.

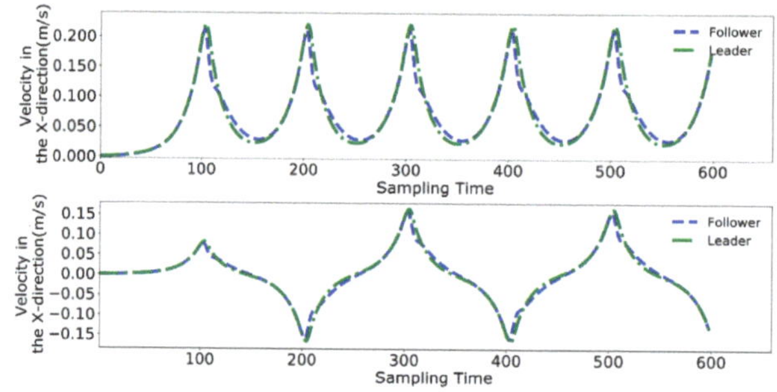

Figure 27. Post-improvement dual-robot centroid velocity.

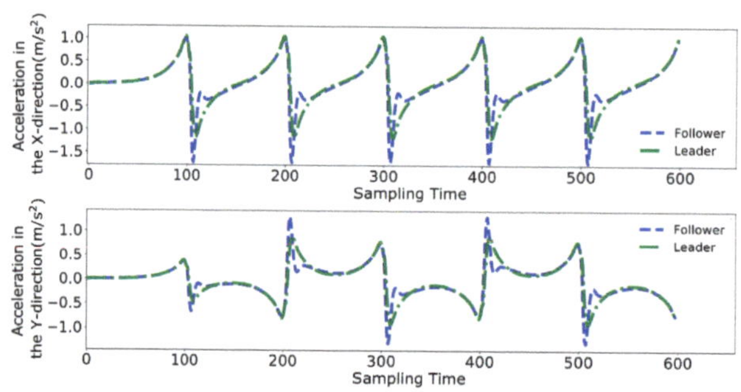

Figure 28. Post-improvement dual-robot centroid acceleration.

Figures 29–31, respectively, present the centroid displacement, velocity, and acceleration errors of the two robots based on the improved DMPC collaborative control proposed in this paper. Figure 29 displays the centroid displacement error, with a maximum error of 0.004 m in the X direction, which not only enhances control precision compared to the displacement error results shown in Figure 23 but also maintains the error around zero as time increases. Figure 30 shows the centroid velocity results of the two robots at the same moment, with a maximum error of 0.02 m/s in the X direction and 0.015 m/s in the Y direction. Compared to the maximum errors of 0.045 m/s and 0.03 m/s shown in Figure 24 for the X and Y directions, respectively, there is a significant improvement. Figure 31 illustrates the centroid acceleration collaboration results in the dual-robot system, with maximum errors of 0.6 m/s^2 and 0.45 m/s^2 in the X and Y directions, respectively. In comparison to Figure 25, there is an enhancement of 0.4 m/s^2 in the X direction and 0.3 m/s^2 in the Y direction.

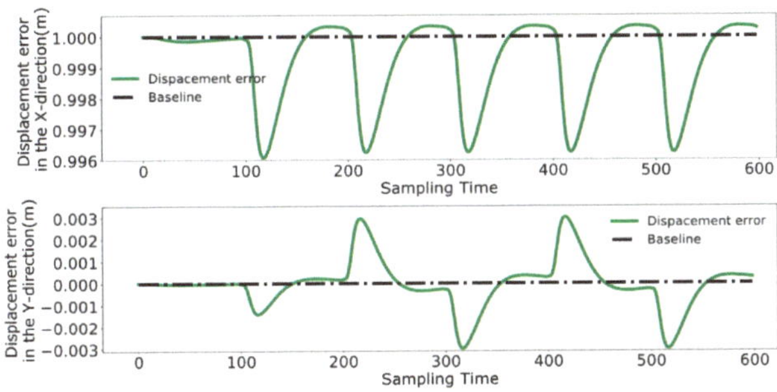

Figure 29. Post-improvement dual-robot centroid displacement error.

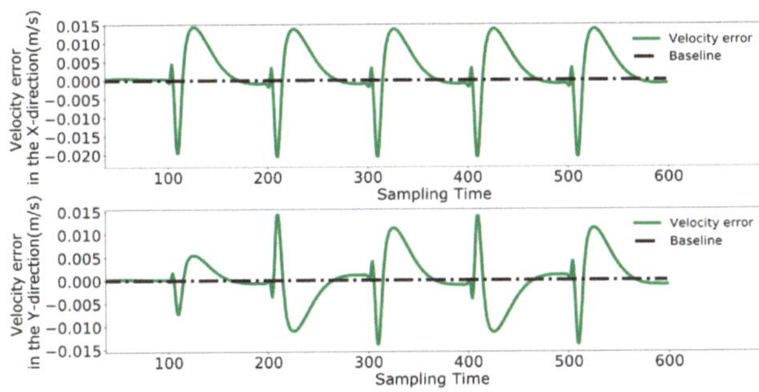

Figure 30. Post-improvement dual-robot centroid velocity error.

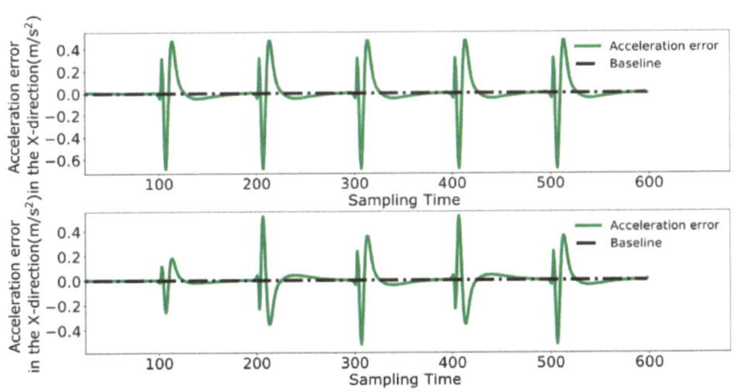

Figure 31. Post-improvement dual-robot centroid acceleration error.

Through a comparative analysis employing two methods, in this paper, certain improvements in control precision for both leader control and dual-robot collaborative control compared to the method presented in reference [24], enhancing the steady-state performance of the system, are shown. Additionally, by validating the errors in centroid displacement, velocity, and acceleration of the two robots using both methods, it is proven that the leader-follower collaborative control based on DMPC designed in this study is effective and feasible.

Figure 32 illustrates the dual-robot collaborative transport process, where they initially move simultaneously to the right side of the leader robot, circumventing the placement stage. Subsequently, they proceed in the positive X direction of the leader robot for transportation. During this process, the follower robot walks backward, continuously receiving positional information from the leader robot to ensure that the center of mass for both robots remains aligned in the X and Y directions within the leader robot's coordinate system.

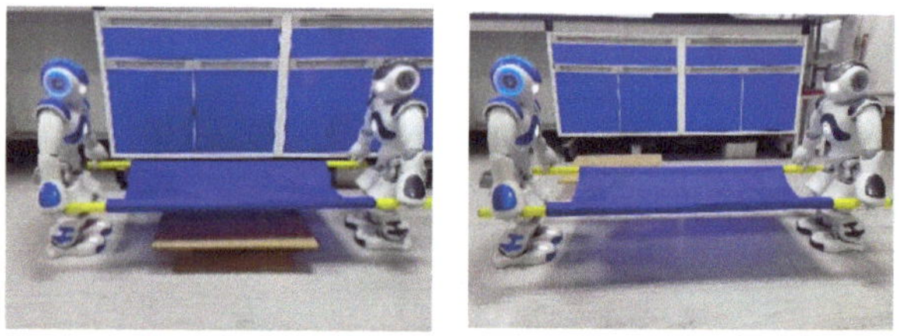

(a) (b)

Figure 32. Collaborative transportation of dual humanoid robots: (**a**) two robots lifting the stretcher simultaneously; (**b**) robots cooperatively transporting a stretcher.

5. Conclusions

This paper presents a dual bipedal robot DMPC (Dynamic Motion Primitives Control) collaborative control system for carrying stretchers using two NAO robots against the backdrop of rescue robots at casualty accident sites. In the dual-robot collaborative control system, a dual bipedal robot consistency communication protocol is constructed to address the issue of consistency in collaborative control. Concurrently, to mitigate the significant trajectory tracking error of the dual-robot system, a DMPC control method considering cumulative error effects is proposed. Compared with existing methods, the proposed control approach demonstrates superior steady-state performance and enhanced control accuracy, with a tracking error within 0.01 m, whether in the individual control of leaders and followers or in the collaborative control of both robots. Furthermore, the dual-robot control system effectively coordinates the velocity and acceleration of the two robots, confirming the effectiveness of the established collaborative control system for dual bipedal robots.

Given that the environment at accident sites is generally complex, characterized by uneven terrain and numerous obstacles, this paper does not delve deeply into research on this aspect. Therefore, future work will focus on how to implement collaborative control of dual robots in complex environments. This could involve upgrading and adding sensor equipment to the robots, enabling them to better perceive their surroundings. Additionally, neural networks such as CNNs (convolutional neural networks) [27] and BP (backpropagation) can be utilized to enhance the intelligence of the robots, thereby strengthening their adaptability to various conditions.

Author Contributions: S.W. and Z.S., contributed to conception and design of the study; methodology, Z.S. and S.W.; simulation, Z.S.; validation, Z.S.; formal analysis, H.L. and S.W.; investigation, S.W.; resources, S.W.; data curation, Z.S.; writing original draft preparation, Z.S.; writing review and editing, S.W. and H.L.; visualization, H.L.; supervision, S.W.; project administration, S.W.; funding acquisition, S.W. All authors have read and agreed to the published version of the manuscript.

Funding: This research was funded by the Natural Science Foundation of Shandong (ZR2023MF024), Support Plan for Science and Technology Innovation Teams in Higher Education Institutions of Henan Province (24IRTSTHN024) and Key Research and Development Projects in Henan Province (231111221600).

Institutional Review Board Statement: Not applicable.

Data Availability Statement: Data are contained within the article.

Conflicts of Interest: The authors declare no conflicts of interest.

References

1. Zhao, X.; Tao, B.; Ding, H. Multimobile Robot Cluster System for Robot Machining of Large-Scale Workpieces. *IEEE/ASME Trans. Mechatron.* **2022**, *27*, 561–571. [CrossRef]
2. Li, B.; Liu, H.; Xiao, D.; Yu, G.; Zhang, Y. Centralized and optimal motion planning for large-scale AGV systems: A generic approach. *Adv. Eng. Softw.* **2017**, *106*, 33–46. [CrossRef]

3. Yu, X.; Ma, J.; Ding, N.; Zhang, A. Cooperative target enclosing control of multiple mobile robots subject to input disturbances. *IEEE Trans. Syst. Man Cybern. Syst.* **2021**, *51*, 3440–3449. [CrossRef]
4. Hu, Z.; Shi, P.; Wu, L. Polytopic event-triggered robust model predictive control for constrained linear systems. *IEEE Trans. Circuits Syst. Regul. Pap.* **2021**, *68*, 2594–2603. [CrossRef]
5. Ning, B.; Han, Q.L.; Zuo, Z.; Ding, L.; Lu, Q.; Ge, X. Fixed-time and prescribed-time consensus control of multiagent systems and its applications: A survey of recent trends and methodologies. *IEEE Trans. Ind. Inform.* **2023**, *19*, 1121–1135. [CrossRef]
6. Zhang, H.; Feng, T.; Liang, H.; Luo, Y. LQR-based optimal distributed cooperative design for linear discrete-time multiagent systems. *IEEE Trans. Neural Netw. Learn. Syst.* **2017**, *28*, 599–611. [CrossRef] [PubMed]
7. Wei, L.Z.; Gong, I.W.; Chen, H.Y.; Li, Z.; Gong, C. Tracking and aiming adaptive control for unmanned combat ground vehicle on the move based on reinforcement learning compensation. *Acta Armamentarii* **2022**, *3*, 1947–1955.
8. Li, C.Y.; Guo, Z.C.; Zheng, D.D.; Wei, Y.L. Multi-robot Cooperative Formation Based on Distributed Model Predictive Control. *Acta Armamentarii* **2023**, *44*, 178–189.
9. Shen, C.; Shi, Y. Distributed Implementation of Nonlinear Model Predictive Control for AUV Trajectory Tracking. *Automatica* **2020**, *115*, 1626–1640. [CrossRef]
10. Mohseni, F.; Frisk, E.; Nielsen, L. Distributed Cooperative MPC for Autonomous Driving in Different Traffic Scenarios. *IEEE Trans. Intell. Veh.* **2020**, *6*, 299–309. [CrossRef]
11. Cao, Y.; Wen, J.; Ma, L. Tracking and Collision Avoidance of Virtual Coupling Train Control System. *Future Gener. Comput. Syst.* **2021**, *120*, 76–90. [CrossRef]
12. Wei, H.; Sun, Q.; Chen, J.; Shi, Y. Robust Distributed Model Predictive Platooning Control for Heterogeneous Autonomous Surface Vehicles. *Control Eng. Pract.* **2021**, *107*, 533–542. [CrossRef]
13. Dai, L.; Hao, Y.; Xie, H.; Sun, Z.; Xia, Y. Distributed Robust MPC for Nonholonomic Robots with Obstacle and Collision Avoidance. *Control Theory Technol.* **2022**, *20*, 32–45. [CrossRef]
14. Pan, Z.; Sun, Z.; Deng, H.; Li, D. A Multilayer Graph for Multiagent Formation and Trajectory Tracking Control Based on MPC Algorithm. *J. IEEE Trans. Cybern.* **2021**, *52*, 13586–13597. [CrossRef]
15. Zhou, J.J.; Shi, Z.F. Research on multi-UAY formation flying control method based on improved leader-follower algorithm. In Proceedings of the 4th China Aeronautical Science and Technology Conference, Beijing, China, 14–19 September 2018; China Aviation Publishing & Media Co.: Shenyang, China, 2019; Volume 11.
16. Munir, M.; Khan, Q.; Ullah, S.; Syeda, T.M.; Algethami, A.A. Control Design for Uncertain Higher-Order Networked Nonlinear Systems via an Arbitrary Order Finite-Time Sliding Mode Control Law. *Sensors* **2022**, *22*, 2748. [CrossRef] [PubMed]
17. Ullah, S.; Khan, Q.; Zaidi, M.M.; Hua, L.G. Neuro-adaptive Non-singular Terminal Sliding Mode Control for Distributed Fixer-time Synchronization of Higher-order Uncertain Multi-agent Nonlinear Systems. *Inf. Sci.* **2024**, *659*, 120087. [CrossRef]
18. Zhang, H.S.; Zhang, H.; Wang, C.S. Collaborative transportation for bulky items based on multi-robot formation control. *J. Shandong Univ.* **2023**, *53*, 157–162.
19. Liu, Q.; Gong, Z.; Nie, Z.; Liu, X.-J. Enhancing the terrain adaptability of a multirobot cooperative transportation system via novel connectors and optimized cooperative strategies. *Front. Mech. Eng.* **2023**, *18*, 38. [CrossRef]
20. Wu, X.; Yu, W.; Lou, P.H. Coordinated Guidance Control for Multi-robot Cooperative Transportations of Large-sized objects. *China Mech. Eng.* **2022**, *33*, 1586–1595.
21. Ding, Y.P.; Zhu, X.J.; Sun, X.Q. Kinematics simulation and control system design of robot. *Control. Eng. China* **2021**, *28*, 546–552.
22. Liu, X.; Li, S.; Liang, T.; Li, J.; Lou, C.; Wang, H. Follow control of upper limb rehabilitation training based on Kinect and NAO robot. *J. Biomed. Eng.* **2022**, *39*, 1189–1198.
23. Mohamed, S.A.; Maged, S.A.; Awad, M.I. A performance comparison between closed form and numerical optimization solutions for humanoid robot walking pattern generation. *Int. J. Adv. Robot. Syst.* **2021**, *18*, 17298814211029774. [CrossRef]
24. Meng, Y.; Zhou, F.N.; Lu, Z.Q.; Wang, P.P. Implementation Method of Predictive control of Five-Centroid Model for Biped Robot. *Mach. Des. Manuf.* **2022**, *3*, 254–257.
25. Neri, F.; Forlini, M.; Scoccia, C.; Palmieri, G.; Callegari, M. Experimental Evaluation of Collision Avoidance Techniques for Collaborative Robots. *Appl. Sci.* **2023**, *13*, 2944. [CrossRef]
26. Zhou, C.; Wang, J.R. Distributed collision avoidance control of UAV formation based on navigation function and model predictive control. *Electron. Opt. Control* **2023**, *30*, 100–104.
27. Wang, Y.; Cang, S.; Yu, H. A Survey on Wearable Sensor Modality Centred Human Activity Recognition in Health Care. *Expert Syst. Appl.* **2019**, *137*, 167–190. [CrossRef]

biomimetics

MDPI

Article

Verification of Criterion-Related Validity for Developing a Markerless Hand Tracking Device

Ryota Suwabe [1], Takeshi Saito [1,2] and Toyohiro Hamaguchi [1,*]

[1] Department of Rehabilitation, Graduate School of Health Sciences, Saitama Prefectural University, Saitama 343-8540, Japan; 2681309e@spu.ac.jp (R.S.); saitotakeshi@tdc.ac.jp (T.S.)
[2] Department of Rehabilitation, Tokyo Dental College Ichikawa General Hospital, Chiba 272-8513, Japan
* Correspondence: hamaguchi-toyohiro@spu.ac.jp; Tel.: +81-48-973-4125

Abstract: Physicians, physical therapists, and occupational therapists have traditionally assessed hand motor function in hemiplegic patients but often struggle to evaluate complex hand movements. To address this issue, in 2019, we developed Fahrenheit, a device and algorithm that uses infrared camera image processing to estimate hand paralysis. However, due to Fahrenheit's dependency on specialized equipment, we conceived a simpler solution: developing a smartphone app that integrates MediaPipe. The objective of this study was to measure hand movements in stroke patients using both MediaPipe and Fahrenheit and to assess their criterion-related validity. The analysis revealed moderate-to-high correlations between the two methods. Consistent results were also observed in the peak angle and velocity comparisons across the severity stages. Because Fahrenheit determines finger recovery status based on these measures, it has the potential to transfer this function to MediaPipe. This study highlighted the potential use of MediaPipe in paralysis estimation applications.

Keywords: MediaPipe; hand tracking; paralysis estimation

1. Introduction

The human "hand" contains a unique set of features within its complex assembly of joints, finely tuned for its specialized role as a delicate manipulative organ [1]. Understanding that hand function relies on specialized areas within the central nervous system to monitor the environment via sensory receptors and control fine movements through both exogenous and endogenous muscles is crucial [2]. Therefore, impairments in finger function following stroke are significant issues that directly affect the activities of daily living (ADL) and quality of life (QOL), making the improvement of finger function a key goal in the rehabilitation of stroke survivors [3].

To rehabilitate paralyzed fingers, it is important to optimize the quality and quantity of tasks. Quality refers to the appropriate level of difficulty, and quantity refers to the appropriate amount of practice [4,5]. To improve the quality of training, it is necessary to grade training according to its severity [6], and accurate evaluation is required to make this possible.

Typical measures of motor function in stroke include the Fugl-Meyer assessment [7], Brunnstrom stage [8], and Stroke Impairment Assessment Set (SIAS) [9], all of which score motor function based on the expert's visualization of motor onset and end. Since motor function assessment is performed and judged by experts, it can be performed in a short time without special equipment and has been widely used. However, visual evaluation of the starting and stopping positions of movements does not include factors related to the speed and trajectory of hand movements. In addition, the accurate evaluation of complex and high-speed hand motions using human vision is limited.

The development of image analysis in recent years has been remarkable, and various three-dimensional (3D) motion analyzers have been applied to provide temporal and spatial analyses of human motion. Our research team has also demonstrated that a binocular camera can be used to determine the severity of paralysis [10,11]. This device, called the Fahrenheit development code, is a method for recording hand motions and estimating hand functions by estimating joint centers and calculating motion features from different images captured by binocular cameras. Fahrenheit can help both the rehabilitation specialist and the patient record and analyze the motor functions of the patient's hand to determine the effectiveness of training. However, Fahrenheit requires a dedicated infrared camera and computer terminal running an application for analysis, making its use at the bedside somewhat labor-intensive. To widely apply this system in clinical practice, we conceived the idea of implementing a dedicated application on a smartphone using a camera. MediaPipe [12] is an open-source AI solution that can estimate joint centers from videos of fingers captured using an optical monocular camera. In fact, accuracy comparisons between infrared and optical cameras have been made using pose discrimination accuracy [13,14]; however, because they do not consider the temporal component, they do not evaluate the equivalence of motion velocity and trajectory, which is our goal. To replace MediaPipe with the binocular camera imaging function of Fahrenheit, we must verify the equivalence of the hand motion estimation.

Therefore, in this study, we examined the temporal and spatial criterion-related validity of the Fahrenheit and MediaPipe functions. If the equivalence of both functions is ensured, a simple smartphone-based paralysis estimation application can be developed. This will not only assist physicians and therapists in their clinical evaluations but also serve as a basis for the development of independent training devices for patients.

2. Materials and Methods

2.1. Research Design

This study is a retrospective observational design.

2.2. Target

Patients with post-stroke sequelae who participated in the Fahrenheit Development Project at Saitama Prefectural University were included in this study. The participants were patients aged 20 years or older who were admitted to the Tokyo Dental University Ichikawa General Hospital between 1 June 2016 and 31 March 2021 and were recruited from among all inpatients who had suffered a stroke. The patients were classified into 6 groups, stages I–VI, according to the Brunnstrom Recovery Stage (BRS) criteria, with reference to previous studies [10,11]. Using G*power to construct models for F tests and ANCOVA, encompassing fixed effects, main effects, and interactions (with $\alpha = 0.05$, $1 - \beta = 0.8$, effect size $f = 0.5$, and six groups), the minimum sample size was calculated to be 34. The exclusion criteria were as follows: (1) history of multiple cerebral infarctions or transient ischemic attacks; (2) loss of one or more fingers; (3) severely limited range of motion (ROM); (4) difficulty in understanding verbal instructions due to impaired consciousness, dementia, or aphasia; (5) a history of stroke or transient ischemic attack; (6) patients who needed bed rest; (7) patients who had difficulty maintaining a sitting position for at least 30 min, and (8) patients who had difficulty maintaining the required position during measurements even with assistance. Data in which the patient's fingers were not included in the angle of view from which the image was taken were excluded from the analysis.

2.3. Device

2.3.1. MediaPipe

MepipPipe is an open-source machine learning (ML) solution framework provided by Google Inc. There are several solutions available for MediaPipe; in this study, we used MediaPipe Hands, a hand-tracking solution released in 2019. MediaPipe Hands uses machine learning to infer 21 3D hand–finger landmarks from a single frame captured by a monocular camera, providing highly accurate hand and finger tracking. According to Google AI, Blogs [15], the average accuracy of hand tracking is 95.7%. One of the features of this system is that it runs in mobile and powerful desktop environments [12].

2.3.2. Fahrenheit

Fahrenheit, developed by Hamaguchi et al., (code name, Fahrenheit; patent number, 6375328; Saitama Prefectural University, Japan) contains an infrared sensor, the Leap Motion Controller (LMC), and obtains the coordinates of the joint centers using Unity game software (https://unity.com/ja) (accessed on 30 June 2024). Fahrenheit has 24 joint landmarks and can measure the ROM of the hand based on conventional joint range-of-motion measurements. Fahrenheit was operated on a laptop computer with a 64-bit Windows 8 (Microsoft, Kobe, Japan), and the generated data were stored on the hard drive of the same computer. It was implemented with a resolution of 0.001 mm and a sampling rate of 60 frames per second (fps) [10].

2.3.3. Basic Performance of Fahrenheit and MediaPipe

To track delicate hand movements, including dexterity movements, in real time, the processing frame rate of hand tracking must be sufficiently fast. In terms of frame rate, Fahrenheit is superior because processing is performed on the device, whereas MediaPipe is considerably affected by the frame rate of the data from the camera used and the performance of the terminal running the processing algorithm (Table 1).

Table 1. Comparison of basic performance.

	MediaPipe	**Fahrenheit**
Frame rate	30	60 * detune to 30
Number of joint landmarks	21	24
Observable joint range of motion	180	90

Values are expressed as real numbers. * Fahrenheit was downsampled and analyzed at 30 frames per second (fps).

Regarding joint landmarks, Fahrenheit includes the metacarpals of each finger, whereas MediaPipe does not. Therefore, it is difficult to obtain accurate flexion angles of finger MP joints using conventional ROM measurement methods. Fahrenheit is limited to 90° in terms of the estimated range of motion of the hand MP joints, and values exceeding 90° are corrected to 90° or less.

2.4. Protocol

The patients were seated in a chair or wheelchair with the shoulder joint in a resting position, elbow joint in approximately 90° of flexion, forearm in a rotated position, and wrist in an intermediate position, and their hand movements were measured. The subjects were instructed to assume the starting posture of each task at approximately 20 cm above Fahrenheit and to maintain the starting posture for 5 s from the starting cue while watching the monitor on a laptop computer. After 5 s, on the verbal cue, the participants were instructed to repeat hand flexion and extension for as long as possible for 20 s and then maintain the maximum reaching position for 15 s after the start cue. Participants were provided with virtual hand images obtained from the recordings to show their hand movements in real time. The coordinate data obtained by the infrared camera were used as the Fahrenheit data (Figure 1a,b). The videos for the MediaPipe analysis were captured from a video camera used to capture the entire hand motion during the acquisition of the Fahrenheit data. The video was divided into images at a sampling rate of 30 fps, the coordinates were extracted from each image using MediaPipe keypoints, and the keypointed images were converted to video. The images with keypoints were converted into movies (Figure 1c). In both cases, data from 10 to 20 s after the start of the measurement were used for analysis.

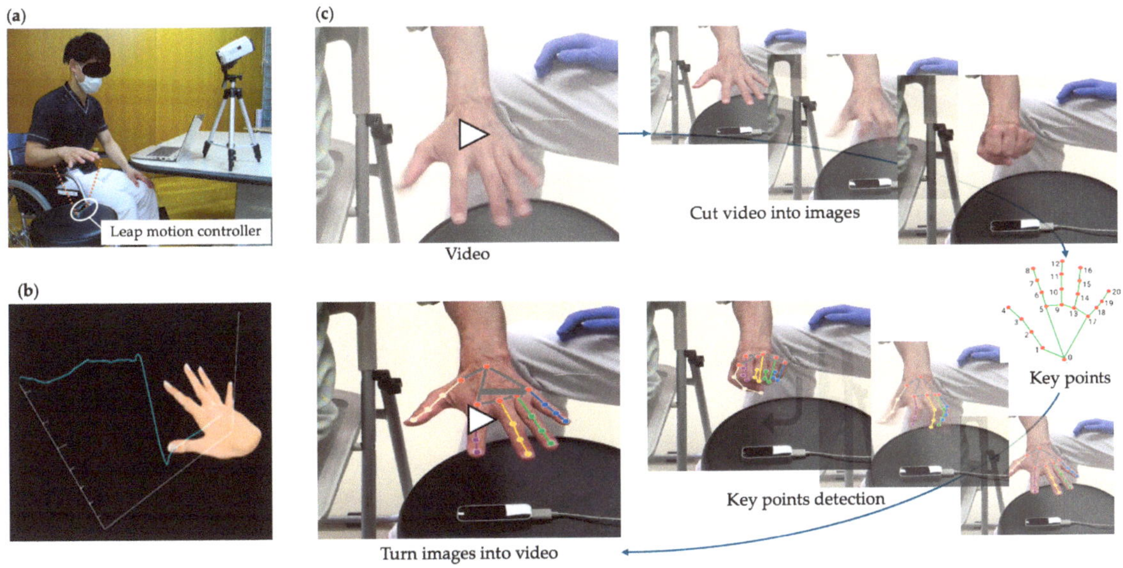

Figure 1. Fahrenheit and MediaPipe measurement method. (**a**) Patient position during measurement. (**b**) Virtual hand motion displayed in real time. The vertical axis indicates the numerical value of finger movement, while the horizontal axis indicates the time of measurement. (**c**) Keypoint detection method of MediaPipe for video data. The video is cut out at 30 frames per second, the keypoints are detected, and the successive images are merged into a video.

2.5. Preprocessing

2.5.1. Data Specification and Angle Conversion

Because both Fahrenheit and MediaPipe were taken from different angles of view, the acquired data were angle-transformed and the evaluations were aligned. In addition, considering the occlusion task of Fahrenheit, in which the DIP joint of each finger is hidden when the hand is maximally flexed, and the task of MediaPipe, in which there are no coordinate data for the metacarpals of each joint, the PIP joint of each finger was targeted for analysis. Fahrenheit's angular transformation method was the same as that used in previous studies [10]. For Mediapipe, a function was programmed to acquire the data. This function is used to calculate the angle of each joint of the body. The angles were calculated using analytic geometry. For example, there are three points, A ($x1$, $y1$, $z1$), B ($x2$, $y2$, $z2$), and C ($x3$, $y3$, $z3$) (Figure 2). Consider A and C as a skeletal structure (two bones). The intersection of line segments A and C lies on point B. The angle between A and C can be found as follows:

Find the inner product of AB and BC:

$$\mathrm{d1} = \overrightarrow{\mathrm{AB}} \cdot \overrightarrow{\mathrm{BC}} \tag{1}$$

Find the norm (distance) of AB/BC:

$$\mathrm{d2} = \|\mathrm{AB}\| \cdot \|\mathrm{BC}\| \tag{2}$$

Convert the cosine of the angle formed by $\overrightarrow{\mathrm{AB}} \cdot \overrightarrow{\mathrm{BC}}$:

$$\mathrm{angle} = \mathrm{arc\ cos}\left(\frac{\mathrm{d1}}{\mathrm{d2}}\right) \left(\text{At this time}, -1 < \frac{\mathrm{d1}}{\mathrm{d2}} < 1\right) \tag{3}$$

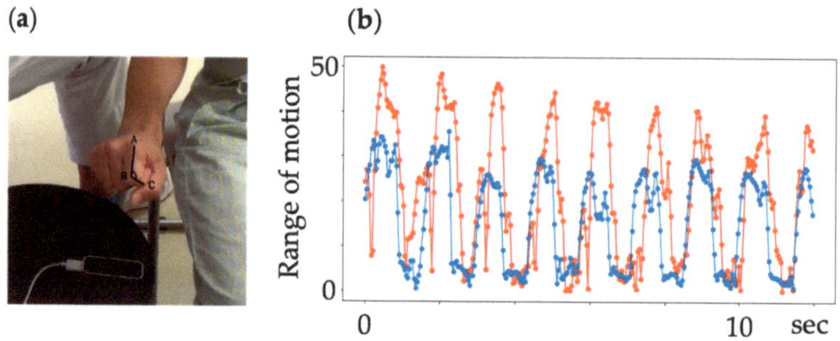

Figure 2. Method of angle conversion. (**a**) Calculation method for joint angles. (**b**) An example of the resulting motion trajectory. The motion trajectories of Fahrenheit and MediaPipe are shown as blue and red lines in the plots, respectively. The horizontal and vertical axes represent time and joint angle, respectively. (**a**) Demonstrates that the maximum achievable range of motion for the patient is realistic, and the flexion of the finger joint appears to be close to 90 degrees. (**b**) Illustrates the range of motion measured during the experiment, which is less than 90 degrees for Fahrenheit and even smaller angles for MediaPipe. These results may differ from the actual angles observed in the captured images because Fahrenheit reconstructs images from two cameras, while MediaPipe estimates angles from two-dimensional images.

2.5.2. Smoothing Process

A low-pass filter (Butterworth 2nd order, cutoff 12 Hz) was used for the first processing of the angle-transformed data. This method is commonly used for marker-based motion capture data, and the optimal cutoff values have been reported [16]. Second, optimal ensemble Kalman smoothing was performed [17]. The ensemble Kalman filter involves a two-way pass to determine the optimal state estimation for a given keypoint trajectory. Hyperparameters, including measurement and transition noises, were optimized using a grid search and cross-validation (Figure 3).

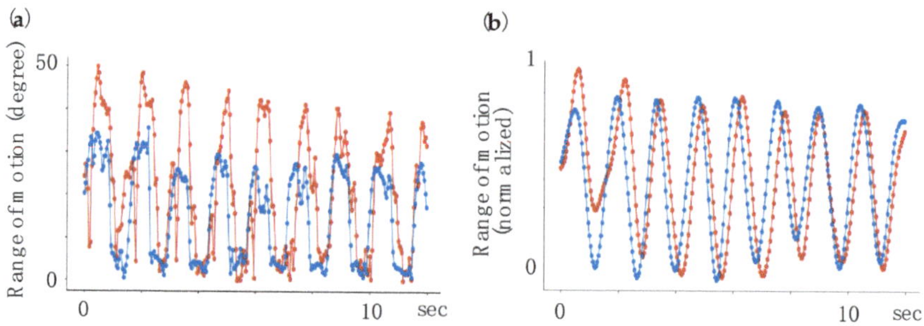

Figure 3. Smoothing of kinematics data obtained by MediaPipe and Fahrenheit. (**a**) Raw data. (**b**) Data after smoothing. Blue and red plots indicate the motion trajectories of Fahrenheit and MediaPipe, respectively.

2.5.3. Detection and Completion of Misestimated Frames

Based on previous studies, misestimation of the coordinate data at frame i was detected by comparing the absolute value of the difference between the previous and subsequent frames [18]. The absolute value of the difference is defined as:

$$d_x = (|x_2 - x_1|, \ldots, |x_i - x_{i-1}|, \ldots) \quad (4)$$

The mean value and standard deviation were calculated for dx, and the acquired xi values were considered as misestimated frames and excluded if any value exceeding the mean ± 3SD was detected. Each frame was then linearly interpolated, and spline smoothing was subsequently applied to obtain new data (Figure 4).

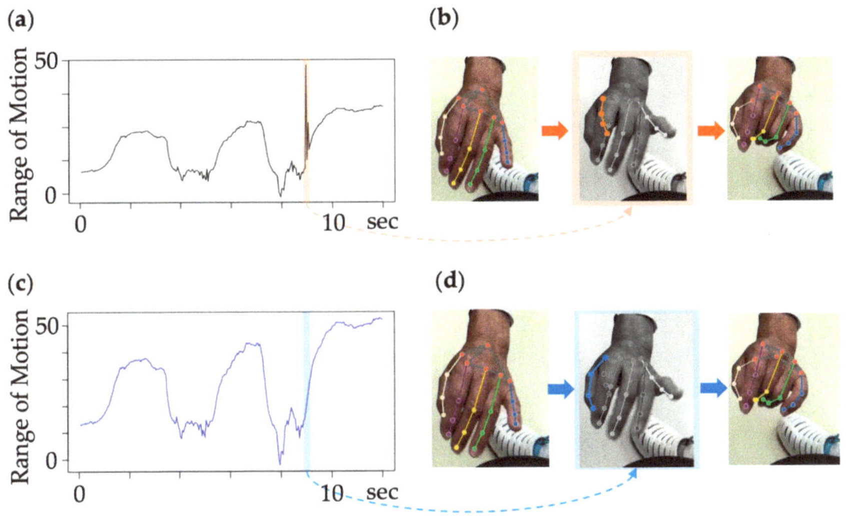

Figure 4. Detection and completion of misestimated frames and completion method. (**a**) The absolute difference dx in the mother finger PIP. Values exceeding the mean ± 3SD are detected in the area circled in red. (**b**) The time frames. (**c**) Implementation of spline smoothing, excluding the misestimated frames. (**d**) Excluded frames.

2.6. Statistical Analysis

2.6.1. DTW Distance

For the preprocessed data, the maximum flexion angle, maximum angular velocity, and average angular velocity were compared for each BRS phase. The DTW distance was calculated for each subject, and the degree of agreement was calculated using cross-correlation analysis.

DTW is a method for calculating the similarity between two time series data with different phases [19]. DTW calculates the distance between each element of two time series data on a round-robin basis and then maps the two time series data to minimize the total distance (DTW distance) between the two time series data. Even if the lengths of the two time series data are different, a DTW correspondence is possible. The DTW distance $D(A, B)$ of two time series data, $A = \{a1, a2, \ldots, aM\}$ and $B = \{b1, b2, \ldots, bN\}$, is defined as follows:

$$D\,(A, B) = d\,(M, N) \tag{5}$$

Let C be the distance function between two time series data and $C(am)$ be the distance between element bn of one time series data and element bn of the other time series data, then $d\,(m, n)$ is calculated recursively as follows:

$$d\,(m,n) = C(am,bn) + min\left\{\begin{array}{c} d(m-1,n) \\ d(m,n-1) \\ d(m-1,n-1) \end{array}\right\} \tag{6}$$

The conditional equation for this case is as follows:

$$d\,(0,0) = 0, d\,(m,0) = d\,(0,n) = \infty \\ 1 \le m \le M, 1 \le n \le N, m, n \in \mathbb{N}\,(2.4) \tag{7}$$

A drawback of DTW is its vulnerability to noise. When noise occurs, the actual distance between elements that should correspond to each other may differ from the distance between elements that should correspond to each other, resulting in inaccurate correspondence. The Euclidean distance between two elements is often used as the distance function of DTW. In this case, DTW is robust to the expansion and contraction or shift of time series data in the time direction but weak against the expansion and contraction or shift in the space direction. In other words, when two time series data have the same trajectory but move in the same spatial direction, accurate correspondence is not possible [19]. An example of mapping using DTW is shown in Figure 5.

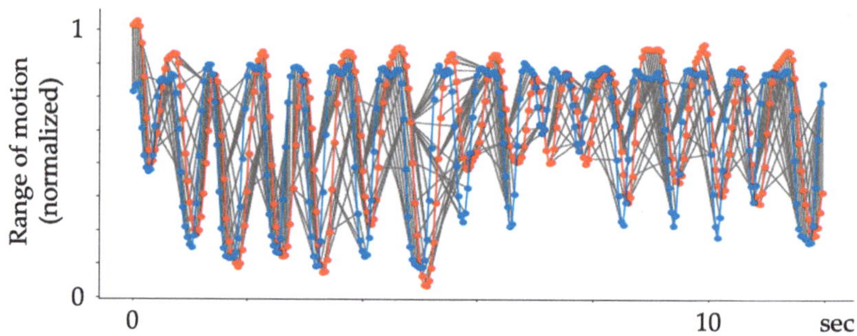

Figure 5. Dynamic Time Warping (DTW) distance calculation. Blue and red indicate the motion trajectories of Fahrenheit and Mediapipe, respectively. The line connecting them indicates the DTW distance.

2.6.2. Cross-Correlation Analysis

Using Lin's cross-correlation coefficients (CCC) [20], we evaluated the criterion-related validity of Mediapipe and Fahrenheit. Fleiss proposed CCC limits that were used for agreement interpretation: very good (CCC > 0.9), acceptable (0.71 < CCC0.9), moderate (0.51 < CCC < 0.7), poor (0.31 < CCC < 0.5), or no agreement (CCC < 0.31) [21].

3. Results

3.1. Subjects

Patients were classified into six groups according to disease severity per the BRS: stage I (2 patients), stage II (8 patients), stage III (6 patients), stage IV (6 patients), stage V (9 patients), and stage VI (9 patients) (Table 2).

Table 2. Demographic data of the subjects.

Attribute	Classification	Values
Age		71 ± 11
Gender	Male/Female	30/10
Paralyzed side	Right/Left	23/17
Day of a patient's illness		13.5 ± 14.8
Diagnosis	Cerebral hemorrhage/cerebral infarction	9/31
Brunnstrom stage (finger)	I/II/III/IV/V/VI	2/8/6/6/9/9

Categorical data are presented as real values, and numerical data are presented as mean ± standard deviation.

3.2. Comparison of Corrections

False estimation frames were detected in three patients using MediaPipe and in nine patients using Fahrenheit. The variation in values in patients with BRS I and II without joint motion was considered noise, and the amount of noise for each patient was evaluated using the Mean Absolute Deviation (MAD). The average values for MediaPipe and Fahrenheit were 2.46 and 1.42, respectively. After smoothing and correction, the MAD improved to an average of 0.81 for MediaPipe and 0.02 for Fahrenheit (Table 3).

Table 3. Comparison of corrections.

	MediaPipe	Fahrenheit
Subjects with mis-estimated frames detected	3/40	9/40
MAD of raw data	2.46 ± 0.87	1.42 ± 0.46
MAD after preprocessing	0.81 ± 0.32	0.02 ± 0.01

Categorical data are presented as real values, and numerical data are presented as mean ± standard deviation.

3.3. Comparison of the Results

The three therapists agreed on the results of the BRS for all subjects on the day of the experiment. The histograms of the middle finger PIP recorded during the 10 s exercise period for each BRS are shown in Figure 6. The histograms show that the percentages of extension (approximately 0–20°) and flexion (approximately 80–100°) increased as the BRS increased, and the distribution became closer to bimodal. This trend was common for both MediaPipe and Fahrenheit (Figure 6).

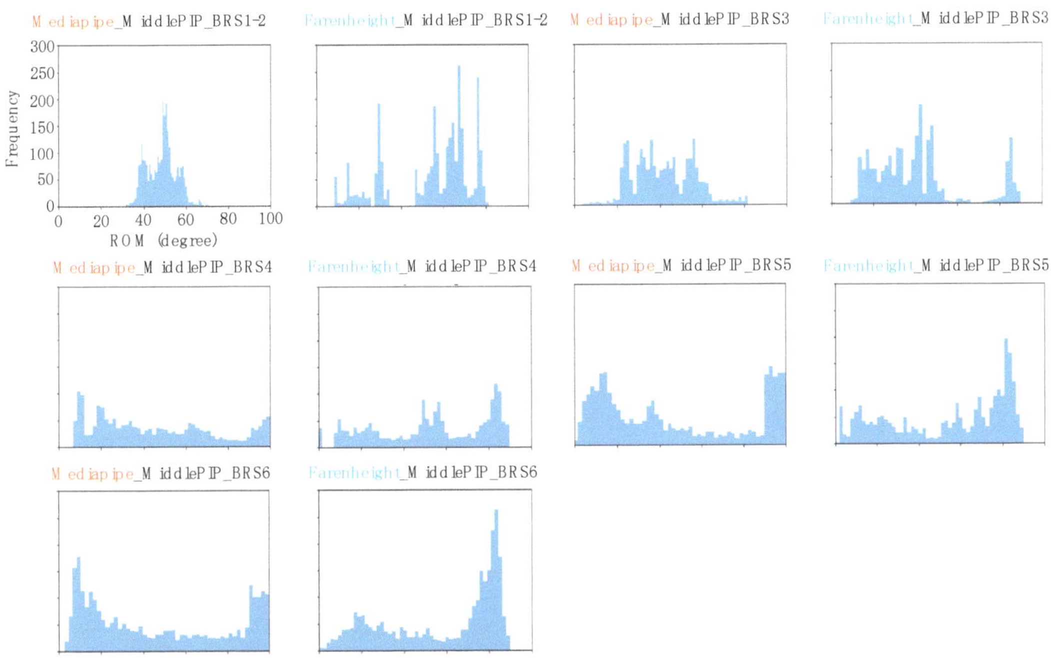

Figure 6. Histograms for each Brunnstrom Recovery Stage (BRS) phase. The *x*-axis indicates the observed range of motion (ROM), divided into intervals of 2° each. The *y*-axis indicates the number of data observed in the interval. Each histogram includes the data of all the subjects divided into BRS phases.

The maximum flexion angle of each finger increased with the BRS stage (Spearman's rank correlation coefficient; MediaPipe: r = 0.723, $p = 0.002$; Fahrenheit: r = 0.691, $p = 0.04$) (Table 4). The maximum angular velocity of each finger increased during the BRS phase (Spearman's rank correlation coefficient; MediaPipe: r = 0.784, $p = 0.0001$; Fahrenheit: r = 0.796, $p = 0.001$) (Table 5). The mean angular velocity of each finger increased during the BRS phase (Spearman's rank correlation coefficient; MediaPipe: r = 0.816, $p < 0.0001$; Fahrenheit: r = 0.832, $p < 0.001$) (Table 5).

Table 4. Maximum flexion angle.

Peak_Flexion (deg)	BRS I–II (n = 10)	BRS III (n = 6)	BRS IV (n = 6)	BRS V (n = 9)	BRS VI (n = 9)
MediaPipe					
Thumb	35.1 ± 14.4	34.9 ± 14.5	55.6 ± 16.3	41.3 ± 12.6	42.3 ± 8.1
Index	54.5 ± 15.6	45.5 ± 26.6	75.9 ± 25.9	93.5 ± 10.7	99.2 ± 0.8
Middle	52.8 ± 9.4	56.0 ± 20.6	87.6 ± 11.3	99.5 ± 0.6	98.6 ± 3.2
Ring	60.8 ± 8.8	60.7 ± 18.2	91.5 ± 7.2	99.5 ± 1.1	99.4 ± 1.1
Pinky	52.8 ± 10.2	59.9 ± 21.2	68.5 ± 15.7	98.9 ± 1.3	97.4 ± 6.0
Fahrenheit					
Thumb	53.3 ± 24.0	52.1 ± 22.0	68.3 ± 17.7	31.5 ± 6.0	40.2 ± 15.4
Index	60.1 ± 19.4	52.9 ± 22.4	73.6 ± 20.0	86.4 ± 3.1	89.1 ± 1.3
Middle	60.5 ± 18.1	51.6 ± 20.6	73.3 ± 20.9	87.1 ± 2.6	88.1 ± 2.2
Ring	59.7 ± 17.7	51.3 ± 21.0	72.4 ± 19.7	86.8 ± 3.1	88.0 ± 2.5
Pinky	18.2 ± 11.6	14.2 ± 22.2	37.8 ± 15.7	87.2 ± 2.3	88.1 ± 2.7

Values are presented as mean ± standard deviation.

Table 5. Comparison of angular velocities.

Peak_Velocity/Average_Velocity (deg/s)	BRS I–II (n = 10)	BRS III (n = 6)	BRS IV (n = 6)	BRS V (n = 9)	BRS VI (n = 9)
MediaPipe					
Thumb	87.9/6.2	111.2/14.7	274.7/55.3	667.6/110.1	780.8/186.9
Index	82.6/9.8	236.1/29.7	505.0/101.6	1296.6/231.1	1400.5/365.9
Middle	61.1/9.7	240.0/26.8	544.8/108.6	1309.0/239.9	1425.9/380.4
Ring	58.5/9.4	256.2/27.6	573.6/110.6	1233.8/227.6	1392.0/365.5
Pinky	104.2/9.6	187.8/25.8	574.8/96.6	1198.8/177.8	1231.8/282.3
Fahrenheit					
Thumb	43.7/7.9	126.0/10.8	167.9/49.6	271.7/99.4	672.0/207.5
Index	72.7/12.3	214.0/28.3	506.4/114.2	1295.8/214.7	1412.8/355.7
Middle	69.2/13.2	225.8/27.0	501.6/102.6	1240.1/217.7	1448.2/373.8
Ring	77.5/14.9	216.8/24.7	587.2/100.2	1300.6/200.2	1403.1/351.1
Pinky	79.9/13.3	165.7/22.5	501.6/74.4	1225.1/205.3	1208.5/345.7

Each cell shows the maximum or average angular velocity. Values are shown as mean ± standard deviation.

3.4. Comparison of Agreement between Measurements

The DTW distance between MediaPipe and Fahrenheit for each subject and the results of the cross-correlation analysis are shown as box plots for each BRS (Figure 7). The results showed a "moderate" or "better" correlation for all data, except for patients with BRS I and II (DTW = 0.092–0.579, CCC = 0.374–0.986). In particular, the middle and ring fingers exhibited a strong average correlation (DTW mean = 0.216, CCC = 0.758). The data of the 10 patients with BRS I and II were highly variable (DTW = 0.106–0.535, CCC = 0. 251–0.963), and two patients showed a weak correlation (CCC < 0.3).

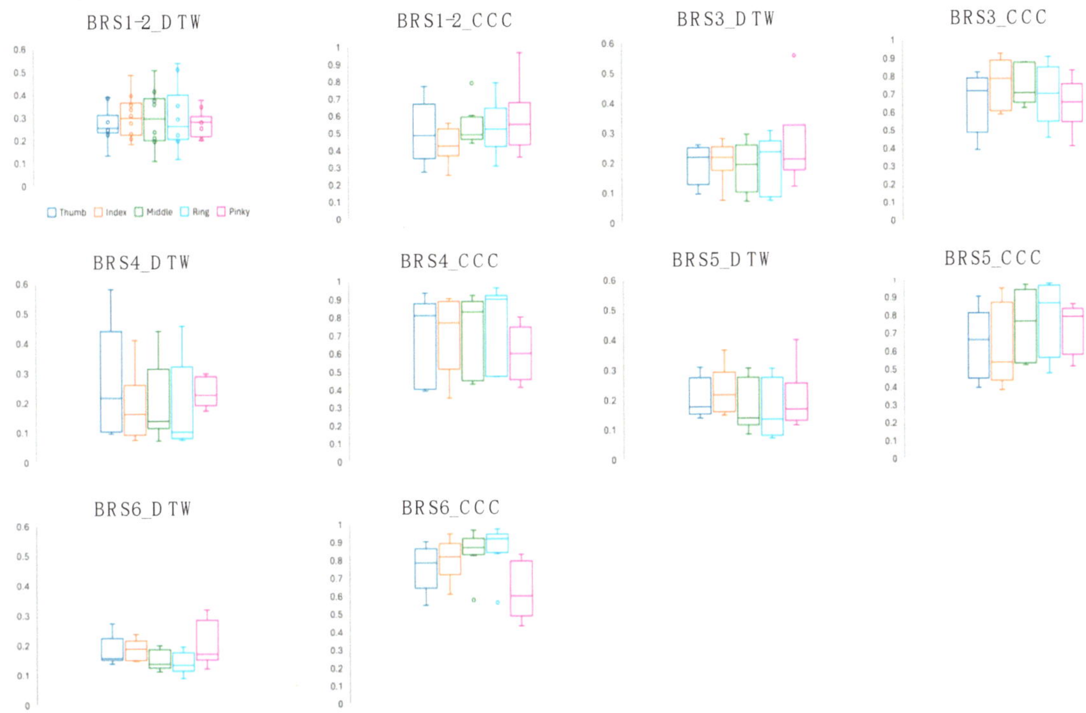

Figure 7. Comparison of the agreement of the measurements. Dynamic Time Warping (DTW) distance between MediaPipe and Fahrenheit for each subject is shown on the left and cross-correlation analysis is shown on the right for each Brunnstrom Recovery Stage (BRS). The bottom line of the error bar indicates the minimum value, the center line indicates the median value, the × mark indicates the mean value, and the top line indicates the maximum value. Boxes indicate interquartile ranges.

4. Discussion

This study verified the criterion-related validity of MediaPipe and Fahrenheit. For the data preprocessed by angle transformation and smoothing, the CCC results and DTW distance showed more than moderate correlations. The results for the peak angle and peak angular velocity were also in general agreement. In a previous study on the development of Fahrenheit, the peak angle and peak angular velocity indicated by the LMC were evaluated to distinguish the recovery state of the finger, and identifiers were created [10]. These identifiers were eventually developed into Fahrenheit as an application for evaluating the recovery stage of paralysis. Based on the concordance of the measurements shown in this study, we believe that the transfer of the Fahrenheit functionality to MediaPipe, which can be used on smartphones and tablet devices, will enhance the effectiveness of hand-movement analysis in clinical practice. This motion analysis system integrated with a camera for capturing video images can analyze minute changes and movement patterns of the subject and will be a powerful tool for sharing practice goals between therapists and patients in evaluation and treatment.

Another advantage of moving from Fahrenheit to MediaPipe is that MediaPipe is a simple motion analysis system. In general, image processing methods, such as MediaPipe, have a wider effective distance for hand tracking than detection algorithms based on LMC [12,14]. Fahrenheit's data acquisition was limited by the installation position, subject's posture, and calibration [11]. The MediaPipe data were obtained from a video camera that was used to capture the hand motions from the start to the end of the hand motions during the Fahrenheit data acquisition. However, the position of the camera and the ratio of the hand on the image to that on the entire screen were not specified. The reason why the agreement of the index and little fingers was lower than that of the other fingers is thought to be that the index and little fingers were occluded by the other fingers and did not appear on the screen sufficiently. Even under these conditions, moderate or better agreement suggests the usefulness of MediaPipe.

In the analysis of the time series data, some limitations of both Mediapipe and Fahrenheit were identified. One of the limitations of Mediapipe is its high noise levels. Because both Mediapipe and Fahrenheit use frame processing to detect hand motions, this limitation may be directly related to the high accuracy of MediaPipe. Although patients with BRS I and II are inherently incapable of voluntary hand movements, angle fluctuations were detected in both MediaPipe and Fahrenheit. The MAD of variation was 2.46 for MediaPipe and 1.42 for Fahrenheit. In other words, the raw data for MediaPipe showed variability within 2.4° on average of the mean value. After preprocessing, based on a previous study, the MAD improved to an average of 0.81 for MediaPipe and 0.02 for Fahrenheit. Fahrenheit eliminated most angle fluctuations, whereas those of MediaPipe remained. These factors can be attributed to problems with the detection algorithm and its accuracy. The frame processing in both MediaPipe and Fahrenheit does not retain the time axis information of the previous and subsequent frames; therefore, the detected hand position deviates from the previous and subsequent frames, which results in noise. Previous studies have also shown that accuracy can be improved by improving the ratio of the hand to the entire screen and resolution [15,22]. Although this is not a simple comparison of accuracy because the results have not been compared under the conditions that optimize the performance of the two methods, it should be considered that there is a need to set measurement conditions for MediaPipe as well and that data processing after the measurement alone is not sufficient. This problem may be solved using a machine learning method such as the recurrent neural network, which is also used for processing time series data, and by incorporating an algorithm that preserves information on the time axis before and after measurement.

One limitation of Fahrenheit is the large number of dropped frames. The breakdown of the subjects for whom falsely estimated frames were detected was five patients with BRS I and II, two patients with BRS III, one patient with BRS V, and one patient with BRS VI. The low body surface temperature of the subjects may have influenced the high number of dropped frames detected at lower motion speeds. The infrared sensor also provides quantitative information by measuring the difference in temperature between the fingers and the back of the hand. Patients with BRS I and II who have difficulty with voluntary movements tend to have weaker muscle contractions and lower body surface temperatures [23]. Limitations of infrared sensors in patients with severe peripheral edema have also been reported [24]. The detectable range of LMC is −30 °C to 60 °C [25], but it is possible that the difference in body surface temperature affected the accuracy of detection. In addition, stroke patients without joint motion are prone to peripheral edema in the acute phase [26], which may have also affected the mis-estimation. In other words, MediaPipe could not completely eliminate the noise due to a drop in accuracy, and Fahrenheit showed many false estimations in patients with BRS I and II, which may have affected the variability of data in patients with BRS I and II (DTW = 0.106–0.535, CCC = 0.251–0.963).

Fahrenheit operates at a frame rate of 60 fps, while MediaPipe processes at 30 fps. This disparity in frame rate can result in variations in dynamic motion measurements. To mitigate this, we downsampled the Fahrenheit data to 30 fps for a more accurate comparison. However, despite this adjustment, some variability may still exist, potentially impacting the comparability of the methods.

The exclusion criteria were established to ensure the reliability and accuracy of the measurements. However, we acknowledge that these criteria may limit the applicability of our findings to a broader patient population. Future studies could consider including a more diverse cohort to evaluate the generalizability of these results.

The primary distinction between Fahrenheit and MediaPipe systems lies in the number of joint landmarks tracked. Fahrenheit includes metacarpals of each finger, whereas

MediaPipe does not, affecting the accuracy of MP joint flexion angle measurements. This limitation in MediaPipe's joint landmarks may impact overall accuracy and clinical applicability. To address the constraint of joint landmarks in MediaPipe, particularly the absence of metacarpals which affects MP joint flexion angle precision, the following method is proposed: (1) Enhanced feature extraction, utilizing Feature Engineering + Long Short-Term Memory (FE + LSTM) models to derive more relevant features from existing landmarks. This approach can infer missing metacarpal positions and movements by enhancing data with temporal and spatial contextual information [27,28]. (2) Integration with additional sensors, combining MediaPipe with depth sensors or infrared cameras to enhance joint tracking capability [29]. (3) Machine learning models, incorporating human pose estimation models can help in capturing and predicting the missing joint movements by analyzing the seasonal and trend components of the hand movements [30,31]. This approach can model the temporal dependencies and seasonal variations in joint movements, thereby improving the accuracy of joint position estimations. (4) Algorithmic enhancements, implementing recurrent neural networks (RNNs) and LSTM networks trained on comprehensive datasets to predict missing joint positions [32]. These models effectively handle temporal dependencies and improve the robustness of motion tracking systems. By integrating STF + LSTM and FE + LSTM into future versions of our application, we can achieve more precise motion estimation and potentially overcome the current limitations of MediaPipe's joint landmark detection.

5. Conclusions

The results showed a moderate-to-high correlation, indicating the possibility of transferring the functionality of Fahrenheit to MediaPipe, which can be used on smartphones and tablets. Fahrenheit's weakness, which causes instability in the joint center estimation when a hand grasps an object (occlusion problem), can be solved by image processing using AI. MediaPipe has an Objectron library that can detect objects [12]. By developing an application that can run the MediaPipe hands and MediaPipe Objectron, it is possible to obtain the positional information of the object and hand simultaneously. This enables an accurate evaluation of human tool manipulation based on the positional relationship with an object, even when the hand is occluded. Currently, conventional motion captures only tracks human motion; the development of an application that can simultaneously track the motion of a human and that of an object generated by human motion can revolutionize the world of motion capture. The next task of our research team is to develop such an application and verify its performance in tracking hand motion with an object.

Author Contributions: Conceptualization, T.H.; methodology, R.S., T.S. and T.H.; validation, T.H.; formal analysis, R.S.; investigation, T.S.; data curation, R.S. and T.S.; writing—original draft, R.S.; writing—review and editing, T.S. and T.H.; visualization, R.S.; supervision, T.H.; project administration, T.H.; funding acquisition, T.S. All authors have read and agreed to the published version of the manuscript.

Funding: This project was supported by the Saitama Prefectural University (SPU) Grant (15001) and JSPS KAKENHI (C) (21K11220).

Institutional Review Board Statement: The study was conducted in accordance with the guidelines of the Declaration of Helsinki and approved by the Ethics Committee of Saitama Prefectural University (approval No. 27083) and the Independent Ethics Committee of Tokyo Dental College Ichikawa General Hospital (I15-71).

Data Availability Statement: Study data will be provided upon request from the corresponding authors.

Acknowledgments: The authors would like to thank Masateru Katayama, Hiraku Hotta, Shin Doumae, Sho Yoshimura, and Ayumi Enami, of the Department of Neurosurgery and Rehabilitation at Tokyo Dental College Ichikawa General Hospital, for their contributions and operational approval to conduct this study.

Conflicts of Interest: The authors declare no conflicts of interest.

References

1. Kivell, T.L.; Baraki, N.; Lockwood, V.; Williams-Hatala, E.M.; Wood, B.A. Form, Function and Evolution of the Human hand. *Am. J. Biol. Anthropol.* **2023**, *181* (Suppl. S76), 6–57. [CrossRef]
2. Ji, Q.; Luo, Z.-X.; Yuan, C.-X.; Wible, J.R.; Zhang, J.-P.; Georgi, J.A. The Earliest Known Eutherian Mammal. *Nature* **2002**, *416*, 816–822. [CrossRef] [PubMed]
3. Langhorne, P.; Bernhardt, J.; Kwakkel, G. Stroke Rehabilitation. *Lancet* **2011**, *377*, 1693–1702. [CrossRef] [PubMed]
4. Nakayama, H.; Jørgensen, H.S.; Raaschou, H.O.; Olsen, T.S. Recovery of Upper Extremity Function in Stroke Patients: The Copenhagen Stroke Study. *Arch. Phys. Med. Rehabil.* **1994**, *75*, 394–398. [CrossRef] [PubMed]
5. Norouzi-Gheidari, N.; Archambault, P.S.; Fung, J. Effects of Robot-Assisted Therapy on Stroke Rehabilitation in Upper Limbs: Systematic Review and Meta-Analysis of the Literature. *J. Rehabil. Res. Dev.* **2012**, *49*, 479–496. [CrossRef] [PubMed]
6. Lin, J.H.; Hsiao, S.F.; Chang, C.M.; Huang, M.H.; Liu, C.K.; Lin, Y.T. Factors Influencing Functional Independence Outcome in Stroke Patients after Rehabilitation. *Kaohsiung J. Med. Sci.* **2000**, *16*, 351–359. [PubMed]
7. Fugl-Meyer, A.R.; Jääskö, L.; Leyman, I.; Olsson, S.; Steglind, S. The Post-Stroke Hemiplegic Patient. I. A Method for Evaluation of Physical Performance. *Scand. J. Rehabil. Med.* **1975**, *7*, 13–31. [CrossRef]
8. Lrvine, R.E. Movement Therapy in Hemiplegia: A Neurophysiological Approach. *Gerontologist* **1972**, *12*, 205–208. [CrossRef]
9. Liu, M.; Chino, N.; Tuji, T.; Masakado, Y.; Hase, K.; Kimura, A. Psychometric Properties of the Stroke Impairment Assessment Set (SIAS). *Neurorehabilit. Neural Repair* **2002**, *16*, 339–351. [CrossRef]
10. Hamaguchi, T.; Saito, T.; Suzuki, M.; Ishioka, T.; Tomisawa, Y.; Nakaya, N.; Abo, M. Support Vector Machine-Based Classifier for the Assessment of Finger Movement of Stroke Patients Undergoing Rehabilitation. *J. Med. Biol. Eng.* **2020**, *40*, 91–100. [CrossRef]
11. Saito, T.; Ishioka, T.; Yoshimura, S.; Hamaguchi, T. Predictive Ability of *Fahrenheit*, a Hand Motion Recording System for Assessing Hand Motor Function in Patients with Hemiplegia Post-cerebrovascular Disease—A Pilot Study. *Appl. Sci.* **2021**, *11*, 8153. [CrossRef]
12. Google. Mediapipe. Github. 2023. Available online: https://github.com/google-edge-ai/mediapipe-samples (accessed on 12 December 2023).
13. Asta, A.K.A.A.; Yuniarno, E.M.; Nugroho, S.M.S.; Avian, C. Handwriting Classification Based on Hand Movement Using ConvLSTM. In Proceedings of the 2023 International Seminar on Intelligent Technology and Its Applications (ISITIA), Surabaya, Indonesia, 26–27 July 2023; Volume 2023.
14. Christian, S.-C.; Dan, G.; Alexandra, F.; Adela, P.-P.; Ovidiu, S.; Honoriu, V.; Liviu, M. Hand Gesture Recognition and Infrared Information System. In Proceedings of the 2022 23rd International Carpathian Control Conference (ICCC), Sinaia, Romania, 29 May–1 June 2022; Volume 2022.
15. Bazarevsky, V.; Zhang, F. On-Device, Real-Time Hand Tracking with MediaPipe. In Google AI Blog. 2019. Available online: https://research.google/blog/on-device-real-time-hand-tracking-with-mediapipe/ (accessed on 30 June 2024).
16. Challis, J.H. A Procedure for the Automatic Determination of Filter Cutoff Frequency for the Processing of Biomechanical Data. *J. Appl. Biomech.* **1999**, *15*, 303–317. [CrossRef]
17. Rauch, H.E.; Tung, F.; Striebel, C.T. Maximum Likelihood Estimates of Linear Dynamic Systems. *AIAA J.* **1965**, *3*, 1445–1450. [CrossRef]
18. Sugiyama, Y.; Uno, K.; Matsui, Y. Types of Anomalies in Two-Dimensional Video-Based Gait Analysis in Uncontrolled Environments. *PLoS Comput. Biol.* **2023**, *19*, e1009989. [CrossRef]
19. Nakamoto, K.; Yamada, Y.; Suzuki, E. Fast Clustering for Time-Series Data with Average-Time-Sequence-Vector Generation Based on Dynamic Time Warping. *Trans. Jpn. Soc. Artif. Intell.* **2003**, *18*, 144–152. [CrossRef]
20. Lin, L.I.-K. A Concordance Correlation Coefficient to Evaluate Reproducibility. *Biometrics* **1989**, *45*, 255–268. [CrossRef]
21. Hills, M.; Fleiss, J.L. The Design and Analysis of Clinical Experiments. *J. R. Stat. Soc. Ser. A (Gen.)* **1987**, *150*, 400. [CrossRef]
22. Dill, S.; Rösch, A.; Rohr, M.; Güney, G.; De Witte, L.; Schwartz, E.; Antink, C.H. Accuracy Evaluation of 3D Pose Estimation with MediaPipe Pose for Physical Exercises. *Curr. Dir. Biomed. Eng.* **2023**, *9*, 563–566. [CrossRef]
23. Alfieri, F.M.; Massaro, A.R.; Filippo, T.R.; Portes, L.A.; Battistella, L.R. Evaluation of Body Temperature in Individuals with Stroke. *NeuroRehabilitation* **2017**, *40*, 119–128. [CrossRef]
24. Anderson, M.E.; Moore, T.L.; Lunt, M.; Herrick, A.L. The 'Distal-Dorsal Difference': A Thermographic Parameter by Which to Differentiate between Primary and Secondary Raynaud's Phenomenon. *Rheumatology* **2007**, *46*, 533–538. [CrossRef]
25. Ultraleap. Tracking | Leap Motion Controller. Available online: https://www.ultraleap.com/leap-motion-controller-overview/ (accessed on 1 December 2023).
26. Giang, T.A.; Ong, A.W.G.; Krishnamurthy, K.; Fong, K.N. Rehabilitation Interventions for Poststroke Hand Oedema: A Systematic Review. *Hong Kong J. Occup. Ther.* **2016**, *27*, 7–17. [CrossRef] [PubMed]
27. Shao, X.; Kim, C.-S.; Sontakke, P. Accurate deep model for electricity consumption forecasting using multi-channel and multi-scale feature fusion CNN–LSTM. *Energies* **2020**, *13*, 1881. [CrossRef]
28. Ryumin, D.; Ivanko, D.; Ryumina, E. Audio-Visual Speech and Gesture Recognition by Sensors of Mobile Devices. *Sensors* **2023**, *23*, 2284. [CrossRef]

29. Amprimo, G.; Ferraris, C.; Masi, G.; Pettiti, G.; Priano, L. GMH-D: Combining Google MediaPipe and RGB-Depth Cameras for Hand Motor Skills Remote Assessment. In Proceedings of the 2022 IEEE International Conference on Digital Health (ICDH), Barcelona, Spain, 10–16 July 2022. [CrossRef]
30. Asteriadis, S.; Daras, P. Landmark-based multimodal human action recognition. *Multimedia Tools Appl.* **2017**, *76*, 4505–4521. [CrossRef]
31. Li, R.; Li, Q.; Yang, S.; Zeng, X.; Yan, A. An efficient and accurate 2D human pose estimation method using VTTransPose network. *Sci. Rep.* **2024**, *14*, 7608. [CrossRef]
32. Yu, J.; Qin, M.; Zhou, S. Dynamic gesture recognition based on 2D convolutional neural network and feature fusion. *Sci. Rep.* **2022**, *12*, 4345. [CrossRef]

biomimetics

MDPI

Review

How Multifunctioning Joints Produce Highly Agile Limbs in Animals with Lessons for Robotics

Stuart C. Burgess

School of Electrical, Electronic and Mechanical Engineering, Bristol University, Bristol BS8 1TR, UK; s.c.burgess@bris.ac.uk

Abstract: This paper reviews how multifunctioning joints produce highly agile limbs in animals with lessons for robotics. One of the key reasons why animals are so fast and agile is that they have multifunctioning joints in their limbs. The multifunctioning joints lead to a high degree of compactness which then leads to a host of benefits such as low mass, low moment of inertia and low drag. This paper presents three case studies of multifunctioning joints—the human wrist joint, knee joint and foot joints—in order to identify how multifunctioning is achieved and what lessons can be learned for robotics. It also reviews the multifunctioning nature of muscle which plays an important role in joint actuation. A key finding is that multifunctioning is achieved through various means: multiple degrees of freedom, multifunctioning parts, over-actuation and reconfiguration. In addition, multifunctioning is achieved through highly sophisticated layouts with high levels of integration and fine-tuning. Muscle also makes an important contribution to animal agility by performing multiple functions including providing shape, protection and heat. The paper reviews progress in achieving multifunctioning in robot joints particularly for the wrist, knee and foot. Whilst there has been some progress in creating multifunctioning robotic joints, there is still a large gap between the performance of animal and robotic joints. There is an opportunity to improve the agility of robots by using multifunctioning to reduce the size and mass of robotic joints.

Keywords: multifunctioning; joints; limbs; muscle

1. Introduction

This paper reviews how multifunctioning joints produce highly agile limbs in animals with lessons for robotics. Animals are well known to have superior performance to robots in the area of locomotion and hand skills [1,2]. One reason for the higher performance is superior control and sensing [1]. Another reason is that of multifunctioning materials [3]. This paper shows that multifunctioning skeletal joints are another key reason for the superior agility of animal limbs. The paper gives examples of how multifunctioning leads to high performance, identifies how multifunctioning is achieved and discusses what lessons can be learned for robotics.

Multifunctioning is widespread in nature. The majority of biological materials are remarkably multifunctional. For example, human skin has many functions including being stretchable, tough, self-renewing, self-healing, self-lubricating and insulating [4]. In addition, skin is host to blood vessels, nerves, sweat glands, hairs and multiple sensors [4]. There have been many studies on the multifunctioning nature of biological materials such as muscle [5], bone [6], nacre [7], feathers [8], plant cuticle [9], cactus spines [10] and synovial fluid [11].

Multifunctioning is also widespread in nature at the subsystem level. The mammalian mouth has multiple functions of breathing, eating, drinking and creating sound [12]. The mammalian penis has two major roles of urinating and mating [13]. Seals use their flippers for ground locomotion as well as swimming [14]. Some birds such as gannets use wings for powered flight in air as well as powered underwater swimming [15]. Water birds like ducks generally have feet that are optimised for both walking on ground as well as swimming. Animal blood circulation systems perform many functions including temperature control, clotting and transportation of multiple materials to multiple destinations [16]. Multifunctioning is also prevalent in microbiology such as proteins [17] and cells [18,19].

Multifunctioning is also common in animal joints. Most animal joints are multifunctioning with multiple degrees of freedom. For example, the mammalian vertebral column has important functions of providing flexibility in multiple degrees of freedom as well as carrying heavy loads and providing a protective structure for a complex network of nerves [20]. By being multifunctioning, the vertebral column is able to be remarkably compact and lightweight.

Multifunctionality is a very advantageous design feature because it reduces the number of subsystems and components and produces a compact design. Multifunctioning in joints leads to a high degree of compactness which then leads to a host of benefits such as low mass, low moment of inertia and low drag. It also leads to reduced energy demands and the ability to meet tight dimensional constraints. Multifunctioning joints also have the additional benefit that it often enables the animal or robot to perform multiple high-level functional tasks. There has been a growing desire to have multifunctioning capabilities in robots such as multifunctioning rescue robots [21], multifunctioning arms [22] and multifunctioning rehabilitation gloves [23].

Significant research effort has gone into producing multifunctional materials such as multifunctioning composites [24,25], multifunctioning coatings [26] and multifunctioning battery structures [27,28]. Some multifunctioning materials are especially suitable for robotics. For example, materials have been developed that transform two-dimensional laminate actuators into three-dimensional structural elements [29,30]. These materials can lead to greater compactness and functionality in robotics and other areas. However, there has not been so much work on bioinspired multifunctioning subsystems such as limb joints.

In comparison to nature, robotic joints tend to be bulky with less multifunctionality. For example, human prosthetic joints are relatively bulky with lower functionality when compared with the human joints they are replacing [31,32]. Two of the most well-developed humanoid robots have been the Asimo robot and the Atlas robot, as shown in Figure 1. These robots have some limbs that are reasonably compact, but they only achieve this due to reduced functionality in the joints. For example, the knee joints of the two robots cannot rotate around the tibia axis and the wrists cannot abduct–adduct. A key reason why robot joints are bulky and/or have less functionality is a lack of multifunctioning in the joints.

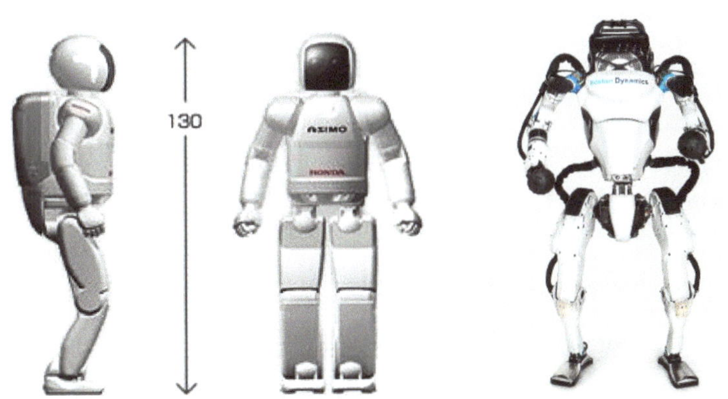

(a) (b)

Figure 1. Two well-known humanoid robots. (**a**) Asimo [33]; (**b**) Atlas [34].

Therefore, there is a gap in the research field in terms of the development of multifunctioning joints and there is potential benefit in learning from the multifunctioning strategies of biological joints. The review seeks to address this gap by analysing three key joints: the human wrist joint, human knee joint and human foot joints. These joints were chosen because they are highly multifunctioning and because they are of particular interest to the field of robotics. It also reviews the multifunctioning nature of muscle which plays an important role in joint actuation.

For each case study, the functions of the joint (or muscle) are defined and the means by which they are achieved is analysed and tabulated. Each section also reviews progress in achieving multifunctioning in robot joints. Key examples of multifunctioning bioinspired joints are given for the wrist, knee and foot. The paper makes recommendations for how to achieve greater multifunctionality in robot joints.

2. The Human Wrist Joint

2.1. Wrist Anatomy

Figure 2 summarises the anatomy of the human wrist joint. The wrist has 8 bones, over 20 articulating surfaces [35] and over 25 ligaments [36]. The ligaments are stiff enough to provide stability under load but flexible enough to allow significant range of movement (ROM) [37]. The radius and ulna bones are connected by an articular disk (an oval plate of fibrocartilaginous meniscus) [38] (Figure 2a). There are six main muscle–tendon units that actuate the wrist (Figure 2b) [1] and these are located in the lower arm.

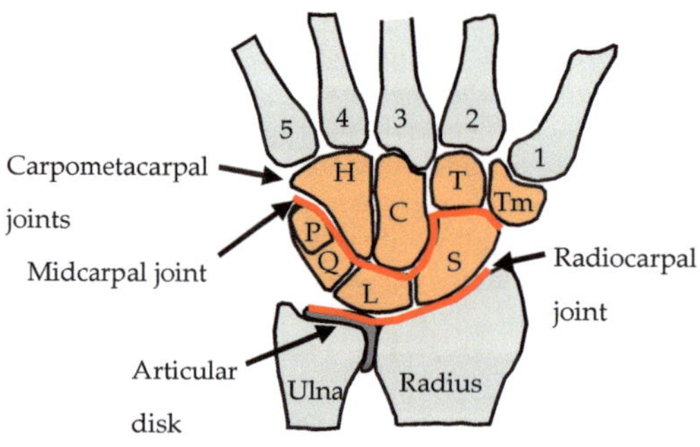

(a) Front palmer view

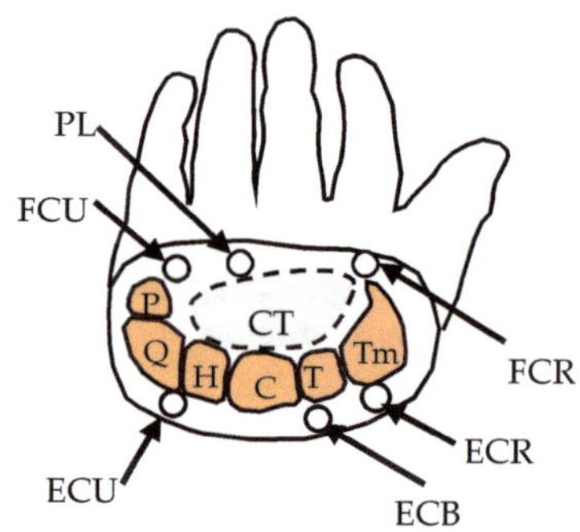

(b) Cross section

Figure 2. Schematic of the right wrist joint based on the anatomy in [35]. The eight wrist bones are highlighted. Key: H = hamate, C = capitate, T = trapezoid, Tm = trapezium, P = pisiform, Q = triquetrum, L = lunate, S = scaphoid. Main muscles: FCR = flexor carpi radialis, PL = palmaris longus, FCU = flexor carpi ulnaris, ECU = extensor carpi ulnaris, ECB = extensor carpi brevis, ECR = extensor carpi radialis, CT = carpal tunnel.

2.2. Requirements and Objectives

Table 1 summarises the four major functions of the wrist and the aspects of subsystem design that make multifunctioning possible. The design objective of the wrist is to form an assembly of parts that creates a biaxial (universal) joint at the same time as creating efficient load paths for in-plane and out-of-plane loads. An additional function of the wrist is to have proprioceptor sensors [39]. The wrist has been described as the most complex joint in the body due to the multifunctioning nature of the joint [40].

Table 1. Summary of wrist functions and how they are achieved.

Main Functions	Flexion–Extension	Abduction–Adduction	Longitudinal Load Paths	Carpal Arch
Schematic	Midcarpal joint Radiocarpal joint 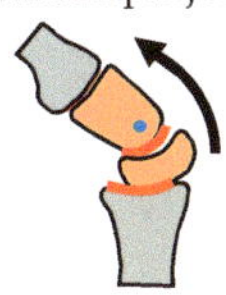	Midcarpal joint Radiocarpal joint 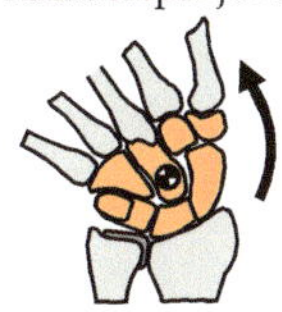	Two load columns Two arches 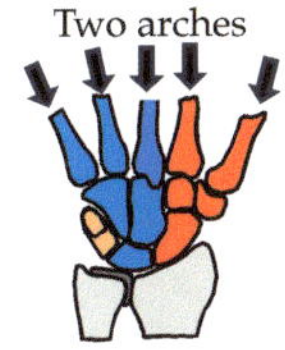	Arch in cross-section of hand 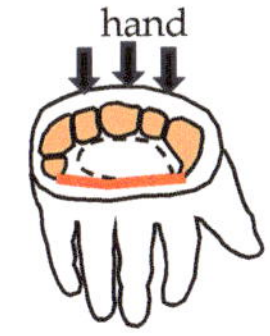
Common parts (wrist bones)	All of top row Three of bottom row 	Three of top row Three of bottom row 	All of top row Two of bottom row 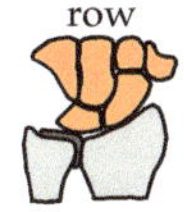	All of top row Two of bottom row 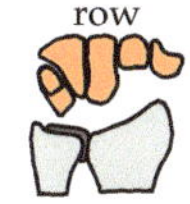
Actuation [41]	Multifunctioning muscles	Multifunctioning muscles		
Fine tuning	Common centre of rotation 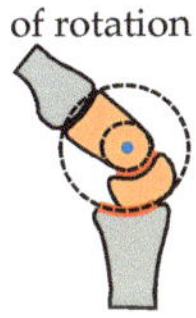	Common centre of rotation 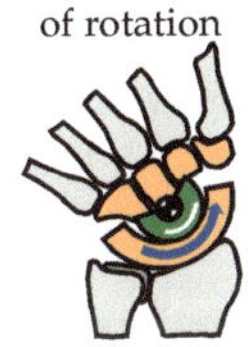	Alignment of metacarpals with top row of wrist bones.	Arch segments aligning simultaneously in two different planes
Integration	Network of 8 bones and >25 ligaments	Multiple joint interfaces (Capitate = 5)	Capitate styloid process	Carpal arch integrated with tendons/nerves
Reconfiguration			Muscle tension to stiffen wrist during loading	Muscle tension to stiffen wrist during loading
Miniaturisation	Sensors, nerves, blood vessels, lubrication	Sensors, nerves, blood vessels, lubrication	Sensors, nerves, blood vessels, lubrication	Sensors, nerves, blood vessels, lubrication

2.3. Wrist Functions and Common Parts

The wrist flexion–extension function is fulfilled with a double joint to produce a significant range of motion. The two joints are the radiocarpal joint and the midcarpal joint, as shown in Figure 2. The radiocarpal joint and midcarpal joints are ellipsoid in shape to enable abduction–adduction to occur as well as flexion–extension [35]. The third function of longitudinal load paths involves six of the wrist bones being aligned to create two compression load paths. The carpal arch has an important function of forming a protective passageway for the delicate tendons and nerves of the hand [42].

It is remarkable how the eight wrist bones perform four major functions. The six bones used in abduction–adduction are all used in flexion–extension. The capitate in particular is a key bone that plays a major role in all four functions. The axis of rotation for all wrist motion passes through the capitate. It is also particularly impressive that the carpal arch is perpendicular to the two arches in the plane of the hand.

2.4. Actuation and Fine-Tuning

The location of the four main wrist muscles (ECU, ECR, FCU and CRF) at the four corners of the wrist (Figure 2) means that they all can multifunction for both flexion–extension and abduction–adduction. For example, the ECU produces both extension and adduction and the FCU produces both flexion and adduction. The layout and curvature of the bones are such that there is an approximate common centre of rotation from the two joints for flexion and extension [43]. To produce a common centre of rotation, the radius of curvature at the radiocarpal joint has to be larger than the radius of curvature at the mid-carpal joint, with a unique proportion that gives a common centre of rotation. This results in a smooth reproducible joint motion. The geometry of the joint is also such that there is an approximate common centre of rotation for abduction–adduction [43]. The wrist bones are also well aligned with the metacarpals to produce a clear longitudinal load path from the metacarpals through the wrist bones to the lower arm with two main compression load columns [44]. This is beneficial because bone is stronger in compression than shear and tension [45].

2.5. Integration, Reconfiguration and Miniaturisation

The capitate is the most highly integrated part with at least five interfaces with adjacent bones, including a shear key (styloid process) where it interfaces with the third metacarpal bone in order to transfer shear loads. Reconfiguration occurs when the wrist joint is tensioned for increasing strength and stiffness under load. Examples of miniaturisation are found in sensors, nerves, blood vessels and synovial fluid. In the case of synovial fluid there is a special form of lubrication that creates a lubricant film between the sliding surfaces through an intricate combination of biomechanical and biomolecular factors acting at a range of scales [11]. The layout of wrist bones represents a unique solution where the layout performs multiple functions with multiple aspects of fine-tuning and integration.

2.6. A Multifunctioning Robotic Wrist

Robotic wrists are relatively bulky compared to the human wrist. In a review of 62 state-of-the-art prosthetic wrists, very few had the three degrees of freedom of the human wrist and those that did were all relatively very bulky [31]. One recent bioinspired wrist design that has improved compactness in the cross-section is showed conceptually in Figure 3 [46]. The design consists of a flexible tube of particles that can be deformed by three cables that surround the tube. The particles fill an annulus and create a stiffness that is low enough to be deformed but high enough to enable wrist functions. The tube can be made to bend in any direction by adjusting the relative tensions of the three cables. The cables are analogous to the tendons that surround the human wrist (Figure 2).

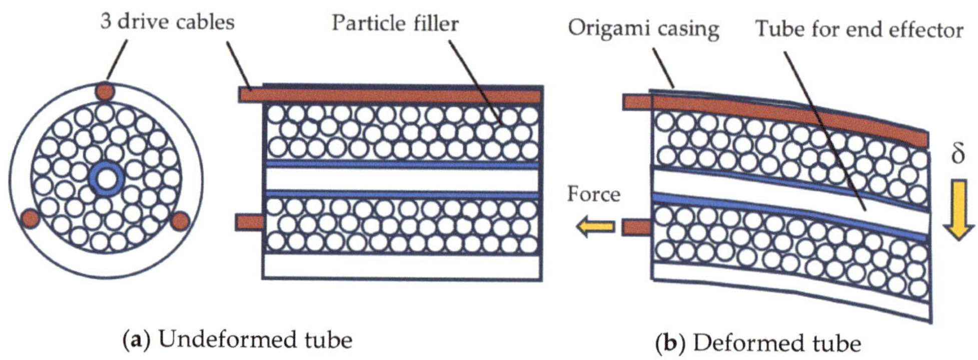

Figure 3. Bioinspired wrist concept using particles [46].

3. The Human Knee Joint

3.1. Knee Joint Anatomy

The main articulation in the knee occurs between the femur and tibia bones, as shown in Figure 4. The femur has two separate condyles with a gap in between to give space for the two cruciate ligaments that join the bones together and guide motion. There is also a sliding joint between the patella and a groove in the femur. The patella forms a low-friction joint that reacts the loads generated from the quadriceps muscle–tendon group. The patella also has a function of increasing the mechanical advantage of the quadricep muscles.

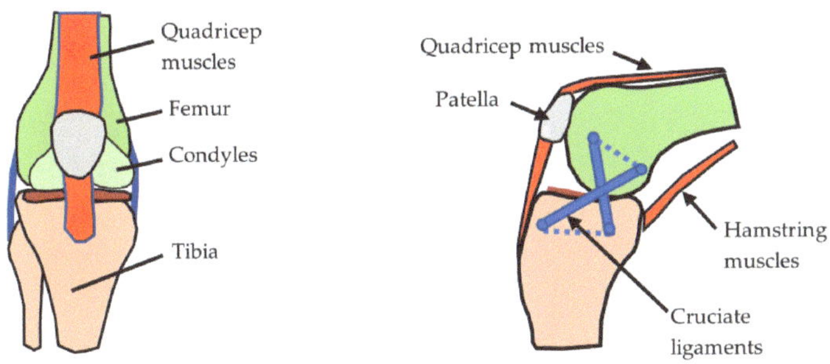

(a) Front view (b) Side view

Figure 4. Anatomy of knee joint.

3.2. Requirements and Objectives

Table 2 summarises the four major functions of the knee joint and the aspects of sub-system design that make multifunctioning possible. The design objective of the knee is to create a joint with a large amount of flexion–extension movement but also with some movement in rotation [47]. The knee also has a function of locking in the fully extended position which helps reduce the effort involved in standing [48]. The knee must also be able to absorb shock loads during activities such as running and jumping. Other functions include load transfer for various external loads and position sensing [49].

3.3. Knee Functions and Common Parts

The flexion–extension function is carried out by the femur and tibia which form a cam-roller joint (Figure 3b). When the knee is in a position of flexion between around 30 and 90 degrees, the tibia is able to rotate axially (Table 2) relative to the femur with up to approximately 45 degrees external rotation and 25 degrees internal rotation [50]. Lesser rotations of the tibia can be achieved outside of the 30–90 flexion range. The tibia axial rotation function is important because it enables abilities like adjusting to uneven surfaces while running as well as general dexterity.

Table 2. Summary of knee functions and how they are achieved.

Main Functions	Flexion–Extension	Rotation	Locking at Full Extension	Shock Absorber
Schematic	Femur rolls against tibia 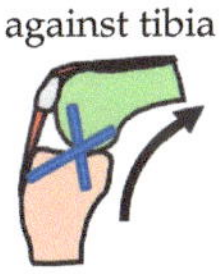	Geometry of joint allows rotation 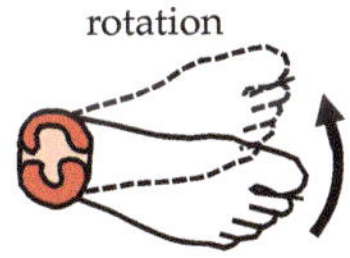	Geometry of joint allows locking 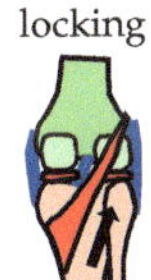	Meniscus absorbs shock 

Table 2. *Cont.*

Main Functions	Flexion–Extension	Rotation	Locking at Full Extension	Shock Absorber
Common parts	Condyles Ligaments Meniscus	Condyles Ligaments Meniscus	Condyles Ligaments Meniscus	Condyles Meniscus
Actuation [51]	Seven muscles over-actuated	Five muscles over-actuated	One muscle for unlocking	
Fine tuning	Geometry of 4-bar linkage	Geometry of tibia plateau. Meniscus geometry	Joint geometry. Popliteus muscle alignment	Meniscus geometry
Integration	Ligaments, muscles and meniscus	Ligaments, muscles and meniscus	Ligaments, muscles and meniscus	Meniscus surrounding condyles
Reconfiguration			Unlocking with popliteus muscle	
Miniaturisation	Sensors, nerves, blood vessels, lubrication	Sensors, nerves, blood vessels, lubrication	Sensors, nerves, blood vessels, lubrication	Sensors, nerves, blood vessels, lubrication

The joint-locking function (sometimes called the screw home mechanism (SHM) [48]) locks the joint in full extension by maximising external rotation (outward rotation of the tibia) and tightening the cruciate ligaments. Locking is important because it gives stability and saves energy when standing. As the knee reaches full extension, external rotation is caused by the geometry of the condyles and by the lateral pull component of the quadricep muscles [47]. The shock absorber function is fulfilled by the meniscus. The meniscus is wedge-shaped on its inner edge and is sandwiched between the femur and tibia during load bearing. This significantly increases the contact area between the two bones, thus reducing contact stresses [52].

Like the wrist, the knee joint has a unique and complex geometry that enables it to perform several functions. The condyles and meniscus play a major role in all four functions. The meniscus performs multiple functions including acting as a dynamic guide for the femur, lubricating layer, nutrition provider and structure that withstands compression, tension and shear loads [52].

3.4. Actuation and Fine Tuning

The knee joint is over-actuated with more actuators than degrees of freedom. Over-actuation allows for fine-tuning of joint movements. In parallel with the cam-roller mechanism there is also a four-bar linkage mechanism [53], as shown in Figure 4b. The linkage mechanism is an inverted four-bar mechanism where the cross links are formed by the anterior and posterior cruciate ligaments (ACL and PCL). The parallel bars are effectively formed by the femur and tibia and these bars are represented schematically by dotted lines in Figure 4b. The four-bar linkage mechanism gives the knee a moving centre of rotation which enables the femur bone to roll around the tibia (Figure 4b) and hence produce large flexion angles [54]. The action of the four-bar mechanism also produces an advantageous increase in mechanical advantage during squatting [54].

The axial rotation function of the tibia is possible due to the geometrical clearance between the condyles on the femur bone and the tibial plateau and because the cruciate ligaments have lengths and orientations that allow a degree of rotation [50]. To maintain stability during flexion and extension, the meniscus translates together with the condyles of the femur bone. Translation of the meniscus is possible because the meniscus is anchored to the tibia by ligaments only on the outer edges of the crescent-shaped rings and not on the inner edges [52].

3.5. Integration, Reconfiguration and Miniaturisation

The parts of the knee joint are highly integrated, especially the meniscus in the way it surrounds the condyles. Joint locking and unlocking represents reconfiguration. To unlock the joint, the tibia is made to rotate internally by flexion muscles, especially the popliteus muscle behind the knee [47], as shown in Table 2. Like the wrist joint, there is miniaturisation in the sensors, nerves, blood vessels and lubrication system that helps to achieve compactness. The layout of the knee joint represents a unique solution where the layout performs multiple functions with multiple aspects of fine-tuning and integration.

3.6. A Multifunctioning Lockable Robotic Joint

A multifunctioning lockable gearbox with analogies to the knee joint has been used on the robotic arm of ESA's METOP satellite launched in 2019 [55,56]. Space satellites have extreme geometrical constraints due to the very limited space in rocket nose cones and so there is a strong desire to create compact designs through multifunctioning. The multifunctioning gearbox is shown in Figure 5b. A standard worm gearbox is shown in Figure 5a for comparison.

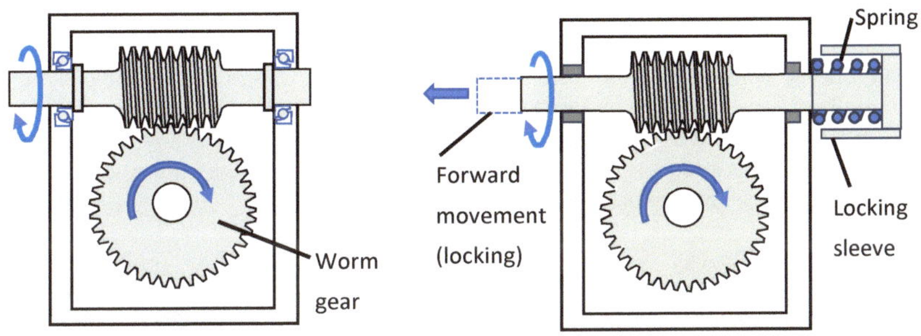

(**a**) Standard worm gearbox (**b**) Double-action worm gearbox

Figure 5. Multifunctioning double-action worm gearbox [55].

The double-action worm gearbox is able to perform three kinematic functions: (i) compliance during launch vibration like a rack and pinion gear; (ii) deployment of the appendage like a worm gear; and (iii) locking function at the end of deployment like a screw-nut. The locking sleeve is able to lock at the end of deployment because when the worm wheel can no longer rotate, the worm shaft screws forward until the locking sleeve comes into contact with the gear casing and engages with a high friction surface. The three gear functions are analogous to the three knee functions of tibia rotation, flexion–extension and locking, respectively. The locking function is analogous because like the femur, the worm wheel comes to the end of travel, and like the femur, there is a screw-home locking action.

The gearbox is called double-acting because the worm shaft has two degrees of freedom—rotation and translation. Interestingly, as with the knee joint, the spacecraft hinge is under-constrained, such that during deployment there are two degrees of freedom and the position of the hinge is a function of the spring reaction as well as the motor torque. This gearbox shows that multifunctioning can lead to extreme levels of compactness. It also shows that innovative ideas can create multifunctioning.

4. Human Foot Joints

4.1. Foot Anatomy

The foot contains approximately 26 bones, 33 joints and over 100 tendons, muscles and ligaments [57]. The main joints used in locomotion are the ankle joint, subtalar joint and MTP joint (Figure 6). The ankle joint is used to produce plantarflexion and dorsiflexion during walking and running. The subtalar joint enables pronation which helps with reducing shock absorption [58]. The MTP joints enables the foot to stand on the toes which is required for pushing off in walking and running.

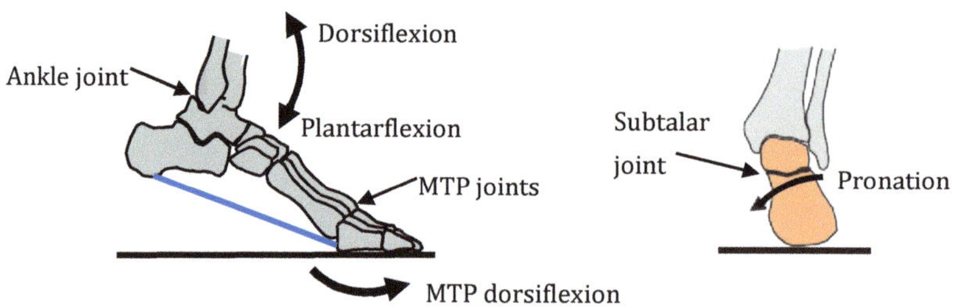

Figure 6. Main joints in the foot.

There are three arches in the foot [57], as shown in Figure 7. The medial arch goes through the three biggest toes to the heel bone (Figure 7a). The lateral arch goes through the two smallest bones to the heel bone (Figure 7b). The transverse arch goes across the foot at the midfoot and also at the MTP joints (Figure 7c). The arches give three-point contact with the ground when standing. The three points of contact are at the bottom of the heel, the base of the big toe and the base of the little toe (Table 3).

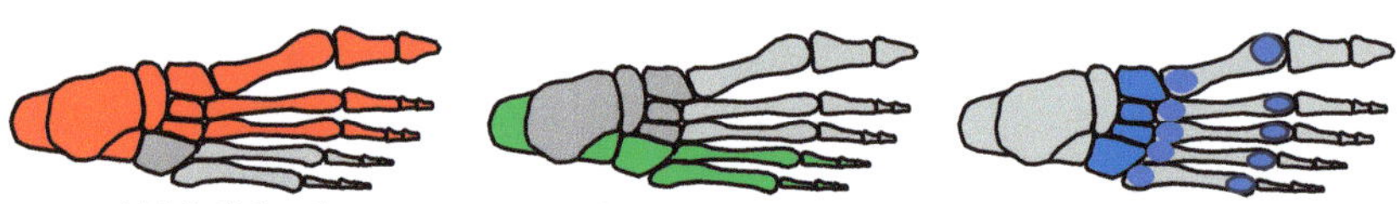

(**a**) Medial arch (**b**) Lateral arch (**c**) Transverse arch

Figure 7. Arches of the foot.

Table 3. Summary of foot functions and how they are achieved.

Main Functions	Stiff Lever (for Push-Off)	Flexible Lever (for Landing)	Stand on Ball of Feet	Three-Point Contact (for Standing)
Schematic	Medial arch 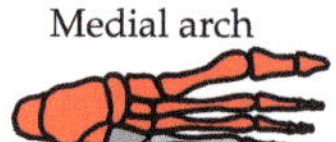	All arches Pronation 	Five MTP joints 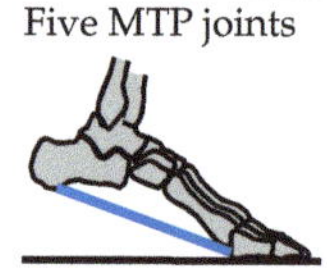	All arches 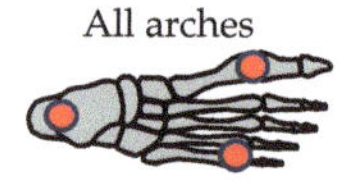
Common parts	Medial arch	Medial arch Lateral arch Transverse arch	Medial arch Lateral arch Transverse arch	Medial arch Lateral arch Transverse arch
Actuation [59]	Seven muscles over-actuated	Four muscles over-actuated	Twenty-two muscles over-actuated	n/a

Table 3. *Cont.*

Main Functions	Stiff Lever (for Push-Off)	Flexible Lever (for Landing)	Stand on Ball of Feet	Three-Point Contact (for Standing)
Fine tuning	Alignment of talus bone with medial arch	Spring ligament alignment	Alignment of five MTP joints	Maximised spacing of contact points
Integration	Bones of medial arch	Integration of three arches	Integration of three arches	Integration pf three arches
Reconfiguration	Tightening of muscles and ligaments			
Miniaturisation	Sensors, nerves, blood vessels, lubrication	Sensors, nerves, blood vessels, lubrication	Sensors, nerves, blood vessels, lubrication	Sensors, nerves, blood vessels, lubrication

4.2. Requirements and Objectives

Table 3 summarises the four major functions of the foot and the aspects of the subsystem design that make multifunctioning possible. The design objective of the foot is to have a compact design that can perform the four functions shown in Table 3. This objective is challenging because stiffness and flexibility are contradictory requirements.

4.3. Foot Functions and Common Parts

The foot needs to form a stiff lever during push-off to transmit the large forces required for walking and running [57]. In one study, the force in the Achilles tendon was found to be over eight times the body weight for jumping [60]. During push-off the medial arch is very stiff with a stiff big toe which points forwards. When landing, the foot transforms into a flexible lever in order to avoid large shock loads [61]. The subtalar joint is a key mechanism for reducing loads by producing a pronation movement which increases the time of landing. Loads are also reduced through the flexibility of the ligaments and tendons in all three of the arches. For example, the spring ligament in the medial arch is capable of absorbing significant energy during landing [62].

A third function of the feet is to be able to stand on the ball of the feet by dorsiflexion of the MTP (metatarsophalangeal) toe joints (Figure 6). Whereas running involves pressing on the base of the big toe, standing on the ball of the feet involves pressing on two points—the base of the little toe as well as the base of the big toe. Standing on the ball of the feet is important not just for reaching up high but also for changing direction during locomotion. The fourth function of three-point contact creates a clear interface with the ground during standing and aids balance. For example, it makes it possible to stand on one leg by placing the centre of gravity of the body through the triangular footprint. The three arches of the foot give three points of contact which are (i) the bottom of the heel; (ii) the base of the big toe and (iii) the base of the little toe. The triangle footprint is maximised by maximising by the distance between the three points of contact. The medial arch plays an important role in all four functions.

4.4. Actuation and Fine-Tuning

The bottom of the foot senses local pressures during locomotion [63] so that the nervous system can quickly make corrective adjustments to leg and foot positions to maintain stability. The medial arch is also closely aligned to the tibia bone via the talus bone so that the power of the leg muscles is focused on the medial arch. It can be seen in Table 3 how the talus aligns exactly with the medial arch. The transverse arch also plays a significant role in producing a stiff lever [64].

4.5. Integration, Reconfiguration and Miniaturisation

The bones, ligaments, muscles and nerve pathways of the three arches are highly integrated. A clear example of reconfiguration occurs when the foot changes from a stiff lever to a flexible lever by changes in muscle tension and changes in ligament tension [65]. Like the wrist joint there is miniaturisation in the sensors, nerves, blood vessels and lubrication system. The layout of the foot joints represents a unique solution that performs multiple functions with multiple aspects of fine-tuning and integration.

4.6. Multifunctioning Robotic Feet

Producing a robotic foot that mimics the human foot is very challenging because of the high complexity involved. Whereas hip and knee joints are commonly replaced with engineered joints, joints of the foot such as the ankle joint and MTP joints are not commonly replaced [66]. However, several researchers have recommended that the design of the human foot be copied to produce better prosthetic joints [67]. There have been several bioinspired designs with recent examples being [68–71]. Figure 8 shows a design with a flexible arch and a spring-loaded MTP joint [68]. Figure 9 shows a design with a tuneable stiffness feature which attempts to mimic the fine-tuning of stiffness in the human foot during the gait cycle [69]. This design has two concentric springs connected by an adapter which can be moved lengthwise by a stepper motor. The longitudinal position of the adapter changes the number of active coils in the outer spring. Figure 10 shows a robot foot concept where the five individual MTP joints are copied to enable close conformity of the foot with the ground [70].

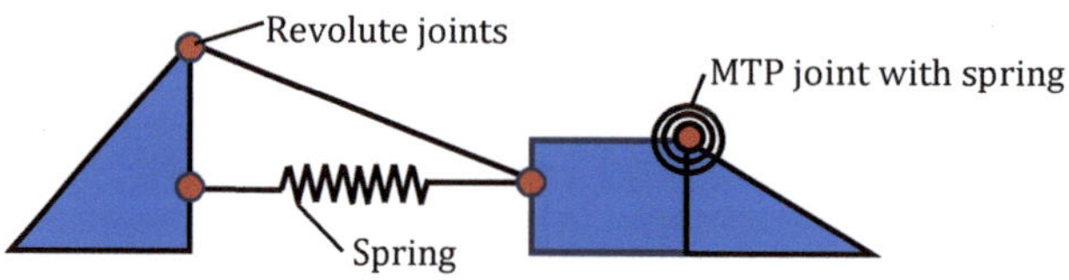

Figure 8. Bioinspired foot with flexible arch and MTP joint [68].

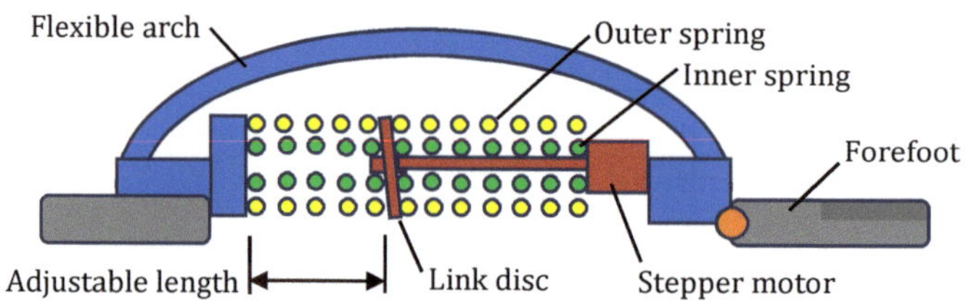

Figure 9. Bioinspired foot with tuneable arch stiffness [69].

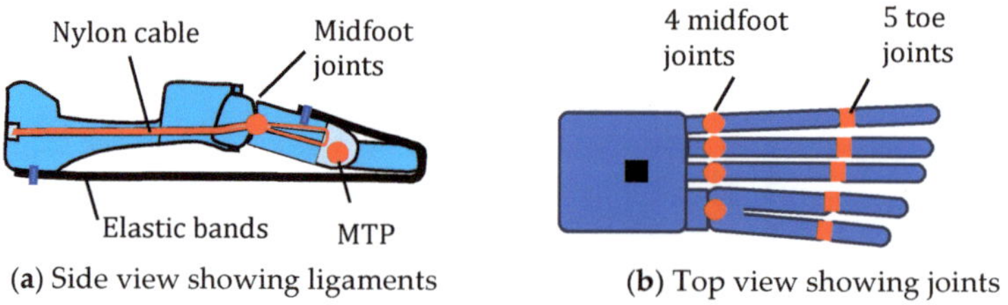

Figure 10. Bioinspired foot with midfoot and MTP joints [70].

Table 4 summarises the functionality of the three robotic foot concepts. Whilst all functions are covered across the three designs, no design performs all the functions.

Table 4. Functionality of three bioinspired robotic feet.

Foot Example	Schematic	Flexible Arch	Tuneable Stiffness	MTP Joint	Individual Toes
[68]	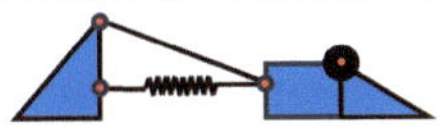	Y	N	Y	N
[69]	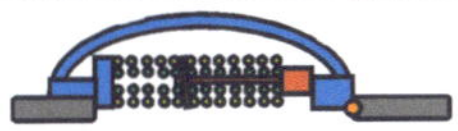	Y	Y	Y	N
[70]	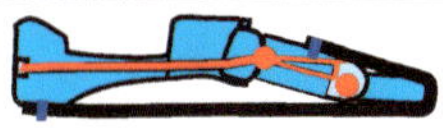	Y	N	Y	Y

5. Skeletal Muscle

5.1. Muscle Anatomy

Skeletal muscle is arranged in parallel bundles of myofibrils that contain many contractile units which give the tissue a striped appearance. Skeletal muscle is usually anchored by tendons to bones and is used to make skeletal movements for posture and locomotion. In skeletal muscle, contraction is stimulated by electrical impulses transmitted by the motor nerves.

5.2. Requirements and Objectives

Table 5 summarises the four major functions of muscle and the aspects of system design that make multifunctioning possible. Muscle–tendon units have been long recognised as a remarkable multifunctioning tissue [5]. The main design objective is to provide the right level of actuation (force and extension) for each joint whilst simultaneously providing the ideal shape for the animal. Additional functions include providing protective barriers in the right places and allowing for efficient heat transfer with blood vessels. As well as these major functions, there are other minor functions such as forming an elastic energy storage system through tendons [72] and providing a place for position sensors [73].

Table 5. Summary of muscle functions and how these are achieved.

Main Functions	Actuator (e.g., Move Joints)	Shape (e.g., Optimise Hydrodynamics)	Cushioning (e.g., Protect Organs & Bones)	Heat Source (for Blood Vessels)
Schematic	Muscle contraction 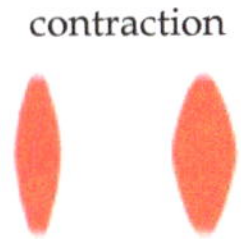	Muscle defines animal shape 	Protective barrier Thigh section 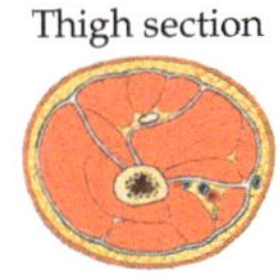	Heat transfer 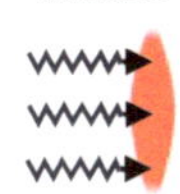
Common parts	Muscle and tendon	Muscle and tendon	Muscle and tendon	Muscle
Fine-tuning	MUs recruited from smallest to largest three types of muscle	Animals: aerodynamic shape Fish: hydrodynamic shape	Distribution Self-healing	Maximal heat transfer

Table 5. *Cont.*

Main Functions	Actuator (e.g., Move Joints)	Shape (e.g., Optimise Hydrodynamics)	Cushioning (e.g., Protect Organs & Bones)	Heat Source (for Blood Vessels)
Integration	Nerves for each muscle unit	Muscles with bones and organs	Muscles with bones and organs	Blood vessels
Reconfiguration				Shivering for generating heat [9]
Miniaturisation	Sensors, nerves, blood vessels, lubrication	Sensors, nerves, blood vessels, lubrication	Sensors, nerves, blood vessels, lubrication	Sensors, nerves, blood vessels, lubrication

5.3. Actuator Functions and Common Parts

The primary function of muscle is actuation. Muscle is well known to be an efficient high strain actuator with high power density [74]. Animal skeletal muscle comes in two main types: red slow twitch and white fast twitch. Red muscle produces forces required for slower and sustainable movements whilst white muscle produces the forces required for rapid movements, such as sprints. Many fish species also have an intermediate pink muscle which is between the two types of muscle.

Muscle also plays a crucial role in providing shape to the body of an animal. This is often an unappreciated function. In the case of land animals, the muscles form a shape that is ideal for locomotion such as an aerodynamic shape for fast animals (Figure 11a). In the case of fish, the muscles often form a precise hydrodynamic fusiform shape (Figure 11b). These shape objectives are met at the same time as creating a muscle mass that is ideal for actuating joints. Skeletal muscle is the main constituent of most animal bodies. Muscle accounts for up to 70% of the mass of fish [75] and 45–55% of the mass of most land mammals [76]. In the case of healthy adult humans, skeletal muscle accounts for 33–43% of body mass [77]. Muscles are also located so as to minimize the moment of inertia of limbs, especially legs. Tendons also play an important role in placing muscle away from joints, such as with the Achilles tendon.

(**a**) Aerodynamic shape (**b**) Fusiform shape

Figure 11. The shape of animal bodies is mainly defined by muscles and tendons.

Cushioning and providing a protective barrier are also important functions. For example, the muscles for the knee joint provide a thick cushion around the complete circumference of the femur bone (Table 5). When an animal falls to the ground, the muscles give a degree of protection to bones and organs. For example, when a human falls, they instinctively put their arms out and land on the thick thenar and hypothenar mounds on the palms of the hand. The thenar mound is formed at the base of the thumb by the intrinsic muscles of the thumb. The hypothenar mound is formed by muscles associated with the little finger. These muscle mounds act as shock absorbers. Even though muscles bruise, this is an injury that can be fully recovered from.

Skeletal muscle also plays a major role in homeostasis by heating the blood [78]. As muscle contracts, there are chemical reactions that generate heat which is transferred to the blood. The tendons function as an energy storage system for fast movements or oscillating movements such as flapping [79]. During fast movements, tendons buffer the work done on muscle by temporarily storing elastic energy then releasing this energy to assist the movement. The tendon deforms in a linear fashion due to the inter-molecular sliding of collagen triple helices.

5.4. Fine-Tuning

The nervous system has fine control over muscle force levels and force take-up because a muscle bundle is typically broken down into dozens of muscle units (MU). Muscle units are usually recruited by the nervous system from the smallest to largest which results in a smooth increase in force [80]. Another aspect of fine-tuning is the way that each muscle size is simultaneously optimal for actuation and shape.

5.5. Integration, Reconfiguration and Miniaturisation

Skeletal muscle is remarkably integrated with each muscle containing dozens of muscle units, each one with individually integrated nerve pathways for control and sensing. In addition, there is integration of the blood vessels. The shivering response to cold is an example of multifunctioning through reconfiguration. When a person gets cold, the body will go into shiver mode, causing muscles to contract and thus creating heat that will be transferred to the blood [81].

5.6. Bioinspired Muscle

There has been significant progress in the development of bioinspired muscle-like actuators [82–85] which are sometimes close to matching the performance of animal muscle in terms of the power density. However, bioinspired muscles are not yet multifunctional in providing shape or a protective barrier [74]. Future developments should investigate developing bioinspired muscle with these multifunctioning properties.

A recent development in bioinspired muscle actuators is that of bilateral dielectric elastomer (DE) actuators assembled in back-to-back configurations, as shown in Figure 12 [86]. This concept has been inspired by the antagonistic pairs of muscles used in animal joints. One configuration involves a star-shaped wheel where each pair of plates can be individually opened out to create asymmetry in the wheel (Figure 12b). When a voltage is applied to the electrodes on both sides of a pair of plates, the DE film is subjected to Maxwell stress and shrinks in the direction of the electric field and expands in the direction perpendicular to the electric field. The star wheel can be used to create motion of a robotic vehicle by sequentially moving each pair of plates in turn [86].

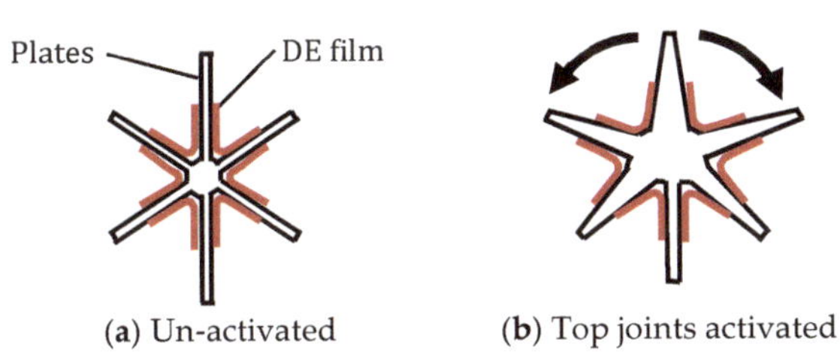

Figure 12. Bioinspired bilateral muscle in star configuration [86].

6. Discussion

6.1. The Importance of Multifunctioning in Engineering Systems

There is growing recognition that multifunctioning can play an important role in the optimisation of human-engineered systems. Multifunctioning has been identified as particularly desirable in aerospace systems where there are tight limits on the mass and size of sub-systems [87]. Recently a multifunctioning morphing wing has been developed that creates optimal lift-to-drag ratios for different speeds [88]. Miniaturisation is seen as a crucial goal for many medical devices [89–91] and multifunctioning is a key means for achieving this goal. Multifunctioning is now seen as an important means of achieving miniaturisation in electronics [92].

The benefits of multifunctioning can be seen in the development of mobile phones which have transitioned from bulky assemblies to compact designs with far more functionality. Modern smart phones pack a large amount of functionality in a small volume, partly by miniaturisation, but also by using multifunctioning materials and sub-systems. Multifunctioning is found in screens that are touch sensitive as well as providing a visual display and protective shield. Sensors are often multifunctioning, such as magnetometers that act as metal detectors as well as a compass. Smart phone processors are multitasking with multiple cores. One of the lessons of mobile phones is that multifunctioning and miniaturisation come at a stage of mature product development and after significant financial investment. At present, the field of robotics has not reached that stage of maturity.

6.2. Energy Benefits of Multifunctioning Limb Joints

Reductions in joint mass not only reduces the overall mass of a robot but also the moment of inertia of the limbs. This has a significant effect on the energy demands of locomotion. Since the legs move backwards and forwards during locomotion, minimising the moment of inertia of the legs helps minimise energy demands. The torque, T, required to cause angular acceleration of the legs is proportional to the moment of inertia ($T = I\alpha$). The moment of inertia of the leg, I, is most influenced by masses, m_n, furthest from the hip joint because the radius of gyration term, k_n, is squared in the equation for I ($I = \Sigma m_n k_n{}^2$). This means that the masses of the joints in the foot have a large effect on the torque required. In one study it was found that a third of the energy required for walking goes into moving the legs backwards and forwards [93].

The proportion of energy required to swing the legs is higher for faster walking and running speeds because the torque, T, required to swing the legs is also proportional to the angular acceleration, α. In one experimental study it was found that adding just 100 g to the mass of each running shoe increased the cost of transport (kJkg^{-1}km^{-1}) by up to 7.4% for high intensity running at 85% of the second ventilatory threshold [94]. The extra 200 g represented a small increase in the mass of the adult of around 0.3%. If a mass of 200 g was added to the torso it would be expected that the increase in energy for locomotion would be only 0.3% since the energy required for running is approximately proportional to the body mass [95]. This confirms that the leg inertia has a large influence on the energy demands of locomotion, especially at high speeds.

As well as reductions in the moment of inertia, a compact joint also results in a low frontal area and therefore lower drag. Therefore, there are very large energy benefits to be gained from minimising the mass and volume of limb joints through multifunctioning.

6.3. Comparison between Animal and Robot Joints

As shown in Sections 2–4, all three joints considered—wrist, knee and foot joints—combine extreme compactness with high functionality. This helps to explain the highly impressive agility of animals. None of the joints have a fully bioinspired engineering counterpart. However, there are some aspects of the joints that have been copied, as illustrated in the relevant sections. The knee joint has far fewer bones than the feet and hands (3 compared to 26/27) which makes bioinspired design more feasible. However, the knee joint has a very complex geometry and the meniscus itself is a very complex component. Much of the functionality and design of the knee joint has been copied with the exception of the rotation function around the tibia axis. If future robots are required to move with agility such as running on uneven ground, then copying the rotation function will be required. Wrist joints and foot joints are a long way off from being fully copied. Bioinspired feet are particularly challenging because of the need to tune stiffness during the gait cycle. A recent study has shown the vertebrate limb pattern to be highly optimal, with a key reason being the multifunctioning capability of the joints [96].

6.4. Strategies for Producing Multifunctioning

Biological joints reveal detailed strategies that can be employed to create multifunctioning systems. These can be considered for bioinspired systems. Some of these strategies are detailed here.

(i) Innovative and finely tuned designs. Multifunctioning in joints requires innovative solutions in order to create multiple degree-of-freedom joints. Therefore, it can be expected that significant innovation is required to produce bioinspired multifunctioning joints. The requirements of multifunctioning are so exacting that fine-tuning of design is also generally required, such as the common centre of rotation in the wrist (Section 2), the geometry of the cruciate ligament 4-bar linkage in the knee (Section 3) and the precise alignment of the medial arch with the talus bone in the foot (Section 4). Therefore, it can be expected that fine-tuning is required for robotic multifunctioning joints. Indeed, this was the case for the three bioinspired designs presented in this paper for the wrist, knee and foot.

(ii) Under-constrained and over-actuated joints. All three joints considered have a degree of under-constraint. For example, the knee joint is free to rotate during most of the flexion range and the wrist joint is free to abduct–adduct throughout most of the flexion range. This freedom means that fine control of actuators is required for stability. The advantage of under-constraining is that it creates more freedom of movement. All three case studies considered also have joints that are over-actuated with more actuators than degrees of freedom. The advantage of over-actuating is that it creates a greater range of possible movements. All three of the bioinspired examples have a degree of under-constraining.

(iii) Multifunctioning actuators. All three joints benefit from the multifunctioning capabilities of animal muscle. For example, the muscles for all three joints provide a protective barrier around the bones of the joints. Despite much progress in the development of soft actuators, they are usually not optimised for additional functions such as shape and protective barriers [97].

(iv) Reconfigurable designs. Reconfiguration can be an effective means for multifunctioning by actively changing the properties of a sub-system such as stiffness or shape. This is particularly seen in the human foot which becomes stiff for push-off by active stiffening by the muscles and by tensioning of the plantar ligament (Section 4). Reconfiguration is used in electronics [98] and active suspension systems [99] but is not commonly used in robotic joints. So, there is an opportunity for using this strategy.

(v) Segmented joints with high DoF. The feet and hands of vertebrates have segmented layouts with multiple bones and many joints. This layout allows for complex geometries and movements with high degrees of freedom. Copying such designs is very challenging because of the difficulty in copying ligament-type connections. However, the benefits of achieving are so high that it is worth the effort in attempting such assemblies. There has been a recent bioinspired robotic foot concept that does copy some of the individual bones of the midfoot [70]

6.5. Trade-Offs for Multifunctioning

Multifunctioning inevitably involves design trade-offs [100] and it is not possible to achieve the maximum performance for every function. For example, in the case of the wrist joint it is not possible to simultaneously have maximum flexion and abduction [37], and the knee joint cannot have maximum rotation and flexion at the same time [47]. The freedom of the knee joint means it can be moved in many directions. However, if the muscles get out of balance the knee can become unstable [101]. Therefore, there is a compromise between stability and functionality [101].

Radar trade-off charts can be used to visualise the trade-offs of functions [102]. To illustrate the trade-offs involved in multifunctioning, Figure 13a presents an illustrative trade-off chart for six aspects of performance for the human knee joint. In terms of compactness, mass and DoF, the knee performs very highly. However, this comes at the cost of complexity of parts, complexity of control and lower stability. The knee joint has a highly complex geometry in the bones and meniscus and also has complex control requiring extensive feedback of muscle positions. The under-constrained nature of the knee is a mixed blessing. On the one hand, it enables complex movements but, on the other hand, it can lead to instability. In terms of overall performance, multifunctioning is the right solution for the knee joint, but the design illustrates the trade-offs involved in multifunctioning.

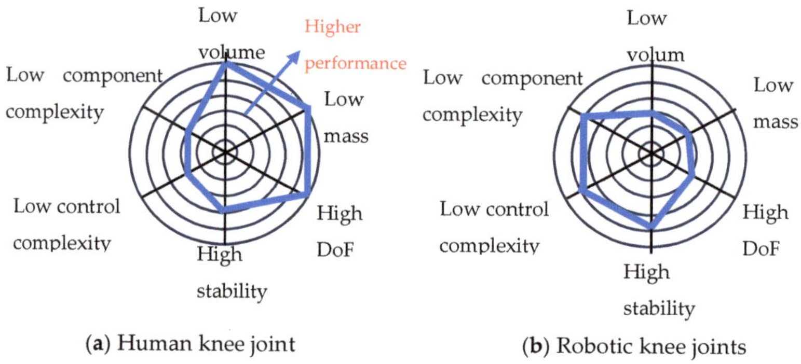

Figure 13. Illustrative radar trade-off charts for knee joints.

The equivalent trade-off chart for robotic knee joints is shown in Figure 13b. In contrast to animals, robotic knee joints perform less well in terms of compactness, mass and DoF but better in terms of complexity and stability [2]. For robots to achieve high levels of compactness and light weight, it may be inevitable that there will be some compromises in terms of complexity and/or stability.

6.6. Design Process for Multifunctioning

Traditional design methods generally do not encourage multifunctioning design. For example, morphological charts imply that all the individual subfunctions should have their own individual function carrier because separate boxes are allocated for each subfunction [103–105]. Whilst it is normal for each component to have 'secondary' functions such as having a protective coating, it is assumed they have one primary function. There have been recent attempts to develop design methods specifically for achieving multifunctioning systems [106–108]. These methods encourage the designer to investigate the possibility of multifunctioning solutions at the conceptual design stage through abstract visualization of function carriers that are multifunctioning.

6.7. The Emergence of Multifunctioning in Biology

Complexity in biological systems is sometimes labelled as an emerging property [109]. However, it is very difficult to explain how a multifunctioning system could emerge from an initially single-functioning system because the first single-functioning system would have to be one of the very few solutions that could lead to a later multifunctioning system [106]. When discussing the origin of mechanical linkage mechanisms in animal joints, Muller has stated that it is very difficult to see how complex mechanical linkage systems can be developed in a bottom-up step-by-step process [53]. Therefore, multifunctioning in biological systems such as limb joints presents a major challenge of irreducible complexity for evolutionary biologists [110].

7. Conclusions

The joints of the wrist, knee and foot contain remarkably sophisticated multifunctioning designs that outperform the best robotic devices. Multifunctioning in robotic joints can produce very important benefits of compactness, low mass, low moment of inertia and low drag. A key finding of this study is that multifunctioning in animal joints is achieved through various means: a high number of degrees-of-freedom, under-constrained assemblies, over-actuation, multifunctioning parts, miniaturisation and reconfiguration. Muscle is a remarkable actuator which performs other important functions including providing shape and protection and a heat source. Whilst there has been some progress in creating multifunctioning robotic joints, the progress is limited and there is much potential for bioinspiration.

Funding: Funding was received from the Biologic Institute BI/2021/SB/1.

Data Availability Statement: Data for this review paper can be found within the paper and the references.

Acknowledgments: This research was carried out as part of a Visiting Fellowship at Clare Hall College, Cambridge University in 2021.

Conflicts of Interest: The author declares no conflicts of interest.

References

1. Burden, S.A.; Libby, T.; Jayaram, K.; Sponberg, S.; Donelan, J.M. Why animals can outrun robots. *Sci. Robot.* **2024**, *9*, eadi9754. [CrossRef] [PubMed]
2. Tong, Y.; Liu, H.; Zhang, Z. Advancements in humanoid robots: A comprehensive review and future prospects. *IEEE/CAA J. Autom. Sin.* **2024**, *11*, 301–328. [CrossRef]
3. Egan, P.; Sinko, R.; LeDuc, P.; Keten, S. The role of mechanics in biological and bio-inspired systems. *Nat Commun.* **2015**, *6*, 7418. [CrossRef] [PubMed]
4. Meglio, P.D.; Perera, G.K.; Nestle, F.O. The Multitasking Organ: Recent Insights into Skin Immune Function. *Immunity* **2011**, *35*, 857–869.
5. Székely, G. A perfect design: The multifunctional muscle. *Behav. Brain Sci.* **1989**, *12*, 668–669. [CrossRef]
6. Mellis, D.J.; Itzstein, C.; Helfrich, M.H.; Crockett, J.C. The skeleton: A multi-functional complex organ. The role of key signalling pathways in osteoclast differentiation and in bone resorption. *J. Endocrinol.* **2011**, *211*, 131–143. [CrossRef]
7. Al-Maskari, N.S.; McAdams, D.A.; Reddy, J.N. Modelling of a biological material nacre: Multi-objective optimization model. *Mech. Adv. Mater. Struct.* **2019**, *28*, 430–439. [CrossRef]
8. Burgess, S.C. Multi-optimisation and multi-functioning in feathers. *J. Des. Nat.* **2007**, *1*, 1–10.
9. Ziv, C.; Zhao, Z.; Gao, Y.G.; Xia, Y. Multifunctional Roles of Plant Cuticle During Plant-Pathogen Interactions. *Front. Plant Sci.* **2018**, *9*, 1088. [CrossRef]
10. Ju, J.; Bai, H.; Zheng, Y.; Zhao, T.; Fang, R.; Jiang, L. A multi-structural and multi-functional integrated fog collection system in cactus. *Nat. Commun.* **2012**, *3*, 1247. [CrossRef]
11. Guilak, F. The Slippery Slope of Arthritis. *Arthritis Rheum.* **2005**, *52*, 1632–1633. [CrossRef] [PubMed]
12. Iwanaga, J.; Tubbs, R.S. Anatomy of the Oral Cavity. In *Atlas of Oral and Maxillofacial Anatomy*; Springer: Cham, Switzerland, 2021.
13. Jung, J.; Ahn, H.K.; Huh, Y. Clinical and Functional Anatomy of the Urethral Sphincter. *Int. Neurourol. J.* **2012**, *16*, 102–106. [CrossRef]
14. Garrett, J.N.; Fish, F.E. Kinematics of terrestrial locomotion in harbor seals and gray seals: Importance of spinal flexion by amphibious phocids. *Mar. Mammal Sci.* **2015**, *31*, 459–478. [CrossRef]
15. Lock, R.J.; Burgess, S.C.; Vaidyanathan, R. Multi-modal locomotion: From animal to application. *Bioinspir. Biomim.* **2014**, *9*, 011001. [CrossRef]
16. Cherian, V.T. Physiological Functions of Blood. In *Blood Substitutes and Oxygen Biotherapeutics*; Liu, H., Kaye, A.D., Jahr, J.S., Eds.; Springer: Cham, Switzerland, 2022.
17. Chapple, C.E.; Robisson, B.; Spinelli, L.; Guien, C.; Becker, E.; Brun, C. Extreme multifunctional proteins identified from a human protein interaction network. *Nat. Commun.* **2015**, *6*, 7412. [CrossRef] [PubMed]
18. Krystel-Whittemore, M.; Kottarappat, M.; Dileepan, N.; Wood, G. Mast Cell: A Multi-Functional Master Cell. *Front. Immunol.* **2016**, *6*, 620. [CrossRef] [PubMed]
19. Boisvert, F.M.; van Koningsbruggen, S.; Navascués, J.; Lamond, A.I. The multifunctional nucleolus. *Nat. Rev. Mol. Cell Biol.* **2007**, *8*, 574–585. [CrossRef]
20. Mahadevan, V. Anatomy of the vertebral column. *Surgery* **2018**, *36*, 327–332. [CrossRef]
21. Ning, M.; Ma, Z.; Chen, H.; Cao, J.; Zhu, C.; Liu, Y.; Wang, Y. Design and analysis for a multifunctional rescue robot with four-bar wheel-legged structure. *Adv. Mech. Eng.* **2018**, *10*, 168781401774739. [CrossRef]
22. Gobinath, A.; Manjula Devi, C.; Ezhilraj, B.; Sabarinathan, T.; Anandan, M.; Rajeswari, P. Design and Development of Multi-function Robotic Arm. In Proceedings of the 2024 IEEE Third International Conference on Intelligent Techniques in Control, Optimization and Signal Processing (INCOS), Krishnankoil, Virudhunagar District, Tamil Nadu, India, 14–16 March 2024; pp. 1–4. [CrossRef]

23. Jiang, Y.; Chen, D.; Ma, J.; Liu, Z.; Luo, Y.; Li, J.; Li, Y. Multifunctional Robotic Glove with Active-Passive Training Modes for Hand Rehabilitation and Assistance. In Proceedings of the 2021 IEEE/RSJ International Conference on Intelligent Robots and Systems (IROS), Prague, Czech Republic, 27 September–1 October 2021; pp. 4969–4974. [CrossRef]
24. Narayana, K.J.; Burela, R.G. A review of recent research on multifunctional composite materials and structures with their applications. *Mater. Today Proc. Energy Harvest.* **2018**, *5*, 5580–5590. [CrossRef]
25. Wang, S.; Urban, M.W. Self-healing polymers. *Nat. Rev. Mater.* **2020**, *5*, 562–583. [CrossRef]
26. Lee, H.; Dellatore, S.M.; Miller, W.M.; Messersmith, P.B. Mussel-inspired surface chemistry for multifunctional coatings. *Science* **2007**, *318*, 426–430. [CrossRef]
27. Zhang, Y.; Ma, J.; Singh, A.K.; Cao, L.; Seo, J.; Rahn, C.D.; Bakis, C.E.; Hickner, M.A. Multifunctional structural lithium-ion battery for electric vehicles. *J. Intell. Mater. Syst. Struct.* **2017**, *28*, 1603–1613. [CrossRef]
28. Eaton, J.; Naraghi, M.; Boyd, J.G. Evaluating multifunctional efficiency of a structural battery composite via thermo-electro-chemical modelling. *Multifunct. Mater.* **2022**, *5*, 015001. [CrossRef]
29. Ranzani, T.; Russo, S.; Bartlett, N.W.; Wehner, M.; Wood, R.J. Increasing the Dimensionality of Soft Microstructures through Injection-Induced Self-Folding. *Adv. Mater.* **2018**, *30*, e1802739. [CrossRef]
30. Jiang, M.; Gravish, N. Reconfigurable laminates enable multifunctional robotic building blocks. *Smart Mater. Struct.* **2021**, *30*, 035005. [CrossRef]
31. Bajaj, N.M.; Spiers, A.J.; Dollar, A.M. State of the Art in Prosthetic Wrists: Commercial and Research Devices. In Proceedings of the IEEE International Conference on Rehabilitation Robotics (ICORR), Singapore, 11–14 August 2015.
32. Versluys, R.; Beyl, P.; Van Damme, M.; Desomer, A.; Van Ham, R.; Lefeber, D. Prosthetic feet: State-of-the-art review and the importance of mimicking human ankle–foot biomechanics. *Disabil. Rehabil. Assist. Technol.* **2009**, *4*, 65–75. [CrossRef]
33. Shigemi, S. ASIMO and Humanoid Robot Research at Honda. In *Humanoid Robotics: A Reference*; Goswami, A., Vadakkepat, P., Eds.; Springer: Dordrecht, The Netherlands, 2019. [CrossRef]
34. Koolen, T. Design of a Momentum-Based Control Framework and Application to the Humanoid Robot Atlas. *Int. J. Humanoid Robot.* **2016**, *13*, 1650007-1. [CrossRef]
35. Gray, H. *Anatomy of the Human Body*, 20th ed.; Lewis, W.H., Ed.; Lea & Febiger: Philadelphia, PA, USA, 1918.
36. Berger, R.A. The Anatomy of the Ligaments of the Wrist and Distal Radioulnar Joints. *Clin. Orthop. Relat. Res.* **2001**, *383*, 32–40. [CrossRef]
37. Mansfield, P.J.; Neumann, D.A. *Essentials of Kinesiology for the Physical Therapist Assistant*, 3rd ed.; Chapter 6—Structure and Function of the Wrist; Elsevier: Amsterdam, The Netherlands, 2019; pp. 120–140.
38. Palmer, A.K.; Werner, F.W. The triangular fibrocartilage complex of the wrist—Anatomy and function. *J. Hand Surg. Am.* **1981**, *6*, 153–162. [CrossRef]
39. Hagert, E.; Rein, S. Wrist proprioception-An update on scientific insights and clinical implications in rehabilitation of the wrist. *J. Hand Ther.* **2024**, *37*, 257–268. [CrossRef] [PubMed]
40. Eschweiler, J.; Li, J.; Quack, V.; Rath, B.; Baroncini, A.; Hildebrand, F.; Migliorini, F. Anatomy, Biomechanics, and Loads of the Wrist Joint. *Life* **2022**, *12*, 188. [CrossRef] [PubMed]
41. Wrist. Available online: https://med.libretexts.org/Bookshelves/Anatomy_and_Physiology/Anatomy_and_Physiology_(Boundless)/9:_Muscular_System/9.9:_Muscles_of_the_Upper_Limb/9.9B:_Muscles_of_the_Wrist_and_Hand (accessed on 1 August 2024).
42. Xiu, K.-H.; Kim, J.-H.; Li, Z.-M. Biomechanics of the Transverse Carpal Arch under Carpal Bone Loading. *Clin. Biomech. (Bristol. Avon.)* **2010**, *25*, 776–780. [CrossRef]
43. Ayhan, C.; Ayhan, E. Chapter 13—inesiology of the Wrist and the Hand. In *Comparative Kinesiology of the Human Body: Normal and Pathological Conditions*; Angin, S., Şimşek, I.E., Eds.; Academic Press: Cambridge, MA, USA, 2020; pp. 211–282.
44. Rainbow, M.J.; Wolff, A.L.; Crisco, J.J.; Wolfe, S.W. Functional kinematics of the wrist. *J. Hand Surg. Eur. Vol.* **2015**, *41*, 7–21. [CrossRef]
45. Havaldar, R.; Pilli, S.C.; Putti, B.B. Insights into the effects of tensile and compressive loadings on human femur bone. *Adv. Biomed. Res.* **2014**, *3*, 101. [CrossRef]
46. Zhou, J.; Cao, H.; Chen, W.; Cheng, S.S.; Liu, Y.-H. Bioinspired Soft Wrist Based on Multicable Jamming with Hybrid Motion and Stiffness Control for Dexterous Manipulation. *IEEE/ASME Trans. Mechatron.* **2023**, *28*, 1256–1267. [CrossRef]
47. Abulhasan, J.F.; Grey, M.J. Anatomy and Physiology of Knee Stability. *J. Funct. Morphol. Kinesiol.* **2017**, *2*, 34. [CrossRef]
48. Hallén, L.G.; Lindahl, O. The "Screw-Home" Movement in the Knee-Joint. *Acta Orthop. Scand.* **2009**, *37*, 97–106. [CrossRef] [PubMed]
49. Johansson, H. Role of knee ligaments in proprioception and regulation of muscle stiffness. *J. Electromyogr. Kinesiol.* **1991**, *1*, 158–179. [CrossRef]
50. Zarins, B.; Rowe, C.R.; Harris, B.A.; Watkins, M.P. Rotational motion of the knee. *Am. J. Sports Med.* **1983**, *11*, 152–156. [CrossRef]
51. Knee. Available online: https://med.libretexts.org/Bookshelves/Anatomy_and_Physiology/Human_Anatomy_Laboratory_Manual_2021/07:_Joints/7.02:_Knee_Joint (accessed on 1 August 2024).
52. Makris, E.A.; Hadidi, P.; Athanasiou, K.A. The knee meniscus: Structure-function, pathophysiology, current repair techniques, and prospects for regeneration. *Biomaterials* **2011**, *32*, 7411–7431. [CrossRef] [PubMed]

53. Muller, M. A novel classification of planar four-bar linkages and its application to the mechanical analysis of animal systems. *Phil. Trans. R. Soc. Lond. B* **1996**, *351*, 689–720.
54. Etoundi, A.C.; Burgess, S.C.; Vaidyanathan, R. A Bio-Inspired Condylar Hinge for Robotic Limbs. *ASME J. Mech. Rob.* **2013**, *5*, 031011. [CrossRef]
55. Burgess, S.C. A backwards design process for Mechanical Design. *J. Mech. Des. ASME* **2012**, *134*, 031002. [CrossRef]
56. Righetti, P.; De Juana Gamo, J.; Sancho, F. Metop-C deployment and start of three-satellite operations. *Aeronaut. J.* **2020**, *124*, 902–916. [CrossRef]
57. Angin, S.; Demirbüken, I. Chapter 23—Ankle and Foot Complex. In *Comparative Kinesiology of the Human Body: Normal and Pathological Conditions*; Angin, S., Şimşek, I.E., Eds.; Academic Press: Cambridge, MA, USA, 2020; pp. 411–439.
58. Lundberg, A.; Svensson, O.K.; Bylund, C.; Goldie, I.; Selvik, G. Kinematics of the ankle/foot complex—Part 2: Pronation and supination. *Foot Ankle* **1989**, *9*, 248–253. [CrossRef]
59. Foot. Available online: https://med.libretexts.org/Bookshelves/Anatomy_and_Physiology/Anatomy_and_Physiology_(Boundless)/9:_Muscular_System/9.10:_Muscles_of_the_Lower_Limb/9.10C:_Muscles_that_Cause_Movement_at_the_Foot (accessed on 1 August 2024).
60. Burdett, R. Forces predicted at the ankle during running. *Med. Sci. Sports Exerc.* **1981**, *14*, 308–316. [CrossRef]
61. Salathe, E.P., Jr.; Arangio, G.A.; Salathe, E.P. The foot as a shock absorber. *J. Biomech.* **1990**, *23*, 655–659. [CrossRef]
62. Ker, R.F.; Bennett, M.B.; Bibby, S.R.; Kester, R.C.; Alexander, R.M. The spring in the arch of the human foot. *Nature* **1987**, *325*, 147–149. [CrossRef]
63. Alfuth, M.; Rosenbaum, D. Long distance running and acute effects on plantar foot sensitivity and plantar foot loading. *Neurosci. Lett.* **2011**, *503*, 58–62. [CrossRef]
64. Venkadesan, M.; Yawar, A.; Eng, C.M.; Dias, M.A.; Singh, D.K.; Tommasini, S.M.; Haims, A.H.; Bandi, M.M.; Mandre, S. Stiffness of the human foot and evolution of the transverse arch. *Nature* **2020**, *579*, 97–100. [CrossRef] [PubMed]
65. Boonpratatong, A.; Ren, L. The human ankle-foot complex as a multi-configurable mechanism during the stance phase of walking. *J. Bionic Eng.* **2010**, *7*, 211–218. [CrossRef]
66. Shah, N.S.; Umeda, Y.; Suriel Peguero, E.; Erwin, J.T.; Laughlin, R. Outcome reporting in total ankle arthroplasty: A systematic review. *J. Foot Ankle Surg.* **2021**, *60*, 770–776. [CrossRef]
67. Grimmer, M.; Seyfarth, A. Mimicking Human-Like Leg Function in Prosthetic Limbs. In *Neuro-Robotics*; Springer: Dordrecht, The Netherlands, 2014; pp. 105–155.
68. Russo, M.; Dayana, B.; Chaparro-Rico, M.; Pavone, L.; Pasqua, G.; Cafolla, D. A bioinspired humanoid foot mechanism. *Appl. Sci.* **2021**, *11*, 1686. [CrossRef]
69. Qaisera, Z.; Kanga, L.; Johnson, S. Design of a bioinspired tunable stiffness robotic foot. *Mech. Mach. Theory* **2017**, *110*, 1–15. [CrossRef]
70. Burgess, S.; Beeston, A.; Carr, J.; Siempou, K.; Simmonds, M.; Zanker, Y. A Bio-Inspired Arched Foot with Individual Toe Joints and Plantar Fascia. *Biomimetics* **2023**, *8*, 455. [CrossRef]
71. Sun, J.; Song, G.; Chu, J.; Ren, L. An adaptive bioinspired foot mechanism based on tensegrity structures. *Soft Robot.* **2019**, *6*, 778–789. [CrossRef] [PubMed]
72. Gautieri, A.; Buehler, M.J.; Redaelli, A. Deformation rate controls elasticity and unfolding pathway of single tropocollagen molecules. *J. Mech. Behav. Biomed. Mater.* **2009**, *2*, 130–137. [CrossRef]
73. Roach, C.; Love, C.; Allen, T.; Proske, U. The contribution of muscle spindles to position sense measured with three different methods. *Exp. Brain Res.* **2023**, *241*, 2433–2450. [CrossRef]
74. Wang, Y.; Ma, X.; Jiang, Y.; Zang, W.; Cao, P.; Tian, M.; Ning, N.; Zhang, L. Dielectric elastomer actuators for artificial muscles: A comprehensive review of soft robot explorations. *Resour. Chem. Mater.* **2022**, *1*, 308–324. [CrossRef]
75. Ibarz, A.; Felip, O.; Fernández-Borràs, J.; Martín-Pérez, M.; Blasco, J.; Torrella, J.R. Sustained swimming improves muscle growth and cellularity in gilthead sea bream. *J. Comp. Physiol. B* **2011**, *181*, 209–217. [CrossRef] [PubMed]
76. West, T.G.; Toepfer, C.N.; Woledge, R.C.; Curtin, N.A.; Rowlerson, A.; Kalakoutis, M.; Hudson, P.; Wilson, A.M. Power output of skinned skeletal muscle fibres from the cheetah (*Acinonyx jubatus*). *J. Exp. Biol.* **2013**, *216*, 2974–2982. [CrossRef]
77. Janssen, I.; Heymsfield, S.B.; Wang, Z.M.; Ross, R. Skeletal muscle mass and distribution in 468 men and women aged 18–88 yr. *J. Appl. Physiol. (1985)* **2000**, *89*, 81–88. [CrossRef]
78. Periasamy, M.; Herrera, J.L.; Reis, F.C.G. Skeletal Muscle Thermogenesis and Its Role in Whole Body Energy Metabolism. *Diabetes Metab. J.* **2017**, *41*, 327–336. [CrossRef]
79. Roberts, T.J.; Konow, N. How tendons buffer energy dissipation by muscle. *Exerc. Sport Sci. Rev.* **2013**, *41*, 186–193. [CrossRef]
80. Marshall, N.J.; Glaser, J.I.; Trautmann, E.M.; Amematsro, E.A.; Perkins, S.M.; Shadlen, M.N.; Abbott, L.F.; Cunningham, J.P.; Churchland, M.M. Flexible neural control of motor units. *Nat. Neurosci.* **2022**, *25*, 1492–1504. [CrossRef] [PubMed]
81. Haman, F.; Blondin, D.P. Shivering thermogenesis in humans: Origin, contribution and metabolic requirement. *Temperature* **2017**, *4*, 217–226. [CrossRef] [PubMed]
82. Liu, C.; Wang, Y.; Qian, Z.; Wang, K.; Zhao, F.; Ding, P.; Xu, D.; Wei, G.; Ren, L.; Ren, L. Bioinspired actuators with intrinsic muscle-like mechanical properties. *iScience* **2021**, *24*, 103023, ISSN 2589-0042. [CrossRef]
83. Pei, Q.; Rosenthal, M.; Stanford, S.; Prahlad, H.; Pelrine, R. Multiple-degrees-of-freedom electroelastomer roll actuators. *Smart Mater. Struct.* **2004**, *13*, N86. [CrossRef]

84. Hwang, T.; Frank, Z.; Neubauer, J.; Kim, K.J. High-performance polyvinyl chloride gel artificial muscle actuator with graphene oxide and plasticizer. *Sci. Rep.* **2019**, *9*, 9658. [CrossRef]
85. Miriyev, A.; Stack, K.; Lipson, H. Soft material for soft actuators. *Nat. Commun.* **2017**, *8*, 596. [CrossRef]
86. Yang, Y.; Li, D.; Sun, Y.; Wu, M.; Su, J.; Li, Y.; Yu, X.; Li, L.; Yu, J. Muscle-inspired soft robots based on bilateral dielectric elastomer actuators. *Microsyst. Nanoeng.* **2023**, *9*, 124. [CrossRef]
87. Sairajana, K.K.; Agliettib, G.S.; Mania, K.M. A review of multifunctional structure technology for aerospace applications. *Acta Astronaut.* **2016**, *120*, 30–42. [CrossRef]
88. Kilian, L.; Shahid, F.; Zhao, J.S.; Nayeri, C.N. Bioinspired morphing wings: Mechanical design and wind tunnel experiments. *Bioinspir. Biomim.* **2022**, *17*, 046019. [CrossRef] [PubMed]
89. Khandalavala, K.; Shimon, T.; Flores, L.; Armijo, P.R.; Oleynikov, D. Emerging surgical robotic technology: A progression toward microbots. *Ann. Laparosc. Endosc. Surg.* **2020**, *5*, 3. [CrossRef]
90. Kim, C. *Evolution of Advanced Miniaturization for Active Implantable Medical Devices*; Nano-Bio-Electronic, Photonic and MEMS Packaging; Wong, C.P.P., Moon, K.S., Li, Y., Eds.; Springer: Cham, Switzerland, 2021. [CrossRef]
91. South, A. Miniaturization of medical products: The development challenge. *Med. Device Technol.* **1998**, *9*, 30–34.
92. Wu, Y. Multifunctional devices from asymmetry. *Nat. Electron.* **2018**, *1*, 331–332. [CrossRef]
93. Jiro Doke, J.; Donelan, M.; Kuo, A.D. Mechanics and energetics of swinging the human leg. *J. Exp. Biol.* **2005**, *208*, 439–445. [CrossRef]
94. Rodrigo-Carranza, V.; González-Mohíno, F.; Santos-Concejero, J.; González-Ravé, J.M. Influence of Shoe Mass on Performance and Running Economy in Trained Runners. *Front. Physiol.* **2020**, *11*, 573660. [CrossRef]
95. Taboga, P.; Lazzer, S.; Fessehatsion, R.; Agosti, F.; Sartorio, A.; di Prampero, P.E. Energetics and mechanics of running men: The influence of body mass. *Eur. J. Appl. Physiol.* **2012**, *112*, 4027–4033. [CrossRef]
96. Burgess, S.C. Universal optimal design in the vertebrate limb pattern and lessons for bioinspired design. *Bioinspir. Biomim.* **2024**, *19*, 051004. [CrossRef] [PubMed]
97. Hammond, M.; Cichella, V.; Lamuta, C. Bioinspired Soft Robotics: State of the Art, Challenges, and Future Directions. *Curr. Robot. Rep.* **2023**, *4*, 65–80. [CrossRef]
98. Zhu, J.; Dexheimer, M.; Cheng, H. Reconfigurable systems for multifunctional electronics. *NPJ Flex. Electron.* **2017**, *1*, 8. [CrossRef]
99. Yu, M.; Evangelou, S.A.; Dini, D. Advances in Active Suspension Systems for Road Vehicles. *Engineering* **2024**, *33*, 160–177. [CrossRef]
100. Jordan Nickel, P.; Duimering, R.; Hurst, A. Manipulating the design space to resolve trade-offs: Theory and evidence. *Des. Stud.* **2022**, *79*, 101095. [CrossRef]
101. Liu-Ambrose, T. The anterior cruciate ligament and functional stability of the knee joint. *BCMJ* **2003**, *45*, 495–499.
102. Al-Ghuwairi, A.R.; Al-Fraihat, D.; Sharrab, Y.; Alrashidi, H.; Almujally, N.; Kittaneh, A.; Ali, A. Visualizing software refactoring using radar charts. *Sci. Rep.* **2023**, *13*, 19530. [CrossRef]
103. Cross, N. *Engineering Design Methods, Strategies for Product Design*, 5th ed.; Wiley: Chichester, UK, 2021.
104. Hülagü, R.; Timur, Ş. Using Morphological Chart for Analysing Existing Designs. *Arch. Des. Res.* **2024**, *37*, 27–41. [CrossRef]
105. Smith, G.; Troy, T.J.; Summers, J.D. Concept Exploration Through Morphological Charts: An Experimental Study. In Proceedings of the ASME 2006 International Design Engineering Technical Conferences and Computers and Information in Engineering Conference, Philadelphia, PA, USA, 10–13 September 2006; Volume 4a, pp. 495–504.
106. Fröhlich, T.; Klaiber, D.; Türck, D.E.; Vietor, T. Function in a box: An approach for multi-functional design by function integration and separation. In Proceedings of the 29th CIRP Design, Póvoa de Varzim, Portgal, 8–10 May 2019.
107. Fröhlich, T.; Vietor, T. Design Principles and Multi-Level Structures for Multi-Functional and Multi-Material Design. *Proc. Des. Soc.* **2022**, *2*, 395–404. [CrossRef]
108. Svendsen, N.W.; Lenau, T.A. Approaches to analyzing multi-functional problems. In Proceedings of the DS 101: Proceedings of NordDesign 2020, Lyngby, Denmark, 12–14 August 2020.
109. Hébert-Dufresne, L.; Allard, A.; Garland, J.; Hobson, E.A.; Zaman, L. The path of complexity. *NPJ Complex.* **2024**, *1*, 4. [CrossRef]
110. Jantzen, B.C. *An Introduction to Design Arguments. Cambridge Introductions to Philosophy*; Cambridge University Press: Cambridge, UK, 2014; pp. 189–206.

biomimetics

MDPI

Article

Bio-Inspired Motion Emulation for Social Robots: A Real-Time Trajectory Generation and Control Approach

Marvin H. Cheng [1,*], Po-Lin Huang [2] and Hao-Chuan Chu [2]

[1] Department of Mechanical, Material & Aerospace Engineering, West Virginia University, Morgantown, WV 26505, USA

[2] Department of Power Mechanical Engineering, National Tsing-Hua University, Hsinchu 300, Taiwan; s2522238@gmail.com (P.-L.H.); okbe32220205@gmail.com (H.-C.C.)

* Correspondence: hmcheng@ieee.org

Abstract: Assistive robotic platforms have recently gained popularity in various healthcare applications, and their use has expanded to social settings such as education, tourism, and manufacturing. These social robots, often in the form of bio-inspired humanoid systems, provide significant psychological and physiological benefits through one-on-one interactions. To optimize the interaction between social robotic platforms and humans, it is crucial for these robots to identify and mimic human motions in real time. This research presents a motion prediction model developed using convolutional neural networks (CNNs) to efficiently determine the type of motions at the initial state. Once identified, the corresponding reactions of the robots are executed by moving their joints along specific trajectories derived through temporal alignment and stored in a pre-selected motion library. In this study, we developed a multi-axial robotic arm integrated with a motion identification model to interact with humans by emulating their movements. The robotic arm follows pre-selected trajectories for corresponding interactions, which are generated based on identified human motions. To address the nonlinearities and cross-coupled dynamics of the robotic system, we applied a control strategy for precise motion tracking. This integrated system ensures that the robotic arm can achieve adequate controlled outcomes, thus validating the feasibility of such an interactive robotic system in providing effective bio-inspired motion emulation.

Keywords: motion recognition; sensor fusion; humanoid robotic system

1. Introduction

As society continues to evolve, social robots are playing an increasingly crucial role in enhancing human life across various sectors. From healthcare and education to industrial assistance, these advanced machines have the power to revolutionize our daily lives. By mimicking natural human interactions, social robots provide companionship, support, and efficiency in a wide range of tasks, from personal care to collaborative work [1–3]. The rapid advancements in artificial intelligence (AI) and humanoid robotics have significantly contributed to the development of social robots [4]. These innovations enable these machines to recognize, interpret, and replicate human behaviors with remarkable accuracy and responsiveness [5]. This has led to the creation of assistive robotic platforms, also known as exoskeleton devices, specifically designed for the healthcare market.

The development of social robots has been revolutionized by AI technologies such as machine learning and convolutional neural networks (CNNs) [6]. These advancements have enabled social robots to process vast amounts of data, learn from human interactions, and improve their ability to perform complex tasks [7]. Humanoid robots, designed to mimic human form and movements, provide an intuitive interface for interaction, making them more acceptable and effective in social contexts [8]. These technological advancements are crucial for developing robots that can seamlessly integrate into human environments and perform tasks with high precision and adaptability. Over the past decade, researchers have intensively explored various applications of social robots. Assistive exoskeletons, for example, have been developed to aid stroke patients in their recovery and to help seniors with daily activities. Ensuring the safety and efficiency of these robots is crucial and can be achieved by adopting trajectories that mimic common user movements. This approach helps prevent potential injuries caused by unusual movements or sudden loss of force [9,10]. In educational settings, social robots have been developed to provide personalized learning experiences and assist in classroom management. These robots can engage students through interactive lessons and activities, enhancing the overall learning environment and fostering better educational outcomes [11]. As social robots continue to evolve, they have the potential to transform our lives in profound ways, making them an essential component of modern society.

Recent studies have highlighted the potential of AI-driven motion recognition systems in enhancing the interaction capabilities of robots. In this research, a robotic platform featuring two five-degree-of-freedom robotic arms and a depth camera, designed to mimic a human upper body, was developed to investigate human–robot interaction. CNNs and path planning techniques were utilized to accurately predict and replicate human motions by processing time-series data and converting it into meaningful motion trajectories. Additionally, adaptive robust control (ARC) strategies [12] and a linear controller were implemented to compensate for nonlinear factors in robotic systems and improve steady-state operation, ensuring precise tracking and execution of movements.

In the experimental setup, a depth camera and a regular camcorder were used for motion capture, leveraging their extensive programming resources and high accuracy. These devices were integrated into the robotic platform to enable direct visualization and recognition of human motions. The captured image and depth data were processed using a skeletal recognition library to derive joint angular information. This information was then processed by the developed motion prediction model utilizing CNNs to interpret angle-to-angle representations of joint movements, allowing for accurate prediction and replication of human actions.

To generate accurate and safe trajectories for the robotic platform, actual trajectories of targeted motions collected from various users were used to develop reference trajectories. Previous studies have shown that motion profiles are generally similar among healthy adults, although small deviations and varying durations are common even for repeated motions by the same person [13]. Though this study focused on the motion prediction for a single user, the motion prediction model is capable of providing distinct predictions for individual motions. Once the motion type was predicted, the corresponding reference trajectories were generated. To achieve this, temporal alignment of time series, a method investigated by various research groups, was adopted to align collected profiles onto a single timeframe, enabling the derivation of reference trajectories. Dynamic time warping (DTW) was used for spatiotemporal alignment among multiple time series [14], with the generated reference trajectories scaled from 0% to 100%. This approach allows the output timeframe to be adjusted to the required duration, ensuring adequate path synthesis for the robotic platform.

When comparing the results of this study with those from previous research, the proposed method, combining CNN modeling with a DTW-path planning approach, demonstrates substantial advancements over earlier methodologies. This superiority is primarily due to several factors, including enhanced accuracy, scalability, and the ability to handle complex motion patterns.

Traditional path planning for robotic systems has often relied on deterministic algorithms or classical machine learning models. For example, methods like Dijkstra's algorithm, A* algorithms, and recursive inverse kinematics have been commonly used for spatial pathfinding in collaborative robotic systems, as seen in the works by Cheng et al. [15] and Sohn et al. [16]. While these methods are effective for basic end effector movement tasks for robotic arms, these methods often struggle with dynamic environments or complex motion patterns. In contrast, the proposed approach, which integrates CNN-based modeling and DTW synthesis of movement trajectories, leverages deep learning to predict and understand complex trajectories with higher precision. By using convolutional layers to extract spatial features from motion data, our method achieves a more nuanced understanding of the task. The proposed control strategy also provided better tracking and synchronization accuracy.

Scalability and flexibility are critical for path planning in assistive devices, yet previous methods often lacked these qualities, especially when applied to diverse motion patterns across different users [17,18]. The proposed approach, however, allows for the scaling of reference trajectories based on individual user characteristics, such as body size and motion duration. This adaptability represents a significant improvement, enabling the system to cater to a broader population and a wider range of applications.

Moreover, previous research frequently overlooked nonlinear factors affecting robotic motion, such as varying inertia, inconsistent friction, and dead zones in motors. For instance, in Liu et al. [19], while the focus was on optimizing the path planning algorithm, the impact of these nonlinearities was not adequately addressed. In contrast, the proposed approach integrates these nonlinearities directly into the control model using an ARC mechanism, which compensates for both structured and unstructured uncertainties. This integration results in more reliable and stable performance, particularly in tasks requiring fine motor control. Additionally, the combined use of ARC and linear controllers in the proposed control strategy effectively addresses both nonlinearities and steady-state tracking performance. The ARC controller compensates for nonlinear factors such as dead zones and actuator saturation, while the linear controller ensures smoother tracking during steady-state conditions. This integrated approach optimizes the tracking performance of the robotic platform for various motions.

In summary, the proposed method based on prediction and the temporal path planning approach in this study demonstrates significant improvements in handling complex, real-time robotic tasks by incorporating advanced machine learning techniques, robust adaptive control mechanisms, and a scalable framework. These enhancements make it superior to previous research methods, offering a more versatile and reliable solution for motion planning in robotic systems.

The structure of this paper is as follows: Section 2 covers the motion capture setup and data processing techniques utilized in this study. Section 3 discusses the experimental setup of the robotic platform. Section 4 elaborates on the motion recognition system and the development of the CNN-based angle-to-angle representation. Section 5 focuses on the design and implementation of the integrated ARC and linear control strategy. This section also presents the simulated results and compares the tracking performance of the proposed controllers. It also discusses the strategies of the integrated control scheme. Section 6 explains how the robotic platform mimics human motion and addresses motion duration adjustments. Finally, Section 7 concludes the paper with a summary of findings and future research directions.

2. Computer Vision for Human Motion Acquisition

Recently, humanoid robots have gained increasing popularity. These robots not only provide a sense of comfort and familiarity to humans but can also imitate human actions, performing tasks that users typically do themselves. For a social or humanoid robot to effectively mimic human motion, it must accurately understand human biomechanics and the trajectories that individual joints follow during movement. This requires sophisticated motion acquisition and processing systems that can capture and analyze the nuances of human motion in real time. By integrating advanced sensors and algorithms, these robots can learn and replicate complex human activities, enhancing their utility in various applications such as personal assistance, healthcare, and entertainment.

2.1. Motion Acquisition Devices

Over the past decade, numerous motion acquisition techniques have been investigated for capturing body movements. When choosing a suitable motion acquisition method for humanoid or social robots, several factors must be considered, including the integration of sensors into the robotic system. The cost of these systems varies widely depending on the technology used. To balance cost and accuracy, this study tested three acquisition methods:

1. Depth Camera: Depth cameras enable the simultaneous acquisition of individual joint movements in three-dimensional space with minimal limitations. However, they tend to be more expensive than regular camcorders. Both camcorders and depth cameras offer the advantage of non-contact acquisition but require operation within a designated space. Additionally, measurements from these cameras can be distorted by environmental conditions.
2. Standard RGB Camcorder: Camcorders, being a mature technology, are widely integrated into almost all laptop computers, making them a cost-effective option for motion acquisition using existing devices. However, like depth cameras, they need to be used within a designated space and can also be affected by environmental factors.
3. Data Fusion System Combining Low-Cost Wearable Gyroscopic Sensors and Accelerometers: Accelerometers and gyroscopic sensors, which are directly attached to users, can capture motion regardless of the surrounding environment. These attached sensors can be easily integrated into an exoskeleton robotic platform as mechatronic components. They are typically low-cost, easily replaceable, and beneficial for their integration flexibility. However, they have limitations in measuring certain types of motion due to mechanical constraints. This type of sensor requires more preparation time before motion acquisition can begin.

Each method has its advantages and limitations. Depth cameras provide detailed 3D motion data but at a higher cost. Camcorders are more affordable and widely accessible but only capture 2D projections and can be affected by environmental conditions. Sensor fusion systems offer flexibility and low cost but require physical attachment to the user and may face mechanical constraints.

These sensors are deployed based on specific scenarios where humanoid and social robots operate. Modern robots are typically equipped with LiDARs or cameras to perceive their surroundings. Cameras, in some cases, can directly facilitate motion acquisition. However, when cameras alone cannot suffice for capturing motion, additional wearable sensors can fill this gap, ensuring that the robots can accurately capture and replicate human movements in various environments and tasks. This dual approach enhances the versatility and reliability of robotic systems in interacting with and assisting humans.

2.2. Motion Acquisition Process

To adequately plan the movement trajectories of a robotic platform for human-like motions, the control system must include the trajectories of involved joints, particularly for the arms. To enable a social robot to mimic human motions, a library of reference trajectories needs to be compiled in advance. In this study, the robot platform was programmed to respond to three pre-selected self-similar motions to ensure consistent performance of human-like movements. Self-similar motions were targeted to guarantee that the robot could always perform actions with similar trajectories, enhancing the human-like quality of its movement. However, not all motions exhibit self-similarity. Therefore, this study focused on two specific motions: object lifting and right arm raising. The motions were selected to assist human workers in a warehouse environment. In this setting, it is essential for human workers to recognize robots as trustworthy partners rather than merely another piece of installed equipment. These motions were chosen because they are common in everyday interactions, making their recognition and replication crucial for the robot's functionality. Once a motion type was identified, the robot mechanism responded accordingly, using baseline trajectories designed for these motions as reference motions for the experimental robotic platform.

This study employed two different depth cameras for motion acquisition: a Kinect Xbox 360 depth camera manufactured by Microsoft and a RealSense D435 made by Intel. The Kinect depth camera was chosen for its extensive programming library resources, and a motion acquisition interface was developed using the Microsoft Kinect SDK package v1.8. The motion acquisition rate for the Kinect was set to 25 frames per second. The Intel RealSense D435 was selected for its advanced capabilities, and the motion acquisition interface for this camera was developed using the Cubemos library. To ensure effective capture, the depth camera was placed 3 m in front of each test subject, whose heights ranged from 160 cm to 190 cm, as illustrated in Figure 1a. Upon the motion being recorded, the positions of individual joints in 3D space were captured, and joint angular information was subsequently derived from this spatial data. However, a major drawback of this method is that optical signals can be disrupted, compromising accuracy. The motion data was processed using an AMD Ryzen 7 4700U processor. Each test subject completed 50 trials of each motion. During these trials, the positions of shoulder, elbow, and wrist joints for both arms were recorded. These recorded positions were then converted to the angular movements of the joints of the shoulder and elbow joints. Figure 1b demonstrates the joints involved in the motion acquisition process. For the selected motions, only the time series data of the arm joints were recorded. A modeling method developed by Cheng et al. [13] was employed to determine the motion identification model. The model was subsequently used to identify the ongoing motion and determine the corresponding trajectories for the robotic platform.

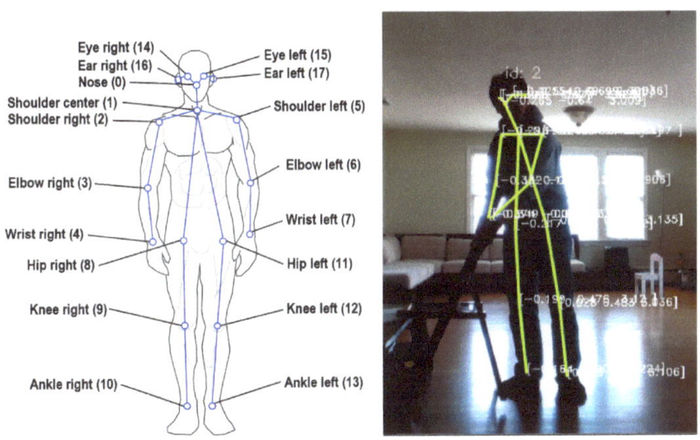

(**a**) (**b**)

Figure 1. (**a**) Recordable joints and the corresponding locations for human motions using the Cubmos library. (**b**) Image of the moving object motion captured with an Intel RealSense D435 camera and processed using the Cubemos framework in C#.

Once established, the model can predict the type of ongoing motion, enabling the generation of joint trajectories tailored to these motions. This capability is valuable in healthcare applications, where it can assist patients by providing personalized support based on their movements. Additionally, in various workspaces, such as manufacturing or rehabilitation settings, this predictive ability can enhance efficiency and safety by automating tasks or providing real-time assistance based on observed motions.

This study specifically focuses on motions that can be performed by a single arm. The decision to concentrate on single-arm movements is motivated by the relative simplicity in processing and modeling these motions. Single-arm actions involve fewer degrees of freedom compared to tasks requiring the coordination of both arms or multiple joints, which in turn reduces the computational complexity.

Modeling and controlling movements that engage both arms or involve multiple joints require significantly more intensive calculations and longer processing times. These factors would demand advanced modeling techniques and computational resources, making them outside the scope of this study. By focusing on single-arm motions, the research aims to develop more efficient and streamlined models, allowing for a more targeted exploration of the proposed methods.

2.3. Prediction Model of Human Motions

Once the motion data acquired from the depth camera are formatted appropriately, the next step involves immediate identification of the corresponding types of motions. Various pattern recognition techniques have been proposed, including Markov chain [20,21], temporal alignment [22], and other methods. Recent advancements in deep learning models have shown outstanding performance in action recognition [23]. The prediction model can also be used to trigger the robotic platform to react to certain selected motions. This section introduces a motion prediction model based on a deep learning approach.

One efficient approach for recognizing arm motions involves utilizing convolutional neural networks (CNNs) with dynamic motions captured by motion acquisition devices. Instead of relying on time-series information, representing the motion information of involved joints through angle-to-angle data has been proven effective and intuitive. This method simplifies the required calculation by transforming individual time-based data into spatial 3D representations. The adoption of angle-to-angle representation offers several advantages. A single portrait can demonstrate profiles with unique characteristics, making it straightforward to distinguish different motions when the positions of multiple joints are presented geometrically. Considering these factors, 3D arm movements can be processed using 2D calculations by projecting a specific combination of joints onto 2D planes. This can dramatically reduce the required calculation using the neural network.

Figure 2 illustrates the angle-to-angle representation of the three selected motions collected from a test subject. The unique characteristics of these motion profiles are evident. Even when the same motion is performed at different times, the major features remain the same, allowing for differentiation from other motions. With enough samples collected, a prediction model can be established based on the profiles of the selected motions.

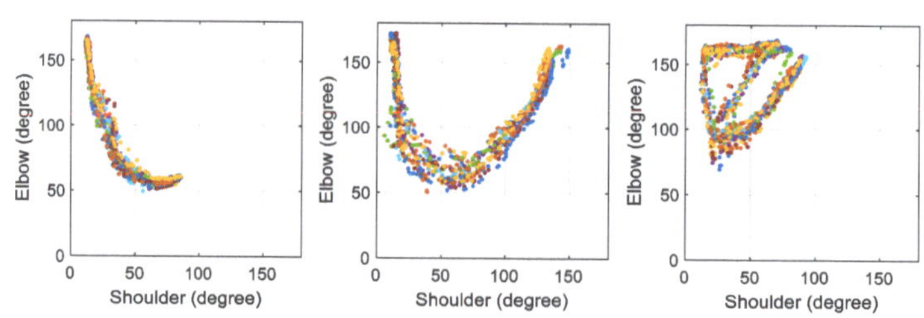

Figure 2. Angle-to-angle plots of three selected motions: drinking water, raising right hand, and object lifting (The different colors represent 10 trials).

The process of using CNNs for motion recognition follows a similar approach to that of regular image recognition. However, before applying CNNs, a preprocessing step is necessary to convert individual time-series information into angle-to-angle data. The transformed data is then utilized for both modeling and prediction purposes. In this study, once the motions were captured by motion sensors, the movements of the elbow and shoulder joints were selected for transformation into angle-to-angle representations.

To ensure consistency, each set of angle-to-angle data was standardized to a size of 256×256 points. To gather a sufficient number of samples for the training process, each motion was recorded 200 to 400 rounds from the test subject. After gathering all the samples, the training process began. This method leverages the unique characteristics of angle-to-angle representations to facilitate accurate and intuitive motion recognition. Figure 3 illustrates the angle-to-angle representation of the three selected motions collected from a test subject, highlighting the distinct features that enable the CNN model to differentiate between different motions effectively. In these figures, 10 trials of each individual motion are displayed. It is evident that the selected motions are consistent and self-similar across all trials performed by the same person.

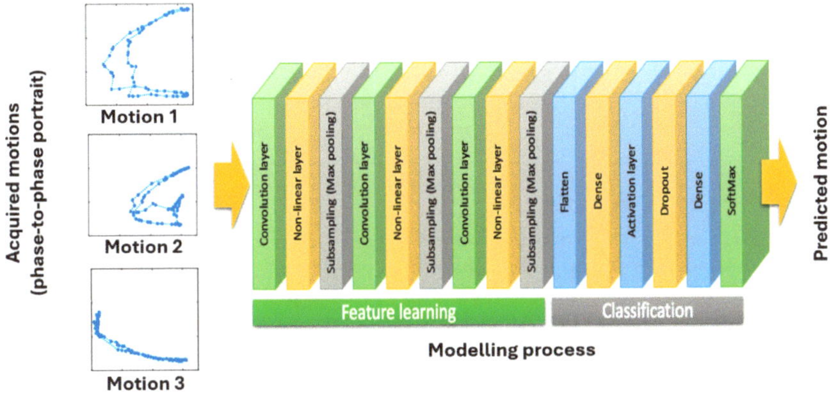

Figure 3. Training process using the framework of CNNs.

The training process using CNNs followed a standardized approach, as illustrated in Figure 3 [6]. The angle-to-angle representation served as inputs for the entire process. Three convolution layers were used to extract the major features from the collected motions. To enhance motion recognition, 3×3 filters in the convolution layers were selected based on the profiles of individual motions. After each convolution layer, a nonlinear activation function identified distinct features in each hidden layer. A subsampling (max pooling) layer was utilized to reduce the resolution of the features from layer to layer, making them more robust against noise and distortion. The total number of layers in the training process was 15. Upon completion, the classifier generated over four million neurons for the targeted motions. The resulting model was then stored and used to detect ongoing motions.

In this study, data collected from a single test subject were used for the training process, totaling 162 datasets. An additional 32 datasets were used for verification. Using a Ryzen 7 4700U CPU, the training time for a single model was less than 300 s. The verification and prediction time for each motion was less than 0.5 s. The derived model, utilizing the CNNs method, achieved a prediction success rate of over 90% for the three selected motions when processing just one-third of the motion cycle. Additionally, the model consistently provided correct predictions to indicate that the performed motion was not among the three selected.

To make the prediction model practical for the robotic platform to react to specific motions, it needs to predict the intended movement at the beginning of a motion rather than waiting for a motion to be completed. In this study, the robotic platform repeated the predicted motion immediately following the prediction. Once the motion was identified, the corresponding joint trajectories were planned. An important objective of using deep learning in robotic platforms is to estimate and generate complete movement trajectories from partial information. This allows the robot to react promptly, providing necessary assistance in healthcare or industrial applications. Therefore, it is critical to determine the minimal portion of a motion required for accurate prediction. This section investigates modeling with partial motion data to ensure timely motion prediction, enabling the humanoid robot platform to deliver the required assistance efficiently.

Previous studies also recommended capturing at least one-third of the total motion to ensure accurate motion prediction with a successful rate greater than 95% [13]. For the path planning of the robotic platform, the robot's movements were programmed to respond to specific human motions. Therefore, trajectories of individual joints were planned based on motion data after one-third of the motion cycle was performed by the user. Table 1 presents the predicted results generated by the proposed model using the selected motions.

Table 1. Estimated results of motions with different targeted duration (100%, 50%, 33%, and 20%) [13].

Type of Motion / Cycle (%)	Drinking	Arm Raising	Object Lifting
100%	100%	100%	100%
50%	100%	100%	100%
33%	97.14%	100%	97.14%
20%	~35%	~35%	~35%

2.4. Validation of Prediction Model

To verify the prediction results obtained from CNNs in this study, we employed the k-fold cross-validation method. This method is particularly effective for assessing the performance of machine learning models for CNNs, as it provides a robust evaluation of the model's ability to generalize across different subsets of the data. The entire dataset was divided into k subsets, known as folds. The value of k is typically chosen based on the size and variability of the data, with common values being 5 or 10. For this study, we selected $k = 6$ to achieve a balance between computational efficiency and a comprehensive evaluation.

During the training and validation phase, the CNN model was trained on $k - 1$ of these folds, while the remaining fold was used for validation. This process was repeated k times, ensuring that each fold served as the validation set once. By averaging the model's performance across all folds, we obtained a more accurate and reliable estimate of its predictive accuracy. Throughout each iteration, key performance metrics such as accuracy, precision, recall, and $F1$-score were calculated for the validation set. These metrics were then averaged across all folds, providing a comprehensive assessment of the model's overall performance.

The k-fold validation process also offered valuable insights into the model's strengths and weaknesses. These insights guided adjustments to hyperparameters, architecture, or training strategies, allowing us to address any observed issues such as overfitting or underfitting by refining the CNN model's configuration. Following the completion of cross-validation, the CNN model was retrained on the entire dataset using the optimized parameters identified during the validation process. This final model was then used to make predictions on unseen data, ensuring that it was both well-tuned and capable of generalizing effectively.

To further enhance the robustness of the model, the k-fold validation process was repeated multiple times with different random seeds for data splitting. This repetition reduced the impact of any particular data split on the final results, reinforcing the consistency and reliability of the CNN's predictions.

By utilizing k-fold cross-validation, this study rigorously verified the CNN model's prediction results, which are 99.6%. The approach ensured that the model performed well not only on the training data but also on new, unseen data, thereby increasing the reliability and applicability of the model in practical scenarios.

3. Configuration of the Robotic Platform

One crucial function of a humanoid robot is to respond to specific human actions. To achieve this, the robotic platform must identify ongoing human motions and react accordingly. This requires the robot's processing unit to determine its movements based on the types of human motions detected. This goal necessitates the integration of machine vision, human motion modeling, path planning, and control strategies for a multi-axial robotic mechanism. The proposed configuration for such a system in this study is illustrated as Figure 4.

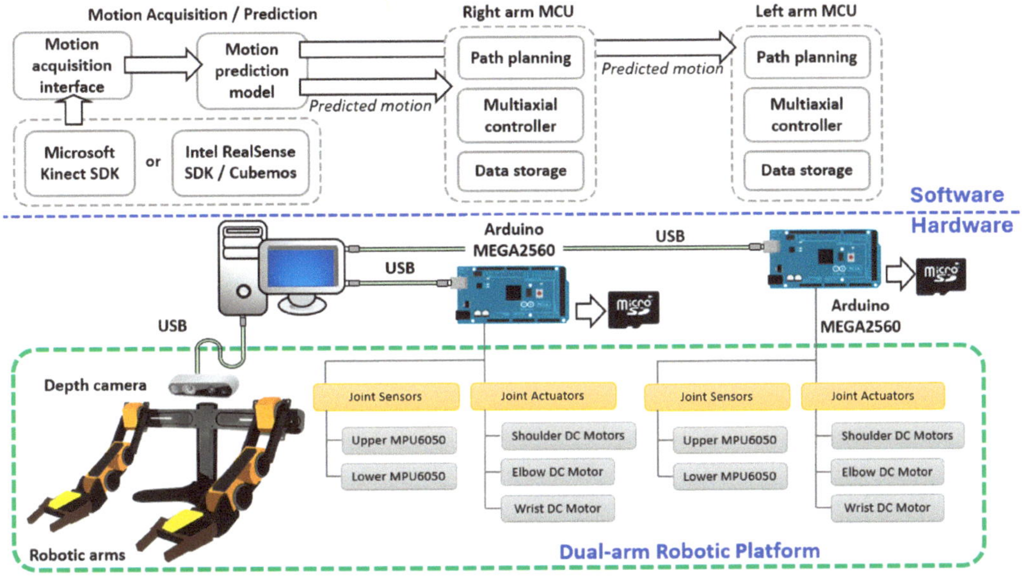

Figure 4. System configuration of the dual-arm robotic platform and the corresponding motion acquisition and control subsystems.

The hardware of the experimental platform was divided into two main components: the motion acquisition/prediction component and the robotic mechanism. The motion acquisition/prediction component was responsible for capturing human movements and determining the motion types. This subsystem integrated an Intel RealSense D435 depth camera and a processing computer. The robotic mechanism included two OWI robotic arms, position sensors, and microcontroller units (MCUs). The predicted motion type from the motion acquisition component was sent to the MCUs of the individual robotic arms to select the corresponding joint trajectories.

With this experimental setup, the robotic platform first predicted the type of ongoing human motions. Once the motion type was determined, the robotic platform performed the corresponding movement in response to the detected human motions. These specific motions for this study were as follows:

(1) Moving an object from a lower position to a higher position.
(2) Raising the right hand.
(3) Drinking a cup of water.

3.1. Mechanism of Robotic Platform

In this study, a dual-arm robotic platform mimicking the upper body of a humanoid robot was developed using two robotic arms. The dual-arm mechanism in this study was optimized by using off-the-shelf, battery-powered robotic arms, each with 5-DOF to mimic the natural motion of a human arm. These arms were carefully balanced on a supporting frame to ensure stability and accuracy in motion replication. The 5-DOF configuration was selected to provide sufficient flexibility while keeping the system efficient, with each joint corresponding to a real human arm's movement, allowing for precise control and operation. This approach streamlined the integration process, leveraging readily available hardware for robust performance. A depth camera was mounted on the upper central section of the mechanism to capture both human motions and the movements of surrounding objects. The mechanism is illustrated in Figure 4. Each arm mechanism in this robotic system included two gyro sensors, two motor driver modules, and a control unit. The shoulder and elbow joints were controlled by a synthesized controller that accounted for cross-coupled dynamics between the included limbs. The trajectories for individual joints were sent by a host controller and stored locally in a SD card.

In each arm's movement, motor speeds at the shoulder and elbow joints were directly related to their applied voltages. MCUs used for these robotic arms produced pulse-width modulated signals as motor controls. These signals required amplification by H-bridge drivers, specifically L298N modules, enabling smooth joint movement. Accelerometer and gyroscope data from MPU6050 sensors provided real-time feedback on the arm's orientation and speed at both joints. Individual Arduino Mega2560 MCUs were responsible for managing data collection, sensor integration, executing control algorithms, communication between arms, and saving information to storage mediums. Each MCU employed an I2C interface with MPU6050 sensors for instantaneous motion tracking. The program incorporated pre-loaded joint trajectory data from a text file located on the SD card corresponding to chosen movements to formulate necessary control signals. For storing detailed movement information and facilitating post-processing, the same integrated SD card was used, with the output transmitted via USB connections back to the host computer for examination or presentation purposes.

To simulate real-life conditions in which objects were manipulated, additional external forces can be introduced into this experimental robotic platform. The overall control system was designed to handle higher-frequency operations; it can process functions at over 200 Hz during single-axis movements that include file management tasks and complex sensor integration processes. However, the operational frequency of the MCUs in charge of controlling movement was deliberately limited to a steady 100 Hz. This conservative sampling rate ensured that adequate time wasallocated for executing control algorithms, which are crucial for accurate and stable joint positioning.

To interact with the users, the dual-arm robotic platform was positioned three meters away from the user, ensuring that the depth camera had sufficient distance to accurately detect human motion. As illustrated in Figure 4, the motion prediction model was triggered once the depth camera detected a human. If the performed motion matched one of the three selected motions, the corresponding reference motion type and necessary parameters were sent to the MCUs of the individual arms. The reference motion trajectories, stored in pre-saved files, were then loaded. These trajectories were adjusted based on the acquired human motion to ensure precise alignment. Finally, the adjusted trajectories were fed to the controllers to compensate for the tracking performance.

3.2. Modeling of Joint Actuators

Developing a controller that can effectively compensate for the joint movements of the robotic platform requires creating theoretical models representing individual joint dynamics. The mechanical components in question are driven by independent DC motors and manifest both linear and nonlinear characteristics in their responses. To simplify the modeling effort, the linear dynamics of each joint were analyzed to compensate for angular movements. The initial step involved determining essential parameters from a motor's response to step inputs to establish transfer functions that characterize the linear aspects of these joints.

Figure 5a illustrates where the DC motors and sensors were located. While individual discrepancies in coefficients due to hardware-induced nonlinearities may occur, all motors driving the robotic structure were homogeneously DC-powered. Nonetheless, variations in inertia among different limbs—the upper and lower ones—emerged from their orientation relative to the mechanism's pose and which motor was active. These dynamic fluctuations were integrated into the modeling phase, allowing for a refined discretization of linear equations within each joint's controller design framework. A simple first-order transfer function was implemented as an approximation for predicting the motion of each joint. It is

$$G_p(s) = \frac{\theta(s)}{U(s)} = \frac{K_0}{s\cdot(\tau s + 1)}. \quad (1)$$

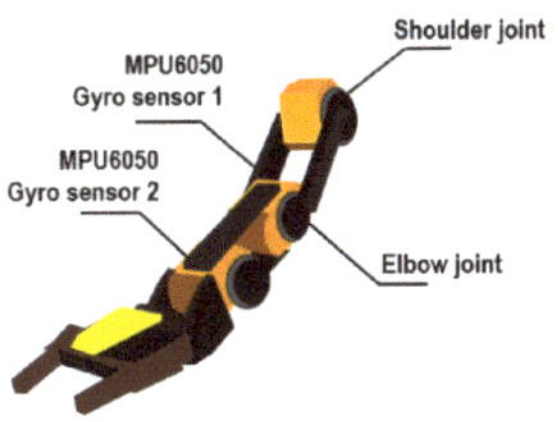

(**a**)

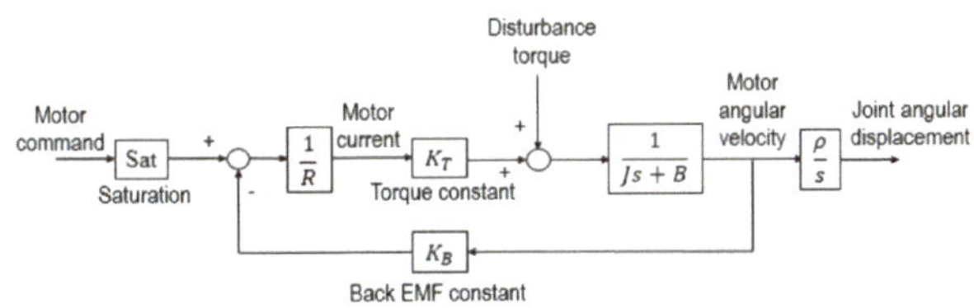

(**b**)

Figure 5. (**a**) Locations of DC motors and sensors. (**b**) Block diagram of a single DC motor used for joint angular movement.

The coefficients in this equation can be derived from the following equations, which are

$$K_0 = \frac{K_T\rho}{BR + K_T K_B} \text{ and } \tau = \frac{JR}{BR + K_T K_B}. \quad (2)$$

Within this framework, each robotic joint's angular displacement over time $\theta(s)$ is directly influenced by an input control signal U, which itself is a product of data collated from sensor fusion techniques to be elaborated upon in subsequent discussions. The research approach diverged from individual parameterization; instead, it adopted a collective method where representative coefficients K_0 and τ were determined through empirical analysis derived from step response experiments.

Figure 5b illustrates a typical block diagram of a single joint, where J represents the effective rotational moment inertia exerted on the DC motor, B is the viscos friction coefficient, R denotes the motor armature resistance, ρ signifies the gear ratio between the DC motor and the physical joint, and K_T and K_B represent the motor torque constant and motor back-EMF constant, respectively.

The coefficients of the involved joints were identified experimentally on an individual basis, which can be derived from previous studies [24,25]. These coefficients were used to synthesize and model the robotic platform. However, after the platform was constructed, these coefficients could change due to mechanical coupling with the attached mechanisms. Additionally, when external loads were applied, some coefficients, such as moments of inertia, deviated from their original values. Consequently, the model needed to be dynamically adjusted to ensure precise tracking performance.

3.3. Sensor Fusion of Position Measurements with the Kalman Filter and Complementary Filter

The design for the robotic platform's feedback mechanism emphasized compactness and cost-efficiency by integrating accelerometer and gyroscope sensors within MPU6050 modules installed at each limb. To enhance measurement precision, sensor fusion was applied, combining inputs from both sensors to leverage the unique attributes of gyroscopes for detecting swift orientation changes and accelerometers for maintaining accuracy over extended periods. Gyroscopes tend to suffer from drift effects, compromising precision, while accelerometers are less adept at capturing high-speed maneuvers and can be prone to linear acceleration distortions. However, the synergy through sensor fusion ensured more reliable and consistent angular position readings. This section evaluates two leading sensor fusion algorithms: the Kalman filter and the complementary filter.

The Kalman filter achieves this through a two-step process: prediction and correction.

- Prediction: The filter projects the current state estimate forward in time, accounting for uncertainties in both the system state and the measurements. This allows for a more accurate prediction compared to relying solely on a single measurement.
- Correction: Once the prediction is made, actual sensor measurements are used to refine the state estimate and its associated uncertainty. The filter calculates a weighted average, where the weights are determined by the Kalman gain. This gain reflects the confidence placed in the new measurement based on sensor reliability.

This continuous update process allows the Kalman filter to learn and improve its estimates with each new data point. The core strength of the Kalman filter lies in its ability to statistically account for both process and measurement noise. This enables it to deliver high-fidelity estimations, making it a valuable tool for applications requiring precise motion detection and control. In this application, the algorithm combined data from gyroscopes and accelerometers, commonly used for motion tracking. Figure 6 shows the updating process of the Kalman filter. By effectively integrating measurements from these various sensors, the Kalman filter significantly enhances the reliability and accuracy of the sensor fusion process.

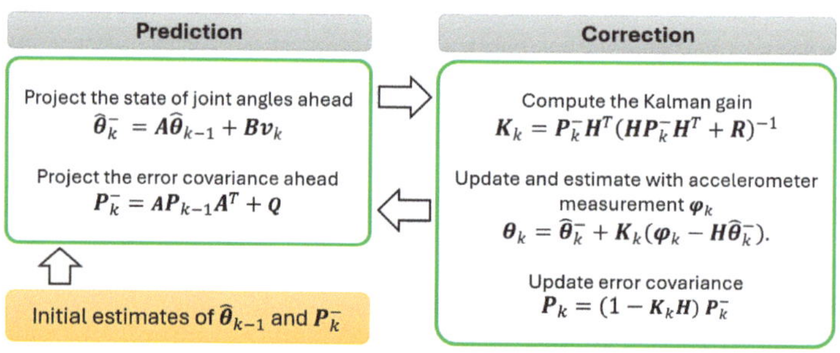

Figure 6. Procedures of sensor fusion utilizing the Kalman filter algorithm, demonstrating the sequential steps for combining accelerometer and gyroscope measurements to estimate pitch angles with reduced noise and enhanced accuracy.

In this prediction step, the matrix $\boldsymbol{H}$ relates the current state variables to the measurement, $\boldsymbol{Q}$ represents the process noise covariance, and $\boldsymbol{R}$ is the measurement noise. As for the variables, $\hat{\boldsymbol{\theta}}_k^-$ represents the posteriori state variables, v_k denotes the control input, $\boldsymbol{P}_k$ signifies the error covariance, $\boldsymbol{K}_k$ is the Kalman filter gain, and $\boldsymbol{\varphi}_k$ is the pitch angle measured by the accelerometer. It is assumed that the system state of the sensors does not change between two steps, so the control input u_k can be zero. In this study, various combinations of $\boldsymbol{R}$ and $\boldsymbol{Q}$ values were experimentally tested to optimize the sensor performance. Specifically, the measurement noise $\boldsymbol{R}$, associated with the accelerometer, was set to 5 to achieve the optimal results. The process noise covariance $\boldsymbol{Q}$, determined through trial and error, was set at a constant value of 0.1. The state variables were adjusted to integrate measurements from both the gyroscope and accelerometer, allowing the Kalman filter to be set up accordingly.

In the Kalman filter algorithm, the parameters included the predicted angle, denoted as $\theta_{1,k}$, and the targeted gyro angle, denoted as $\theta_{2,k}$. The state vector $\hat{\boldsymbol{\theta}}_k$ is a 2×1 vector $\begin{bmatrix}\hat{\theta}_{1,k} & \hat{\theta}_{2,k}\end{bmatrix}^T$. Based on the calibration of sensors, the coefficients of state matrix A were determined, which led to the following equation:

$$\hat{\boldsymbol{\theta}}_k^- = \begin{bmatrix}\hat{\theta}_{1,k}^- \\ \hat{\theta}_{2,k}^-\end{bmatrix} = \begin{bmatrix}1 & a_{12} \\ 0 & 1\end{bmatrix} \cdot \begin{bmatrix}\theta_{1,k} \\ \theta_{2,k}\end{bmatrix}. \quad (3)$$

If the error covariance P_k was defined as

$$\boldsymbol{P}_k = \begin{bmatrix} P_{00} & P_{01} \\ P_{10} & P_{11} \end{bmatrix}_k, \tag{4}$$

the matrix was updated itself by:

$$\boldsymbol{P}_k^- = \begin{bmatrix} 1 & a_{12} \\ 0 & 1 \end{bmatrix} \cdot \boldsymbol{P}_{k-1} \cdot \begin{bmatrix} 1 & 0 \\ a_{12} & 1 \end{bmatrix} + \begin{bmatrix} Q & 0 \\ 0 & Q \end{bmatrix} \Delta t. \tag{5}$$

With this information, the Kalman filter gain, K_k, represented as $\begin{bmatrix} K_{0,k} & K_{1,k} \end{bmatrix}^T$, can be updated using the following equation.

$$\boldsymbol{K}_k = \boldsymbol{P}_k^- \cdot \begin{bmatrix} 1 \\ 0 \end{bmatrix} \cdot \left(\begin{bmatrix} 1 & 0 \end{bmatrix} \cdot \boldsymbol{P}_k^- \cdot \begin{bmatrix} 1 \\ 0 \end{bmatrix} + R \right)^{-1}. \tag{6}$$

Additionally, since the relationship between current state and measurement was one-to-one conversion, the vector of H becomes $\begin{bmatrix} 1 & 0 \end{bmatrix}^T$. Thus, the predicted angle $\theta_{1,k}$ was calculated as

$$\theta_{1,k} = \hat{\theta}_{1,k}^- + K_{0,k} \cdot \left(\varphi_{1,k} - \hat{\theta}_{1,k}^- \right). \tag{7}$$

In this equation, $\varphi_{1,k}$ represents the angle measured by the accelerometer, and the error covariance can be updated using a similar approach.

The Kalman filter is a technique designed to estimate the state of a dynamic system by processing a series of noisy and incomplete data points. On the other hand, the complementary filter offers a simpler alternative. Its core principle is to merge and adjust the proportions of the two signals obtained from the sensors to achieve an improved result. This principle is applied through the following algorithm.

$$\theta_{1,k} = \alpha \cdot \theta_{2,k} + (1 - \alpha) \cdot \varphi_k. \tag{8}$$

In this equation, the coefficient α is constrained to be between 0 and 1. While the Kalman filter can yield precise joint angle estimates, it requires more computational steps. Conversely, the complimentary filter achieves comparable estimates with fewer calculations. Despite the accurate results calculated by the Kalman filter, minor oscillations caused by uncompensated noises can still occur, potentially leading to unwanted vibrations in the driving DC motors. Due to these considerations, the complementary filter algorithm was chosen and used in the experimental platform.

3.4. Nonlinear Factors of Physical Setup

The robotic platform utilized in this study was influenced by various nonlinear factors impacting its operational dynamics. These factors within a single joint include the following: (1) motor dead-zones; (2) inconsistent friction coefficients; (3) varying inertia based on joint positions; (4) backlash in gear mechanisms; and (5) elastic deformations in the twisted-string actuators.

Among these, the dead zones in the DC motors and the saturation of the power supply's output voltage were particularly significant. The joints demonstrated dead zones where the motors would not respond to control commands if the command voltage was within ±0.6 V. To address this, control signals below this threshold were boosted to at least 0.6 V to initiate motor movement. Additionally, power saturation occurred when the control commands generated by the controller approached the voltage limits of the power supply, leading to potential clamping of the output. Moreover, loading effects could arise if the required current exceeded the capacity of the power amplifiers, especially under sudden torque demands or excessive loading. To protect the DC motors, the control commands were restricted to a maximum of ±8 V, with a current limit of 1 A per motor.

Inconsistent friction was noticeable when the motor rotated in opposite directions and fluctuated slightly with changes in the load applied to the joint. Although this friction could be estimated to some extent, variations persisted during actual operation, with different joints exhibiting differing friction levels. This inconsistency made it difficult to compensate accurately using a linear controller. As a result, an adaptive mechanism was employed to approximate the friction's magnitude in the controller and effectively compensate for the angular movement assigned to each specific joint.

Mechanical components could also introduce nonlinearities into the system. Minor misalignments and backlash of transmission created during manufacturing can disrupt smooth operation. As the posture of the robot platform changes, the effective moments of inertia, J, of both upper and lower limb joints changed continuously. Even with lubrication, some friction remained, acting as a constant external disturbance. Additionally, slight imbalances or eccentricities of joints during rotation could lead to periodic changes in both angular speed and position. The frequency of these disturbances varied with the rotational velocity of the joint, requiring adaptive control strategies to minimize their impact on overall system performance. To address these nonlinear issues, the controller for each joint was designed with these nonlinearities explicitly considered.

These nonlinear factors were identified through experimental analysis, during which the behavior of the robotic platform was meticulously observed under various operating conditions. The parameters and ranges of these nonlinear factors, such as dead zones, inconsistent friction, and changing inertia, were derived by systematically varying input commands, loads, and joint positions and measuring the platform's response. These data were then analyzed to establish the specific thresholds and ranges of the nonlinearities affecting each joint. Once these ranges were determined, they were integrated into the simulation model of the robotic platform to ensure an accurate representation of its real-world dynamics. Additionally, tailored compensation strategies were developed and implemented within the control system to mitigate these nonlinear effects, ensuring more stable and precise operation across different scenarios.

4. Identification of Human Motions and Path Planning

4.1. Reference Trajectories of Selected Motions

The primary goal of this experimental robotic platform was programmed to replicate specific tasks performed by the user, focusing on the movement of the right arm. The robot's joint positions were determined by angular trajectories for the elbow and shoulder joints. As mentioned in the previous section, these reference trajectories were derived from pre-derived human reference motions captured through multiple repetitive motion recordings by a test subject. Each trajectory detailed the angular movement and range of the selected joint. Timing for these trajectories was expressed as a percentage of the overall cycle rather than in seconds, allowing for adjustments based on the physical scenario. This required modifying the initial and end positions, timing, and magnitude of the motions to fit the specific application. The motions of the right arm were emphasized because it played a major role in the tasks studied, ensuring detailed and accurate movements that mimic those of the user's dominant arm [13].

To compensate for the joint movements of the robotic platform, it was essential to synthesize reference trajectories of selected motions for the joints involved in specific motions. This process collected sufficient motion data to extract featured characteristics. While motion prediction models using CNNs could effectively identify motions, they fell short in synthesizing the required trajectories due to a lack of precise time information. Even when time information was included, it remained insufficient for creating complete trajectories due to variations in the duration of each motion. Consequently, reference trajectories for the involved joints must be derived from the recorded movements of the same motion, repeatedly performed by the same user or test subject. Instead of using angle-to-angle information, time series of joint movements must be employed to synthesize the trajectories of reference movements. However, since the reference movements derived from pre-recorded data were unlikely to match the duration of the ongoing movement, the trajectories of the involved joints needed to be adjusted based on the acquired ongoing trajectories. This adjustment ensured that the robotic platform can accurately replicate the intended motions, despite variations in individual motion durations and initial positions.

To synthesize reference trajectories for various movements, it was necessary to collect a sufficient number of motion samples. After collecting these samples, a temporal alignment method, dynamic time warping (DTW), was employed to align the collected trajectories. The initial step involved aligning the collected motion samples to a common timeframe, facilitating the comparison among the acquired motions and characteristic extraction for the reference motions. DTW, a widely used tool for temporal alignment, minimized the effects of time shifts and distortions by creating a warping path that detected similar characteristics between signals at different phases [26]. This method adopted a nearest neighbor algorithm with DTW as the distance measure, demonstrating a high level of effectiveness in identifying motion profiles within multivariate series. By aligning the collected data onto a common timeframe, DTW enabled the generation of accurate and consistent reference trajectories from the motion samples.

To align joint trajectories from two recorded motions, individually acquired angular positions presented as two time series, X and Y, and were expressed as

$$X = (x_1, x_2, \ldots, x_{T_1}),\ Y = (y_1, y_2, \ldots, y_{T_2}). \tag{9}$$

The two recorded trajectories X and Y have distinct time durations, which are T_1 and T_2. Calculating a distance metric between two vectors representing the same position is challenging if the datasets are not aligned to the same scale. Therefore, a local cost measure is defined as follows:

$$d : \mathcal{F} \times \mathcal{F} \to \mathbb{R} \geq 0, \tag{10}$$

To determine the best similarity between two recorded datasets presented as distinct time sequences, a cost measure was used. This measure evaluated similarity, with a lower cost indicating more similar characteristics and a higher cost indicating less similarity. The cost was calculated locally as a matrix for each component in the matrix, measuring distances to the four surrounding components. The process began at the initial boundary (x_1, y_1) and continued to the end boundary (x_{T_1}, y_{T_2}). Due to different durations of the two separate time series, T_1 and T_2, the recorded trajectories X and Y did not have direct one-to-one mapping. The cost sought the minimal distance values at each step, storing these costs in a mapping. Once the calculation was complete, the two recorded trajectories could be aligned to the same timeframe. This alignment process standardized the motions onto a uniform timeframe, represented on a percentage scale from 0% to 100%. By converting all sampled data to this consistent timeframe, it became possible to extract motion characteristics for synthesizing a reference trajectory. This method ensured precise alignment and comparison of the recorded datasets.

This method was employed to create reference trajectories for selected motions. When a motion is detected by the motion prediction model, the subsequent trajectories of the involved joints can be retrieved from these established reference trajectories. Figure 7 depicts the outcomes of processing 10 recorded motions involving elbow and shoulder joints (object lifting) using DTW. Originally, the durations of these motions ranged from 5 to 7.5 s. The horizontal axis represents relative time scaled from 0% to 100%, rather than absolute seconds. The average profiles extracted from individual motions were used as their respective reference trajectories. Additionally, Figure 7 displays the average motion trajectories from 30 samples, with boundaries defined at ±3 standard deviations.

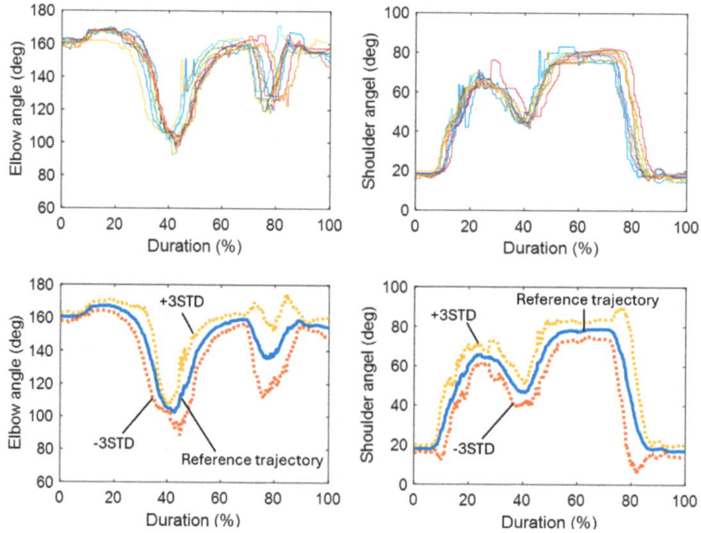

Figure 7. Motion profiles adjusted through temporal alignment and the derived reference trajectories for the elbow and shoulder joints (object lifting). The top two figures display ten recorded motions aligned from 0 to 100%, while the lower figures present the derived reference trajectories for the motion.

4.2. Reference Motions

The three reference motions selected for this study were (1) object lifting, (2) raising the right arm, and (3) drinking. The trajectories for these reference motions, including both elbow and shoulder joints, are illustrated in Figure 8. These trajectories were adjusted and extended based on the parameters obtained from the detected motions.

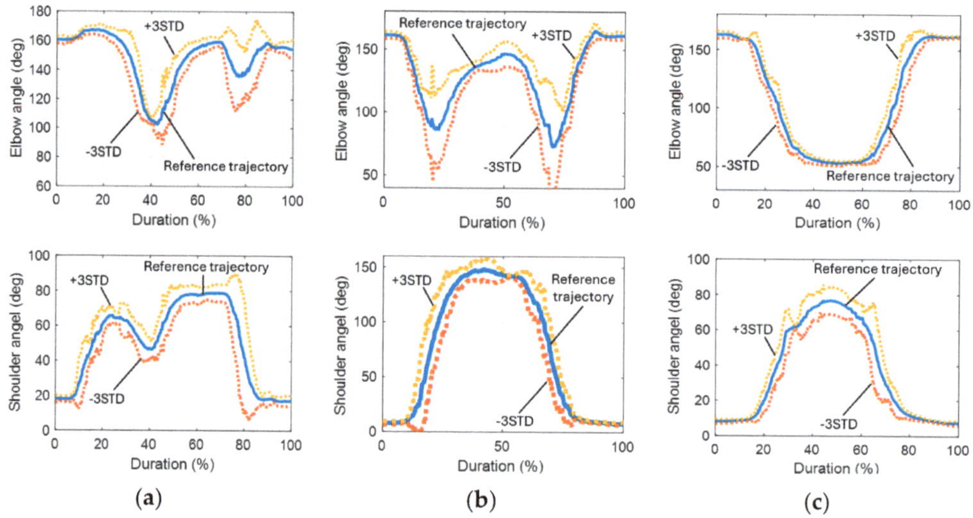

Figure 8. Derived reference trajectories of three selected motions: (**a**) object lifting, (**b**) raising the right arm, and (**c**) drinking.

4.3. Adjustments of Reference Trajectories

Once the reference trajectories of selected motions are established, path planning for the robotic platform's joints can be synthesized according to the real movement acquired from the user. However, as discussed in a preceding section, reliable motion prediction is only feasible after completing at least one-third of the motion. It was observed that arm movements varied from time to time, requiring adjustments to the real trajectories based on the actual situation. While the motion profile remained consistent, deviations could occur due to various factors, such as inconsistent amplitude, drifting initial position, varying total motion duration, and collision with external objects. To accommodate these factors and ensure accurate execution by the robotic platform, adjustments to the reference trajectories of the predicted motions were necessary.

To address these factors, adjustments to the reference trajectories were necessary. The initial positions of involved joints should be aligned to the same starting angular positions to ensure that the robot can generate a similar motion. Joint movement amplitudes were scaled based on data from the first one-third of the predicted motion. This specific time is defined as the motion prediction time, T_P. Additionally, the total duration of the motion was adjusted according to the duration of this one-third motion cycle. Figure 9 illustrates the process of generating synthesized trajectories. Initially, real trajectories of a specific motion were consolidated into a reference trajectory. This reference trajectory was then adjusted based on the actual acquired motion using several required parameters. Once these adjustments were made, minor refinements were applied to ensure that the generated trajectories maintained identical angular positions at both the starting time and T_P when the trajectories were transferred to the robotic platform.

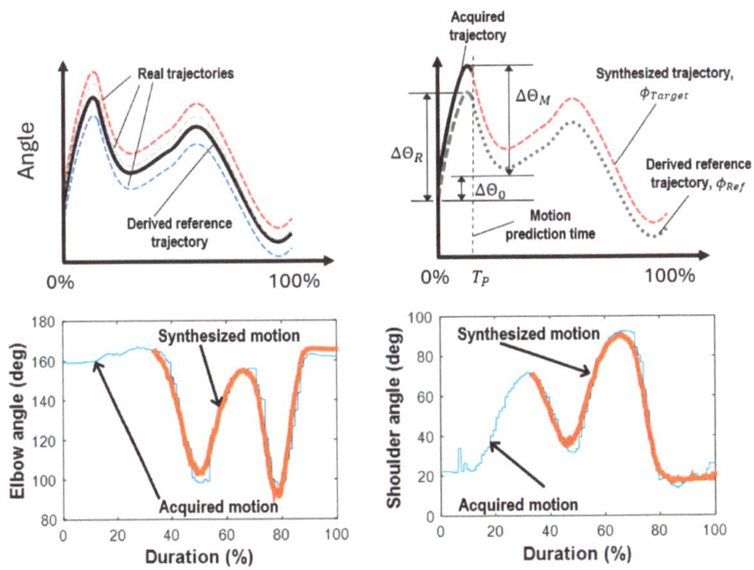

Figure 9. Adjustment of joint trajectory based on reference motions and real-world movements.

The synthesized trajectory was adjusted from the reference trajectory based on the actual movement scenarios. Two major adjustments were required: (1) the deviation in the start position between the real and reference trajectories, denoted as $\Delta\Theta_0$, and (2) the adjusting ratio, $\Delta\Theta_M/\Delta\Theta_R$, which scaled the amplitude of the reference trajectory based on the comparison between the real and reference trajectories. The amplitudes $\Delta\Theta_R$ and $\Delta\Theta_M$ represented critical variations used to determine the scaling between the two trajectories by comparing the peak variation of the joint angular position with the starting angular position. Accurately determining whether the robot's operator had driven the platform to the targeted duration T_P can be challenging, so both angular velocity and angular position were measured to ensure precise adjustments.

The adjustment process began with collecting necessary motion data to select reference trajectories for involved joints. Initial angular positions of the joints were recorded to identify any initial deviations. The user's angular positions and angular velocities were acquired and assessed. When the motion reached the motion prediction time, T_P, which was set to be one-third, the prediction model determined the type of the ongoing motion. After determining the motion type, the reference trajectories of the predicted motion were adjusted and scaled to generate the targeted trajectories of the joints. The angles of the targeted trajectories, ϕ_{Target}, for a single joint were generated using the corresponding angular position of the reference trajectory, ϕ_{Ref}, an adjusting ratio of amplitude $\Delta\Theta_M/\Delta\Theta_R$, and the shifted angle $\Delta\Theta_0$. Thus, a simple linear approximation could be applied to obtain the targeted trajectories for individual joints, expressed as

$$\phi_{Target} = \frac{\Delta\Theta_M}{\Delta\Theta_R} \cdot \phi_{Ref} + \Delta\Theta_0. \quad (11)$$

These targeted trajectories of individual joints can then be applied to the controllers to drive the robotic platform for specific motions. Figure 9 illustrates the synthesized results of the planned paths for the object lifting motion. Observations show that the synthesized path closely matched the real motion profile. However, deviations of up to 20° at the elbow joint and 16° at the shoulder joint were noted. These deviations might be caused by vibration or inconsistency in the subject's arm movements. Further, psychophysical analysis was necessary to determine whether these deviations can be disregarded.

4.4. Scalability of Acquired Reference Trajectories

Building on prior research, it has been observed that the selected motions exhibit both self-similarity and consistency across different individuals [13]. To make these reference trajectories applicable to a broader population, the trajectories can be scaled according to factors such as body size, total duration, and other physical characteristics. By adjusting these parameters, the acquired reference trajectories can be modified—either expanded or contracted—to meet the specific needs of different users and various situations.

The ability to scale these reference trajectories within the reasonable limits of human movement ensures their adaptability. However, extending these trajectories to include unstudied motions presents significant challenges. This would require gathering additional data on various movements, potentially involving the segmentation of individual motions into smaller components. These segments could then be reassembled to form a complete motion. While machine learning techniques can assist in this reassembly process, certain actions, particularly those influenced by cultural practices, professional requirements, or personal habits (e.g., dining or dressing), cannot be easily reconstructed from a motion library. The process of segmentation and reassembly is complex and represents a separate area of research beyond the scope of this study.

The modeling and path planning approach proposed in this study could potentially be applied to other motions, provided the training procedures and sequences are standardized and streamlined for predicting individual motions. However, it is important to note that the model is best suited for self-similar motions. Motions that do not exhibit consistency over time are not well-suited for this path planning approach. Future research might explore the feasibility of segmenting motions into smaller components, but this would likely require a redesigned learning and path planning methodology.

5. Controller Synthesis and Tracking Performances

To ensure that the angular movements of the robotic platform were compensated properly, both a linear controller and a nonlinear controller were designed for motion tracking. These controllers were planned to be implemented using an Arduino mega 2560 MCU, operating at a sample rate of 100 Hz. The controllers were developed based on the nominal dynamics of individual actuators, while the cross-coupled dynamics [27,28] were treated as external disturbances as well as the unsynchronized deviations of individual joints.

5.1. Synthesis of Linear Controller and Adaptive Robust Controller

A linear controller for motion tracking of a single joint was designed using the pole-placement method, based on the characteristic equation identified earlier. Its tracking performance was compared to that of a unity feedback controller. Given that the robotic platform operates both the upper and lower arms simultaneously, cross-coupled dynamics can act as external disturbances. Therefore, an important consideration is that the synthesized controller must effectively attenuate disturbances and operate within the linear region of the power source without saturation.

To evaluate the performance of the synthesized controller, a cycloidal input trajectory was used as the reference trajectory for both the unity feedback system and the compensated system with the pole-placement controller. The comparison of their controlled results is illustrated in Figure 10, showing the transient response of an elbow joint. The comparison clearly demonstrates that the pole-placement controller achieves a better transient response than the unity feedback compensation, with less tracking error. The steady-state error of the compensated result is less than 5% of the total stroke, indicating superior performance of the pole-placement controller.

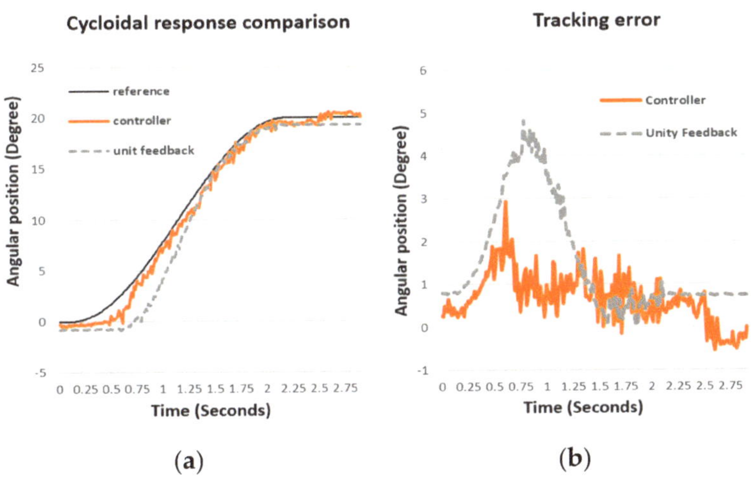

Figure 10. (**a**) Comparison of system response between compensated system and unity feedback control result for with and without an appropriate controller and (**b**) comparison of tracking errors of both control schemes [24].

However, the linear controller can only compensate for the linear response of the robotic platform. The nonlinear phenomena present on the platform cannot be adequately compensated for, resulting in inaccurate position tracking. To address these issues, an adaptive robust control (ARC) scheme was implemented. This ARC scheme was specifically designed to handle identified nonlinear factors, including cross-coupled dynamics, dead zones of individual actuators, and power source saturation. The ARC controller works by continuously adapting to changes and disturbances in the system, ensuring robust performance even in the presence of nonlinearities. Cross-coupled dynamics refer to the interactions between different joints or parts of the robotic system, which can affect each other's movements. Dead zones in actuators are regions where small control signals do not produce any movement, leading to delays and inaccuracies. Power source saturation occurs when the power supply cannot provide sufficient voltage or current, limiting the actuator's performance. The derivation of the rules for ARC controllers designed to enhance both tracking and synchronization accuracy in multiaxial mechanisms is detailed in [12].

Previous studies have explored various methods to compare the performance of different control strategies, such as linear controllers, adaptive controllers, LQR, and ARC [24,29]. These comparisons often focus on metrics like tracking accuracy, system stability, and responsiveness to disturbances. ARC, for instance, has shown superior performance in compensating for nonlinearities and uncertainties in complex systems, whereas linear controllers are typically more effective for simpler, steady-state conditions. LQR, with its optimal control approach, has been widely used in systems requiring a balance between performance and energy efficiency. However, each method has its limitations, and the comparative studies provide valuable insights into selecting the most appropriate controller based on specific system requirements and operational conditions.

In this study, the ARC controller was carefully designed to address the various nonlinearities present in the system, including motor dead zones, inconsistent friction at the joints, varying inertia, and other unmodeled uncertainties. The dead zones in the driven motors were precisely characterized to ensure that the control system could effectively compensate for these non-responsive regions. Additionally, the friction inconsistencies at the joints, which varied with load and rotation direction, were integrated into the controller's design. The controller also accounted for the changing inertia across different joint positions and other unstructured uncertainties that could impact system performance.

The development of the ARC controller followed established methodologies from previous research, incorporating advanced techniques to estimate and compensate for these nonlinear factors in real time. Through rigorous testing and refinement, the ARC controller demonstrated its effectiveness in mitigating the adverse effects of nonlinearities, leading to significantly improved position tracking and overall performance of the multiaxial robotic system.

5.2. Cross-Coupled Dynamics among Joints

One important issue in compensating for the angular positions of the arm mechanism is the coupling of individual dynamics among joints. Cross-coupled dynamics can arise from various factors, such as the continuously changing inertia caused due to the varying poses of the arm, different loads held by the end effector, and vibrations induced by the mechanism's movements. These cross-coupled dynamics are challenging to model and were typically treated as external disturbances. However, compensating for cross-coupled dynamics individually without adequate synchronization can result in inaccurate positioning.

To address this issue, the cross-coupling can be compensated for by considering the asynchronization of the performed motion. This means that the controller needs to consider the angular movements of involved joints as a whole. Deviations in one joint can affect the positional accuracy of the other joints. Therefore, the cross-coupled dynamics adopted in this platform were addressed by focusing on the synchronization error corresponding to individual joints.

5.3. Gain Scheduling

Experimental results from a previous study [12] demonstrate that the ARC controller can effectively compensate for several nonlinear factors, such as the dead zones when initiating movement from a stop and on-the-fly calibration of uncertain system dynamics. Furthermore, the ARC controller can prevent the actuator from operating in the saturated zone. Despite these advantages, the adaptive mechanism of the ARC controller still introduced observable oscillations. In contrast, the linear controller produced smoother tracking results during steady-state operation. However, the designed linear controller also exhibited larger tracking deviation when the control commands fell within the dead zone.

In this study, the gain scheduling control strategy was employed to simulate the compensation for the proposed robotic platform. To ensure accurate tracking of the robotic platform, combining the strengths of both ARC and linear controllers was a feasible approach, which can be implemented through gain scheduling based on the operating conditions of each joint. The situational rules used were as follows:

- Linear Controller: Applied when the required angular velocity is less than 0.35 rad/s, and the required angular acceleration is less than 2.6 rad/s^2.
- ARC Controller: Applied in all other situations.

These rules were used to compensate for both the elbow and shoulder joints of the robotic platform. Such an approach ensured fast response time during transient states as well as smooth tracking performance during steady states.

5.4. Simulation Results

To validate the effectiveness and repeatability of the proposed controllers, the trajectories for the task of object lifting were used as the reference motion for the robotic platform. The original duration of this motion ranged from 6 to 7.5 s for the test subject. For testing purposes, this reference motion was expanded to 15 s to evaluate the tracking performance of the controller. Given the reference motion trajectories, Figure 10 illustrates the simulated tracking performance and corresponding errors of both elbow and shoulder joints for an object-lifting motion. The reference motion for this experiment was obtained from a single test subject. The controllers, based on gain scheduling and the situational rules, demonstrated superior tracking errors, maintaining errors within 5% of the total angular spans for both joints. However, the angular accelerations and velocities of the actuators at the involved joints were constrained by the maximum voltage limits of the DC motors. Consequently, the profiles of generated forces might not perfectly match the profiles of the original motion. If the required speeds or accelerations exceed these limits, delays and fluctuations might still occur while tracking the reference trajectories. To evaluate the repeatability of the proposed control strategy, Figure 11 illustrates the error distribution across 10 simulated trials. In these simulations, the initial angular positions of both elbow and shoulder joints were randomly set to vary within a deviation of 10°. Without significant external disturbance, the majority of the errors in the involved joints were confined within a 12° range. It is evident that the hardware setup exhibited some physical defects at specific angular positions of the shoulder joint.

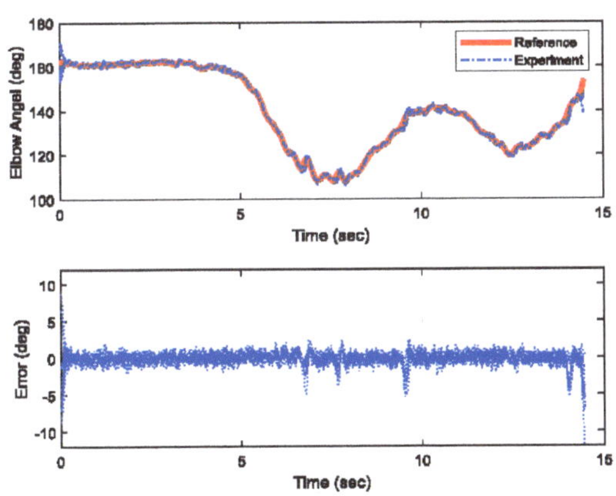

(a)

Figure 11. *Cont.*

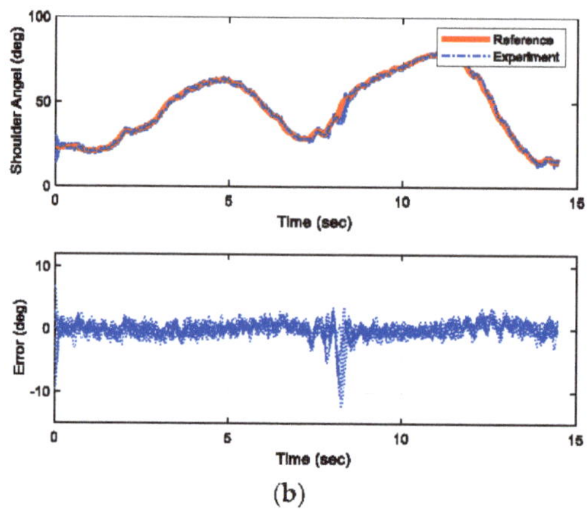

Figure 11. Ten trials of compensated results of elbow and shoulder joint movements (object lifting). (**a**) Tracking performance and tracking error of elbow joint and (**b**) tracking performance and tracking error of shoulder joint.

6. Operation of Proposed Robotic Platform

6.1. Operational Process of the Robotic Platform

The operational sequence of the proposed robotic platform began with motion acquisition using a depth camera. Once the movements of individual joints were recorded, the angular positions of these joints were fed into a motion prediction model to determine the type of ongoing motion. This model utilized a pre-built motion library that contained reference trajectories for various motions.

From the motion library, the system derived the angular trajectories for the involved joints. These reference trajectories were then adjusted based on the acquired motion data from the user, taking into account factors such as amplitude, shifted angles, and the total duration of the motion. This adjustment ensures that the robotic platform can accurately replicate the intended human motions. The adjusted trajectories were subsequently sent to the controllers of the robotic platform. These controllers, which integrated both ARC and linear control strategies through gain scheduling, used the adjusted trajectories to compensate and drive the joints of the robotic platform. The dual control strategy ensures both fast response times during transient states and smooth tracking performance during steady states. Figure 12 illustrates the overall process for a complete cycle of the robot's operations to mimic human arm motions, from initial motion acquisition to the execution of the adjusted trajectories by the robotic platform. This comprehensive approach ensures precise and effective replication of human motions by the robotic system. In this study, the compensated trajectories of the selected motions were simulated.

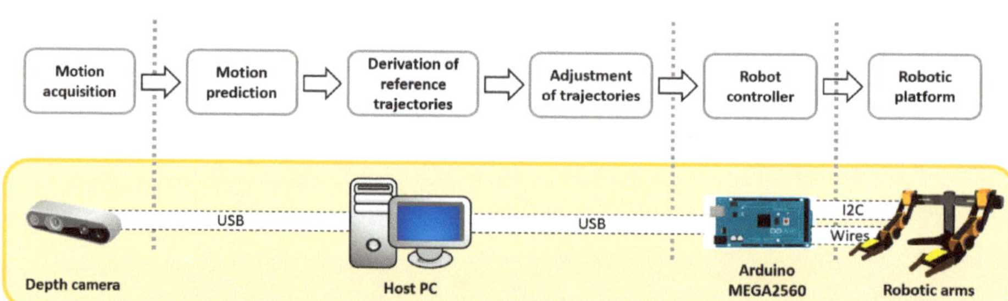

Figure 12. Operational process of the robotic platform to mimic human arm motions.

6.2. Response Time for Adopted MCU

While the control strategy effectively followed the motion when the duration was sufficiently long, it is important to identify the types of motions that the robotic platform can reliably mimic. Given that the time constant of the motors at individual joints ranged from 0.4 to 0.6 s, the platform must be selective about the motions it attempts to replicate to ensure accuracy and responsiveness.

To evaluate the feasible motion cycles that the robot can follow, two specific motions were simulated: object lifting and hand raising with the right arm. These motions were chosen because they are typical and straightforward actions that involve significant joint movements. Normally, these motions can be completed within approximately 6 to 7.5 s. For the purposes of this evaluation, the trajectories for these motions were rearranged to vary between 5 and 15 s while maintaining an identical sample rate. This range allows for the assessment of the platform's performance across different speeds and durations, providing insight into its capability to mimic human motions accurately. By testing these adjusted trajectories, we can determine the limits and optimal conditions for the robotic platform's operation in real-world scenarios.

Figures 13 and 14 display the simulated results of the object lifting motion completed over various durations. The simulations indicated that the deviation between the reference trajectories and the physical outputs for both the elbow and shoulder joints exceeded 10° when the total duration of the motion was less than 8 s. If the duration is reduced below a threshold, the deviation can increase significantly. This discrepancy is likely due to the slow time constants of the actuators used. Therefore, it is recommended that the total duration of a motion for this adopted robotic platform should not be shorter than the original duration to avoid significant deviations due to the limited response time of the actuators.

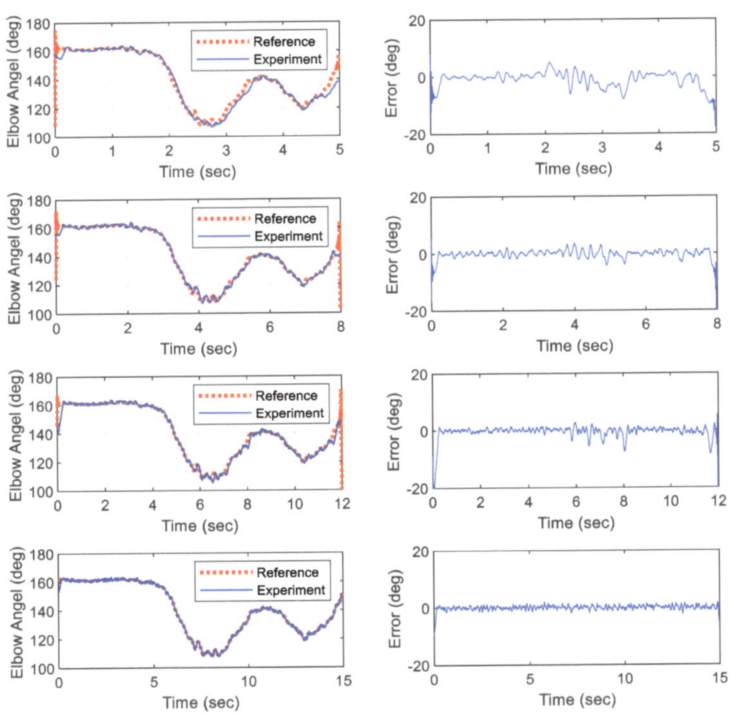

Figure 13. Simulated results of elbow joint with different operation durations for object lifting motion.

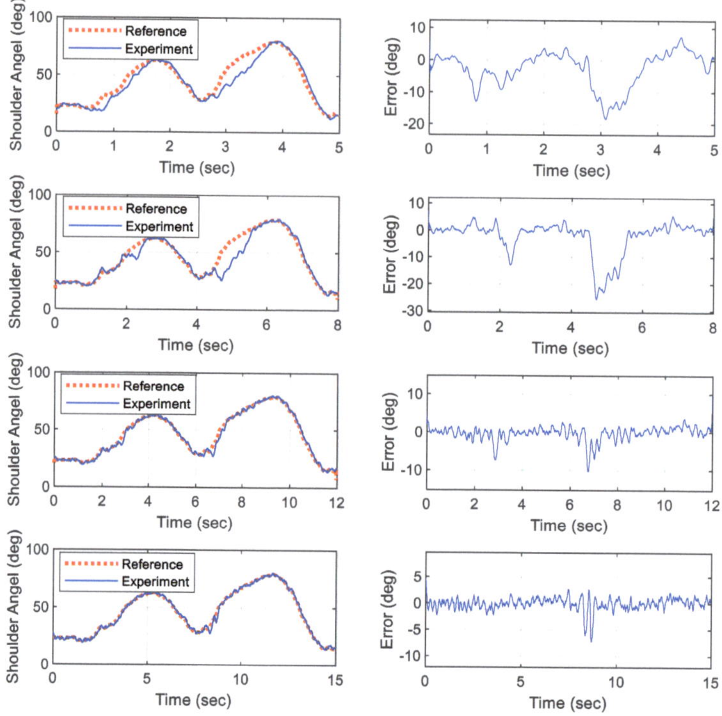

Figure 14. Simulated results of shoulder joint with different operation durations for object lifting motion.

Extending the duration of the motion can offer several advantages. It allows for smoother and more controlled movements, reducing the risk of mechanical stress and wear on the robotic components. Additionally, a slower, more deliberate motion can improve the precision of the robot's actions, making it more suitable for tasks that require fine motor skills and careful handling. On the other hand, performing motions faster than a threshold can lead to several issues. The primary concern is the potential for increased deviations due to the actuators' inability to respond quickly enough. This can result in inaccurate positioning and compromised performance. For the proposed robotic platform, it is recommended to extend the motion cycle by one-third beyond its original duration to ensure accurate tracking. Moreover, rapid movements can introduce additional mechanical stress and potential instability, which might lead to premature wear or damage to the robotic components. Thus, while it is advantageous to extend the motion duration for certain applications, care must be taken not to reduce the duration below the original timeframe to maintain accuracy and protect the mechanical integrity of the robotic platform.

6.3. Experimental Results of Proposed Gain Scheduling Controller

The object-lifting motion was used as the reference trajectory for the robotic platform. Figure 15 displays the tracking performance and corresponding errors for both the elbow and shoulder joints of this selected motion over a 15 s period. With the implementation of the proposed controller and operational rules, tracking errors were maintained within 5% of the total angular spans. However, the angular accelerations and velocities of the actuators at the involved joints were constrained by the maximum voltages that could be applied to the DC motors. When the required speeds or accelerations exceeded these limits, delays and fluctuations were observed in tracking the reference trajectories.

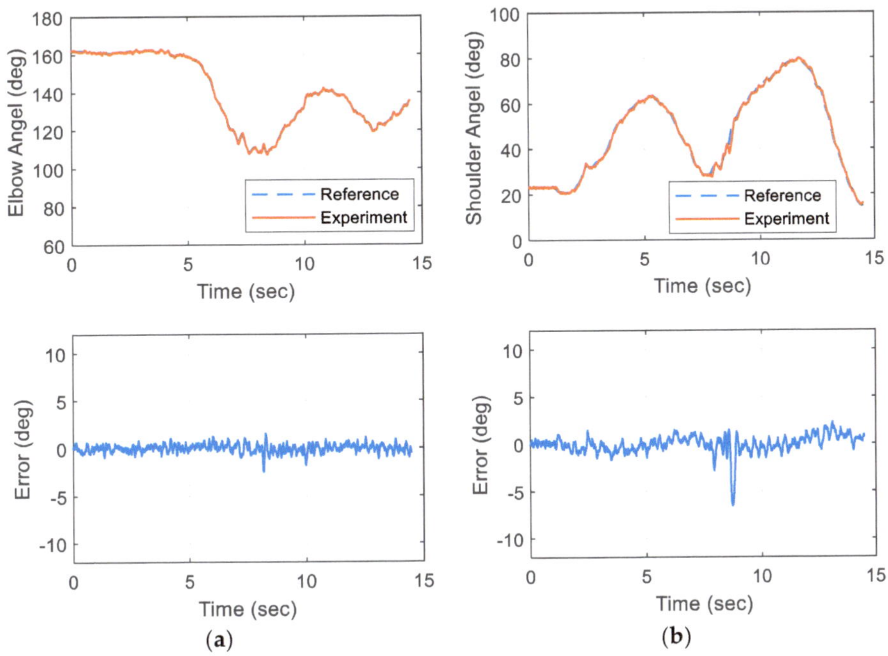

Figure 15. Compensated results of elbow and shoulder joint movements (object lifting). (**a**) Tracking performance and tracking error of elbow joint and (**b**) tracking performance and tracking error of shoulder joint.

To evaluate the repeatability of the proposed control strategy, Figure 16 illustrates the errors distribution over 11 trials of experimental results. Most errors of involved joints were confined within the ±8° range, with over half falling within ±4°. When the ARC controller was integrated with the linear controller, tracking errors were significantly reduced. This indicated that combining both nonlinear and linear controllers resulted in optimal tracking performance.

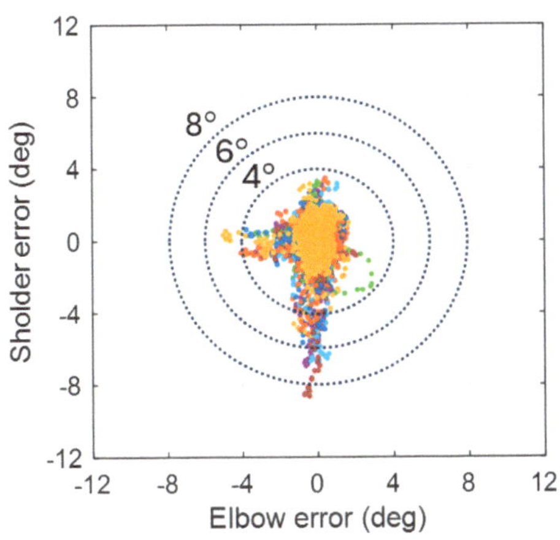

Figure 16. Distribution of tracking errors of the selected motion (10 trials of object lifting).

7. Conclusions

In this study, an integrated control strategy combining both ARC and linear controllers has been developed to address reference tracking and nonlinear factors in a robotic platform. The proposed controller demonstrated adequate tracking performance for the multiaxial robotic system, effectively managing the complexities of motion recognition and trajectory planning. The motion recognition part, which allows CNNs to process angle-to-angle representations of joint movements, proved essential for predicting and replicating human-like motions in real-time. This integration enables the robotic platform to assist human workers by accurately mimicking their movements.

However, the complexity of implementing this controller on an MCU unit remains a challenge. Additionally, the cross-coupled dynamics (CCD) between the two axes were not considered, which is a crucial factor when the robot interacts with human users. In future work, more joints will be integrated into the control strategy, and the perturbations caused by CCD will be thoroughly investigated. Efforts will also be made to simplify the implementation of the controller on the selected MCU unit to achieve a faster sample rate. This will enhance the system's responsiveness and reliability in dynamic, real-world environments.

Author Contributions: Conceptualization, M.H.C.; methodology, M.H.C.; software, P.-L.H. and H.-C.C.; formal analysis, P.-L.H. and M.H.C.; writing—original draft preparation, M.H.C.; writing—review and editing, M.H.C. All authors have read and agreed to the published version of the manuscript.

Funding: This research was funded by the Research Seed Grants from the NASA West Virginia EPSCoR Program.

Institutional Review Board Statement: The research study "Identification of Arm Dynamics in Daily Activities" was reviewed by the West Virginia University Institutional Review Board IRB and was approved in accordance with 46 CFR 46.101b (protocol number 1400179969R001).

Data Availability Statement: The data that support the findings of this study are available from the corresponding authors upon reasonable request.

Acknowledgments: The authors would like to thank Steven Wheeler at West Virginia University for the related discussion about the acquisition of motions.

Conflicts of Interest: The authors declare no conflicts of interest.

References

1. Lemaignan, S.; Warnier, M.; Sisbot, E.A.; Clodic, A.; Alami, R. Artificial cognition for social human–robot interaction: An implementation. *Artif. Intell.* **2017**, *247*, 45–69. [CrossRef]
2. Moniz, A.B.; Krings, B.J. Robots working with humans or humans working with robots? Searching for social dimensions in new human-robot interaction in industry. *Societies* **2016**, *6*, 23. [CrossRef]
3. Alam, A. Social robots in education for long-term human-robot interaction: Socially supportive behaviour of robotic tutor for creating robo-tangible learning environment in a guided discovery learning interaction. *ECS Trans.* **2022**, *107*, 12389. [CrossRef]
4. Stock-Homburg, R. Survey of emotions in human–robot interactions: Perspectives from robotic psychology on 20 years of research. *Int. J. Soc. Robot.* **2022**, *14*, 389–411. [CrossRef]
5. Yang, C.; Yuan, K.; Heng, S.; Komura, T.; Li, Z. Learning natural locomotion behaviors for humanoid robots using human bias. *IEEE Robot. Autom. Lett.* **2020**, *5*, 2610–2617. [CrossRef]
6. Goodfellow, I.; Bengio, Y.; Courville, A. *Deep Learning*; MIT Press: Cambridge, MA, USA, 2016.
7. Nguiadem, C.; Raison, M.; Achiche, S. Motion planning of upper-limb exoskeleton robots: A review. *Appl. Sci.* **2020**, *10*, 7626. [CrossRef]
8. Wonsick, M.; Padır, T. Human-humanoid robot interaction through virtual reality interfaces. In Proceedings of the 2021 IEEE Aerospace Conference (50100), Big Sky, MT, USA, 6–13 March 2021; IEEE: Piscataway, NJ, USA, 2021; pp. 1–7. [CrossRef]
9. Zhou, X. Predictive human-in-the-loop simulations for assistive exoskeletons. In Proceedings of the International Design Engineering Technical Conferences and Computers and Information in Engineering Conference, St. Louis, MO, USA, 17–19 August 2020; American Society of Mechanical Engineers: New York, NY, USA, 2020; Volume 83983, p. V009T09A006. [CrossRef]
10. Wang, Y.; Metcalfe, B.; Zhao, Y.; Zhang, D. An assistive system for upper limb motion combining functional electrical stimulation and robotic exoskeleton. *IEEE Trans. Med. Robot. Bionics* **2020**, *2*, 260–268. [CrossRef]
11. Conti, D.; Cirasa, C.; Di Nuovo, S.; Di Nuovo, A. "Robot, tell me a tale!" A social robot as tool for teachers in kindergarten. *Interact. Stud.* **2020**, *21*, 220–242. [CrossRef]

12. Cheng, M.H.; Li, Y.; Bakhoum, E.G. Controller Synthesis of Tracking and Synchronization for Multi-Axis Motion System. *IEEE/ASME Trans. Control. Syst. Technol.* **2014**, *22*, 378–386. [CrossRef]
13. Cheng, M.H.; Huang, P.-L.; Chu, H.-C. Motion Estimation and Path Planning for Assistive Robotic Devices. In Proceedings of the 2019 ASME International Mechanical Engineering Congress & Exposition (IMECE2019), IMECE2019-12296, Salt Lake City, UT, USA, 8–14 November 2019. [CrossRef]
14. Mohammadzade, H.; Hosseini, S.; Rezaei-Dastjerdehei, M.R.; Tabejamaat, M. Dynamic time warping-based features with class-specific joint importance maps for action recognition using Kinect depth sensor. *IEEE Sens. J.* **2021**, *21*, 9300–9313. [CrossRef]
15. Cheng, M.H.; Liang, C.J.; McKenzie, E.A., Jr.; Dominguez, E.G. Identification of Contact Avoidance Zones of Robotic Devices in Human-Robot Collaborative Workspaces. *IFAC-PapersOnLine* **2023**, *56*, 577–582. [CrossRef]
16. Sohn, J.H.; Oh, S.; Lee, C.H.; Kim, S.S. Recursive inverse kinematic analysis for humanoid robot based on depth camera data. *J. Mech. Sci. Technol.* **2021**, *35*, 3199–3209. [CrossRef]
17. Calaver, D.; Floroian, L.; Grigorescu, S.M. Assistive rehabilitation using a 7-dof robotic arm with self-collision and obstacle avoidance system. In Proceedings of the 2019 E-Health and Bioengineering Conference (EHB), Iasi, Romania, 21–23 November 2019; IEEE: Piscataway, NJ, USA, 2019; pp. 1–4. [CrossRef]
18. Roy, R.; Mahadevappa, M.; Kumar, C.S. Trajectory path planning of EEG controlled robotic arm using GA. *Procedia Comput. Sci.* **2016**, *84*, 147–151. [CrossRef]
19. Liu, Z.; Ai, Q.; Liu, Y.; Zuo, J.; Zhang, X.; Meng, W.; Xie, S. An optimal motion planning method of 7-DOF robotic arm for upper limb movement assistance. In Proceedings of the 2019 IEEE/ASME International Conference on Advanced Intelligent Mechatronics (AIM), Hong Kong, China, 8–12 July 2019; IEEE: Piscataway, NJ, USA, 2019; pp. 277–282. [CrossRef]
20. Luciani, B.; Roveda, L.; Braghin, F.; Pedrocchi, A.; Gandolla, M. Trajectory Learning by Therapists' Demonstrations for an Upper Limb Rehabilitation Exoskeleton. *IEEE Robot. Autom. Lett.* **2023**, *8*, 4561–4568. [CrossRef]
21. Bashir, F.I.; Khokhar, A.A.; Schonfeld, D. Object trajectory-based activity classification and recognition using hidden Markov models. *IEEE Trans. Image Process.* **2007**, *16*, 1912–1919. [CrossRef] [PubMed]
22. Bobick, A.F.; Davis, J.W. The recognition of human movement using temporal templates. *IEEE Trans. Pattern Anal. Mach. Intell.* **2001**, *23*, 257–267. [CrossRef]
23. Simonyan, K.; Zisserman, A. Two-stream convolutional networks for action recognition in videos. In *Advances in Neural Information Processing Systems*; MIT Press: Montreal, QC, Canada, 2014; pp. 568–576.
24. Chu, H.C. Design and Development of an Upper Limb Assistive Robotic System for Stroke Rehabilitation. Master's Thesis, National Tsing Hua University, Hsinchu, Taiwan, 2019.
25. Cheng, M.; Bakhoum, E. Tracking Control Design and Implementation of Multiaxial Controller for Social Robotic Device. In Proceedings of the 2021 ASME International Mechanical Engineering Congress & Exposition (IMECE2021), IMECE2021-70510, Virtual Conference, 1–4 November 2021. [CrossRef]
26. Zhou, F.; De la Torre, F. Generalized time warping for multi-modal alignment of human motion. In Proceedings of the 2012 IEEE Conference on Computer Vision and Pattern Recognition, Providence, RI, USA, 16–21 June 2012; IEEE: Piscataway, NJ, USA, 2012; pp. 1282–1289. [CrossRef]
27. Sun, D.; Shao, X.; Feng, G. A model-free cross-coupled control for position synchronization of multi-axis motions: Theory and experiments. *IEEE Trans. Control. Syst. Technol.* **2007**, *15*, 306–314. [CrossRef]
28. Barton, K.L.; Alleyne, A.G. A cross-coupled iterative learning control design for precision motion control. *IEEE Trans. Control. Syst. Technol.* **2008**, *16*, 1218–1231. [CrossRef]
29. Cheng, M.H.; Chen, C.-Y.; Bakhoum, E.G. Synchronization Controller Synthesis of Multi-Axis Motion System. *Int. J. Innov. Comput. Inf. Control.* **2011**, *7*, 4395–4410.

biomimetics

MDPI

Article

Hydrodynamic Simulation and Experiment of a Self-Adaptive Amphibious Robot Driven by Tracks and Bionic Fins

Minghai Xia [1], Qunwei Zhu [1], Qian Yin [2], Zhongyue Lu [1], Yiming Zhu [1] and Zirong Luo [1,*]

[1] College of Intelligence Science and Technology, National University of Defense Technology, Changsha 410073, China; xiaminghai@nudt.edu.cn (M.X.); zhuqunwei@nudt.edu.cn (Q.Z.); luzhongyue@nudt.edu.cn (Z.L.); yiming.zhu@nudt.edu.cn (Y.Z.)

[2] College of Energy and Power Engineering, Changsha University of Science and Technology, Changsha 410076, China; yqziqiang@mail.sdu.edu.cn

* Correspondence: luozirong@nudt.edu.cn

Abstract: Amphibious robots have broad prospects in the fields of industry, defense, and transportation. To improve the propulsion performance and reduce operation complexity, a novel bionic amphibious robot, namely AmphiFinbot-II, is presented in this paper. The swimming and walking components adopt a compound drive mechanism, enabling simultaneous control for the rotation of the track and the wave-like motion of the undulating fin. The robot employs different propulsion methods but utilizes the same operation strategy, eliminating the need for mode switching. The structure and the locomotion principle are introduced. The performance of the robot in different motion patterns was analyzed via computational fluid dynamics simulation. The simulation results verified the feasibility of the wave-like swimming mechanism. Physical experiments were conducted for both land and underwater motion, and the results were consistent with the simulation regulation. Both the underwater linear and angular velocity were proportional to the undulating frequency. The robot's maximum linear speed and steering speed on land were 2.26 m/s (2.79 BL/s) and 442°/s, respectively, while the maximum speeds underwater were 0.54 m/s (0.67 BL/s) and 84°/s, respectively. The research findings indicate that the robot possesses outstanding amphibious motion capabilities and a simplistic yet unified control approach, thereby validating the feasibility of the robot's design scheme, and offering a novel concept for the development of high-performance and self-contained amphibious robots.

Keywords: amphibious robot; undulating fin; compound drive; computational fluid dynamics; track robot; hydrodynamic simulation

1. Introduction

Natural evolution has rendered biological systems far more advanced than many of the capabilities of modern technology [1]. From the observation of nature and its various biological process, researchers were inspired to design efficient bioinspired systems to solve a wild range of problems in a low-cost and sustainable way [2]. Recent innovations of bioinspired advanced systems have led to the development of biomimetic robotics, including bioinspired structures, materials, actuators, sensors, and control methods.

As an important branch, biomimetic amphibious robots have been attracting increasing attention. Compared with conventional mobile robots that are capable of functioning only in a single medium, amphibious robots exhibit remarkable advantages such as multi-terrain motion maneuverability [3], environmental adaptivity [4], and obstacle climbing capacity [5]. Due to these benefits, amphibious robots hold broad prospects in the fields of industry, defense, and transportation [6]. For example, they can be deployed in scenarios such as environment protection [7], resource exploration [8], and search and rescue [9].

Traditional amphibious robots simply integrate propellers and tracks or wheels to realize aquatic and terrestrial locomotion [10,11]. However, the robots are required to switch among different propulsion mechanisms, resulting in a complex control system. Furthermore, the propellers are fragile in muddy and grassy areas, which renders them unsuitable for unstructured environments. In nature, there exist numerous creatures possessing outstanding amphibious locomotion capabilities, and researchers have drawn inspiration from amphibians. The last two decades have witnessed the advancement of bionic amphibious robots, including bionic crabs [12], turtles [13], lobsters [14], beavers [15], frogs [16], snakes [17], and salamanders [18].

Based on the driving mechanism, the existing bionic amphibious robots can be classified into single bionic amphibious robots and compound amphibious robots [6]. Single bionic amphibious robots imitate natural amphibians' organisms and use the same propulsion components for land and underwater locomotion. This category encompasses three types of robots: snake-like, legged, and fish-like. The robotic snake uses the serpentine movement of its body to achieve propulsion on land and underwater [19,20]. However, the bionic robot snake lags far behind in terms of agility and speed when compared to real snakes, even with the assistance of passive wheels. The legged configuration represents a more comprehensive scheme for amphibious robots. Zhang et al. proposed a hexapod amphibious robot, namely AmphiHex-I [21]. The robot is actuated by six reconfigurable legs that can be controlled to switch between C-shaped legs and straight paddles for land and underwater motion, respectively. However, each leg should be equipped with a driven motor for transformation. In the updated version, AmphiHex-II [22], the robot adopts a variable-stiffness leg for lightweight design. The legs are constructed with rigid fan-shaped frames, working as walking legs, and protect the inner flexible flippers, which propel the robot underwater.

Some researchers used smart materials to build amphibious actuators. Zou et al. used a shape-memory polymer and an elastomeric actuator for reconfigurable limbs in a fin-based amphibious robot. The main problem was that the legs should have been transformed manually and reassembled to the robot [23]. Baines et al. developed a bionic turtle robot based on morphing limbs, which could change stiffness and cross-section area and adopted multiple walking and swimming gaits to adapt to different environments [24]. Wu et al. designed a 3D-printed tortoise-inspired soft robot. With four pneumatic bionic legs capable of bending in three dimensions, the robot could achieve six different gaits [25]. Although soft amphibious robots possess the merits of significant deformation and high adaptability, they are still at a considerable distance from practical applications.

In 2017, the American company issued Velox robot (Pliant Energy Systems, Brooklyn, USA) [26], which features a pair of ribbon fins for multi-terrain locomotion. The robot can adapt to various terrains, including ice, sand, beach, water, and seabed via the unified propulsion of undulating fins. Although similar works were conducted for optimal design [27], the undulating fin was relatively slow for land locomotion.

The compound amphibious robot combines various driving methods to meet practical requirements, showing a better balance between terrestrial and aquatic locomotion performance. Wang et al. [28] designed a leg–fin hybrid crab-like hexapod robot, which integrated two-tail swimming fins for underwater paddle flapping. AQUA2, which was developed based on the RHex series robot [29], adopted wheel–fin hybrid legs for amphibious propulsion [30]. The legs utilized a cage-like circular enclosure to protect the inner flippers when moving on the ground, and the cage worked like a wheel. The maximum velocity on land reached 0.9 m/s. Ma et al. proposed a hybrid amphibious robot named SHOALBOT [31], which could carry out multimode operation using a propeller-leg integrated propulsion device. Under low-speed mode, the robot could run as fast as three body lengths/s on land. When underwater, the robot body could be folded and the four propellers provided vectored thrust force to realize multi-DOF swimming. Yu et al. developed a wheel–leg–fin hybrid amphibious fish that could swim via dolphin-like and fish-like oscillations of its multi-joint body, and move fast on land via a wheel–propeller–fin mechanism [32].

From the above analysis, although the single-bionic-drive method does not require switching of propulsion mechanisms, it tends to perform better in only one environment at the expense of locomotion ability in another medium. The compound amphibious robot adopts optimal schemes for land and underwater scenarios, ensuring good propulsion performance in the two media. Nevertheless, due to the integration of distinct propulsion mechanisms, a greater number of drives are needed. More significantly, the robot must switch control systems, which enhances the difficulty of control.

In this paper, we proposed a track–fin-compound-driven amphibious robot, namely AmphiFinbot-II. The robot depends on bilateral undulating fins for underwater propulsion and tracks for land propulsion, demonstrating remarkable motion performance in both environments. The undulating fins and tracks share a common driving source, thereby moving synchronously. The robot is self-adaptive as it does not need to switch drive sources. The robot possesses identical control logics both underwater and on land, such as forward and backward movement, in-place rotation, and steering motion. Therefore, the same control system can be employed. In this paper, we concentrate on verifying the underwater propulsion performance of the robot through the computational fluid dynamics (CFD) method, and validate the amphibious motion function of the robot via experiments.

The rest of the paper is organized as follows: Section 2 introduces the structure design and locomotion principle of the amphibious robot. Section 3 establishes the simulation model and presents the hydrodynamic performance simulation of the robot. In Section 4, locomotion experiments are carried out in terrestrial and aquatic environments. A conclusion is drawn in Section 5.

2. Locomotion Principle

2.1. Robot Structure

The overall design scheme of AmphiFinbot-II is presented in Figure 1, which consists of a robot body, undulating fins, a track mechanism, and a mobile communication cabin. The features of the robot are as follows: (1) Land driving is achieved through the track mechanism, featuring high speed, excellent maneuverability, and terrain adaptability. (2) Underwater driving is driven by bionic undulating fins, which are safer compared with propellers and can adapt to complex environments. (3) When the robot moves on a riverbed or in a water–land transition zone, the tracks and the undulating fins collaborate to provide thrust. (4) The robot body is designed as a rigid structure, facilitating the installation of sensors and manipulators. (5) Remote control is achieved by a movable sealing cabin. When the robot operates on land and in shallow water, the communication cabin is installed on the robot, and the robot employs wireless communication. When the robot is required to carry out deep-water operations, the communication cabin can be placed on the land and then connected to the robot through cables, and the robot utilizes wired communication. (6) The robot is compact and easily portable. We establish the robot's dimensions based on the optimal space for manual operation, incorporating two handles at either end for convenient handling. While a higher chassis enhances obstacle navigation capabilities, it also results in an elevated center of gravity and increased hydrodynamic resistance. Consequently, the final design specifications for the robot are 0.81 m × 0.68 m × 0.22 m.

Despite the utilization of two sets of propulsion mechanisms, through the gear transmission system, the track mechanism and the undulating fin share a common driving source, enabling simultaneous control for the continuous rotation of the track and the wave-like motion of the undulating fin. Through this design, the robot employs different propulsion methods on land and underwater for optimal performance but utilizes the same drive source, eliminating the need for switching propulsion mechanisms.

The schematic diagram of the compound drive is depicted in Figure 2. The working principle is as follows. The motor's output shaft is linked to spur gear A1, and the power is conveyed to the output shaft via spur gears A2 and A3. One end of the output shaft is connected with the undulating fin, which is responsible for generating underwater propulsion force via the wave motion. The other end of the output shaft is coupled to bevel gear A4 and drives vertically installed bevel gear A5 to rotate continuously. The track wheel, which is fixed with bevel gear A5, serves as the driving wheel, and actuates the track, the support wheel, and the guide wheel to rotate, thereby forming a complete-track walking mechanism.

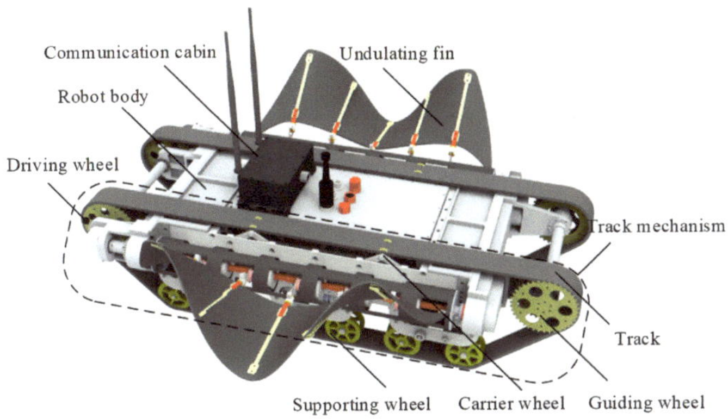

Figure 1. The overall structure of AmphiFinbot-II.

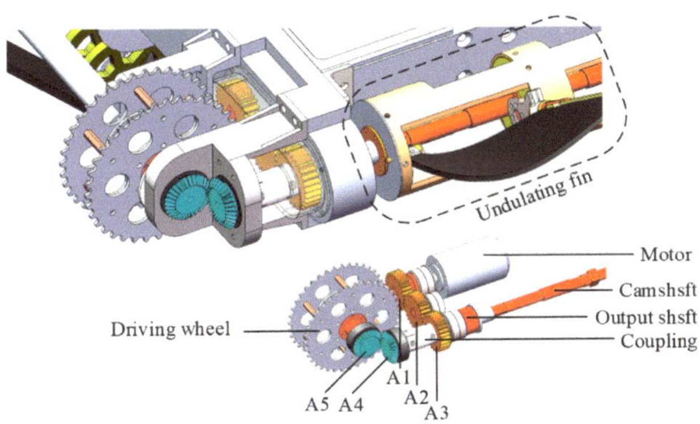

Figure 2. The track–fin compound drive mechanism.

The principle of the modular bionic undulating fin is depicted in Figure 3. Currently, the prevalent approach for undulating-fin-driven robots is to equip each oscillating joint with a drive motor [33–35], which offers the benefit of flexible adjustment of parameters such as wavelength, amplitude, and wave frequency. However, it suffers from several drawbacks, including a large number of motors, complex control, and difficult sealing. More crucially, due to the speed limit of the motor response, the wave frequency cannot be enhanced, and the maximum wave frequency can only reach 3–4 Hz [36]. Hence, we adopt a centralized drive scheme, where a motor drives the camshaft to rotate continuously. The cams connected in parallel to the camshaft installation have a 90-degree phase difference, converting the continuous rotation of the cams into the reciprocating swing of the pendulum rods. Since the swings of the adjacent pendulum rods always have a 90-degree phase difference, a waveform can be formed automatically. By controlling the direction of the camshaft, the transmission direction of the wave can be regulated. Although the waveform of this method is fixed and only the wave frequency and wave deflection angle can be adjusted, its advantages lie in its compact structure, simple control, and reliable sealing, so it is more suitable for field amphibious robots.

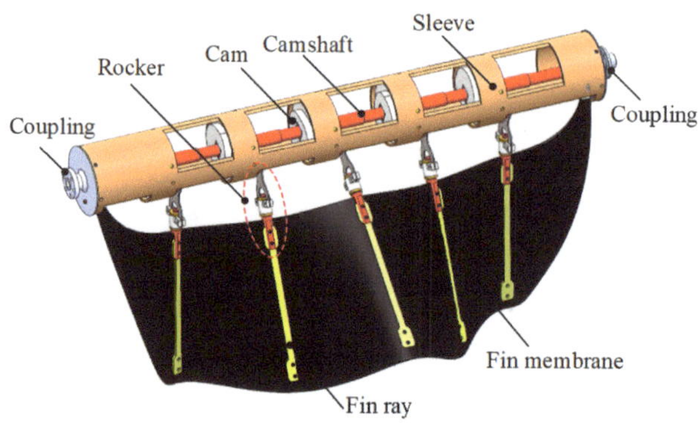

Figure 3. The bionic undulating fin structure.

The overall layout and transmission principle of the robot are depicted in Figure 4, which shows only half of the robot due to symmetry. As explained previously, a speed-regulating motor on the left side is applied for driving the track and the undulating fin. We configure another motor on the right side of the robot, which is also connected to the undulating fin via the gear drive (B1, B2, B3) to control the overall tilting angle of the fin for adjusting the robot's attitude. The robot body is divided into three sealing cabins, i.e., the two motor cabins and the electrical cabin. This design separates the dynamic seal from the static seal to enhance the waterproof performance of the robot.

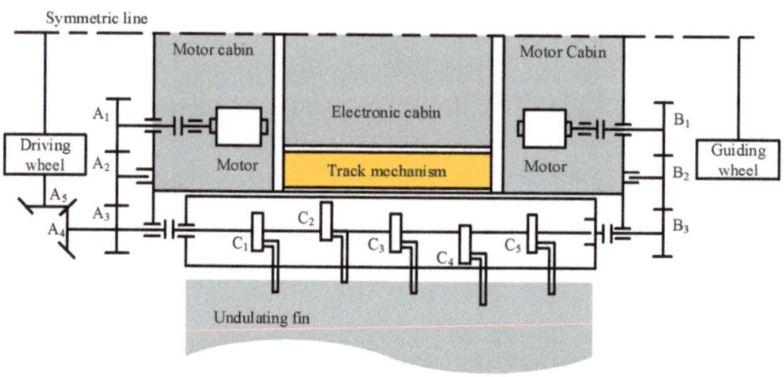

Figure 4. The layout and the transmission principle of AmphiFinBot-II.

2.2. United Operation Mechanism

The existing hybrid amphibious robots operate in underwater, land and land transition environments, often involving distinct moving components and motion mechanisms, thereby necessitating a drive mechanism switch. AmphiFinbot-II employs the same drive source for undulating fins and tracks, making its motion compound and simultaneous. Moreover, the bilateral fins and tracks possess similar differential steering logics, thereby providing AmphiFinBot-II the same maneuvering strategy both underwater and on land, which implies that the robot does not require a switch of mechanical and control systems. The rotational speeds of the left and right tracks are defined as ω_1 and ω_2, with clockwise being positive. The frequencies of the undulating fins are f_1 and f_2, and the underwater propulsion forces generated are F_1 and F_2. The regulations for the robot conducting forward, backward, yaw and in situ steering actions are depicted in Figure 5.

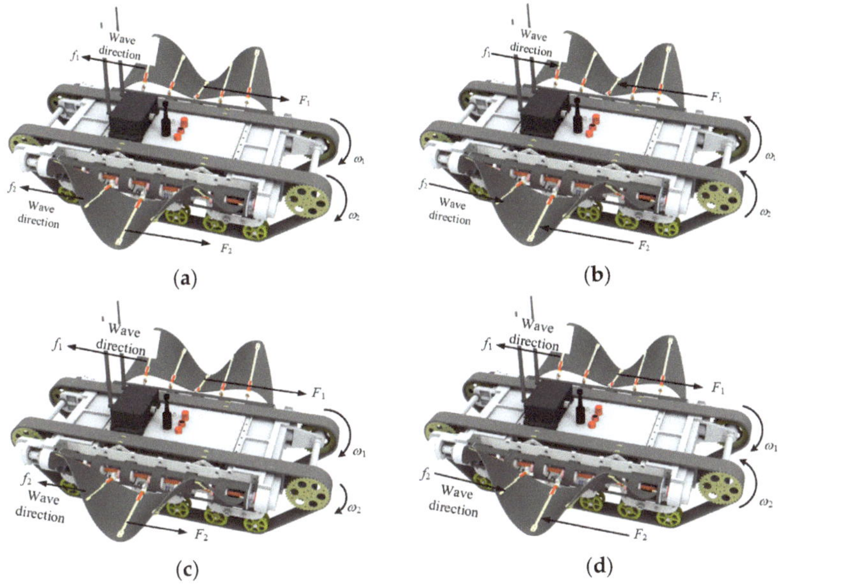

Figure 5. The united operation method of the robot in terrestrial and aquatic environments: (**a**) linear forward motion; (**b**) linear backward motion; (**c**) yaw motion; (**d**) rotation motion.

- Forward movement: When the tracks on both sides rotate clockwise at the same speed ($\omega_1 = \omega_2 > 0$), both tracks offer the same forward speed. Simultaneously, the left and right undulating fins transmit waves to the rear direction with the same frequency ($f_1 = f_2 > 0$), and the undulating fins generate equal forward propulsion force ($F_1 = F_2 > 0$). If underwater, the undulating fins propel the robot to move forward; if on the ground, the tracks drive the robot to move forward.
- Backward movement: As shown in Figure 5b, when the tracks on both sides rotate clockwise at the same speed ($\omega_1 = \omega_2 < 0$), the undulating fins transmit waves to the front direction with the same frequency ($f_1 = f_2 < 0$), and the propulsion force is negative ($F_1 = F_2 < 0$). The robot moves backwards either underwater or on land.
- Yaw motion: Yaw motion can be achieved by regulating the differential rotation of the tracks in the same direction. For example, when $\omega_1 > \omega_2 > 0$, the robot will turn to the right when moving forward on land. The left propulsive force is greater than that on the right side ($F_1 > F_2 > 0$), and the combined force generates additional yaw torque, causing the robot to turn to the right underwater. Similarly, when $\omega_2 > \omega_1 > 0$, the robot performs forward and left-turning locomotion. Due to symmetry, the robot also adopts the same left and right turn control methods when moving backward.
- Rotation motion: Turning in place is a special case of yaw motion, when the two-side tracks' speeds are equal but for opposite directions. As shown in Figure 5d, when $\omega_1 = -\omega_2 > 0$, there is $f_1 = -f_2 > 0$ for undulating fins, so the resultant force is zero. The robot will carry out right-turning motion either on land or underwater. When $\omega_1 = -\omega_2 < 0$, the robot will rotate left.

3. Hydrodynamic Simulation

3.1. Simulation Model

To verify the design feasibility and motion capability of the amphibious robot, the force and motion of the robot in the flow field under various control inputs were analyzed through computational fluid simulation (CFD).

To simplify the grid division and accelerate the simulation calculation, the local particulars of the robot were streamlined. The simulation model was established, as depicted in Figure 6a, and the fluid domain is shown in Figure 6b. The grid of the outer fluid zone was coarsely divided; the robot and the adjacent local area which was shown in color were encrypted. According to the physical prototype's design parameters, the simulation parameters of the robot are given in Table 1.

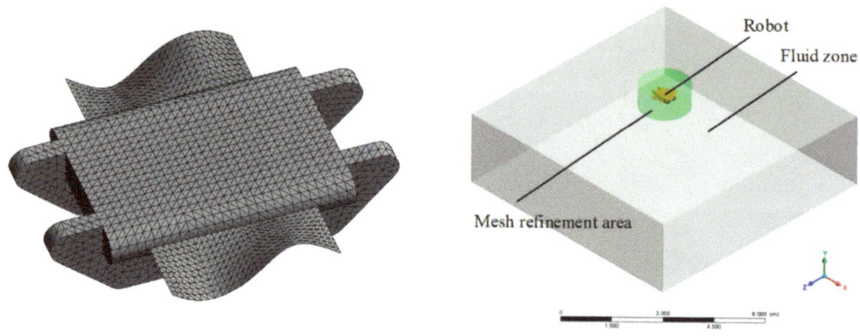

(a) (b)

Figure 6. CFD simulation model: (**a**) the mesh model of the robot; (**b**) the calculation zone.

Table 1. CFD simulation parameters.

Parameter type	Parameters/Unit	Value
Robot body	Body length, L/m	0.775
	Body width, W/m	0.404
	Body height, H/m	0.22
Undulating fin	Fin length, L_f/m	0.38
	Fin width, h/m	0.15
	Fin thickness, t/m	0
	Wave amplitude, $\theta_m/^\circ$	19.8
Control parameters	Left fin frequency, f_1/Hz	0–6
	Right fin frequency, f_2/Hz	0–6
Dynamic parameters	Weight, m/kg	17
	Rotational inertia, J_{xx}/kg·m^2	0.132
	Rotational inertia, J_{yy}/kg·m^2	0.936
	Rotational inertia, J_{zz}/kg·m^2	0.863

3.2. Mathematical Model

3.2.1. Kinematic Model

To describe the motion process of the amphibious robot in the flow field precisely, the kinematic model of the robot and the motion equation of the undulating fin were constructed. The coordinate system during the robot's locomotion was defined, as shown in Figure 7.

- Global coordinate system, $O_eX_eY_eZ_e$: The origin of coordinate O_e was defined by taking any point within the earth; O_eX_e pointed at any horizontal direction, O_eZ_e was vertically downward to the ground, and the direction of O_eY_e was in accordance with the right-hand rule.
- Local coordinate system, $O_bX_bY_bZ_b$: The origin O_b was defined as the mass center of the robot. O_bX_b was along the longitudinal axial direction. O_bZ_b pointed straight down the body. O_bY_b was located to the right of the body.
- The fin coordinate system, $O_fX_fY_fZ_f$: The origin O_f was located at the base point of the undulating fin. O_fX_f pointed forward along the undulating fin's base line. O_fZ_f was in the symmetrical plane of the undulating fin and pointed outwards. O_fY_f was determined by the right-hand rule. The coordinate systems $O_{f1}X_{f1}Y_{f1}Z_{f1}$ and $O_{f2}X_{f2}Y_{f2}Z_{f2}$ were established for the left and right undulating fins, respectively.

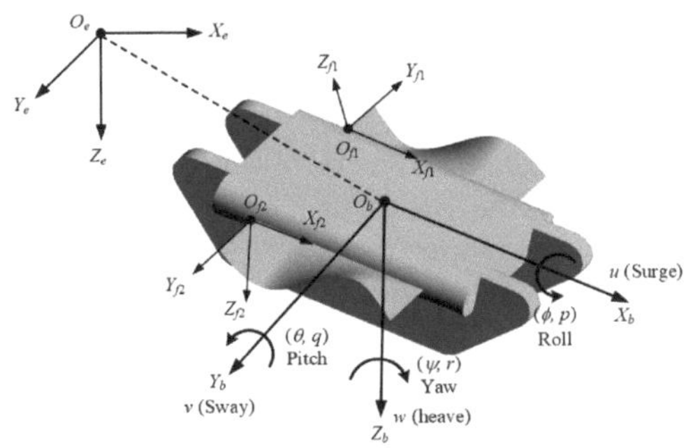

Figure 7. Coordinate system of the robot.

Suppose $\boldsymbol{\eta} = (X, Y, Z, \phi, \theta, \psi)^{\mathrm{T}}$ is the position and attitude vector of the robot in the global coordinate system. $\boldsymbol{\eta}_1 = (X, Y, Z)^{\mathrm{T}}$ is the three-dimensional coordinate, and $\boldsymbol{\eta}_2 = (\phi, \theta, \psi)^{\mathrm{T}}$ is the three-axis attitude angle, corresponding to roll, pitch, and yaw angle. $\boldsymbol{V} = (u, v, w)^{\mathrm{T}}$ and $\boldsymbol{\omega} = (p, q, r)^{\mathrm{T}}$ are the three-axis linear and angular velocities of the robot in the local coordinate system, respectively. The relationship between the local coordinate system and the robot system can be expressed as follows:

$$\begin{cases} \dot{\boldsymbol{\eta}}_1 = \boldsymbol{J}_1(\boldsymbol{\eta}) \cdot \boldsymbol{V}, \\ \dot{\boldsymbol{\eta}}_2 = \boldsymbol{J}_2(\boldsymbol{\eta}) \cdot \boldsymbol{\omega}, \end{cases} \tag{1}$$

where $\boldsymbol{J}_1(\boldsymbol{\eta})$ and $\boldsymbol{J}_2(\boldsymbol{\eta})$ are the rotation matrices, which are solved as follows:

$$\boldsymbol{J}_1(\boldsymbol{\eta}) = \begin{bmatrix} \cos\psi\cos\theta & \cos\psi\sin\theta\sin\phi - \sin\psi\cos\phi & \cos\psi\sin\theta\cos\phi + \sin\psi\sin\phi \\ \sin\psi\cos\theta & \sin\psi\sin\theta\sin\phi + \cos\psi\cos\phi & \sin\psi\sin\theta\cos\phi - \cos\psi\sin\phi \\ -\sin\theta & \cos\theta\sin\phi & \cos\theta\cos\phi \end{bmatrix}, \tag{2}$$

$$\boldsymbol{J}_2(\boldsymbol{\eta}) = \begin{bmatrix} 1 & \tan\theta\sin\phi & \tan\theta\cos\phi \\ 0 & \cos\phi & -\sin\phi \\ 0 & \sec\theta\sin\phi & \sec\theta\cos\phi \end{bmatrix}, \tag{3}$$

The motion law of the undulating fin is described in the fin coordinate system. Let the coordinates of any point on the fin surface in the $O_f X_f Y_f Z_f$ system be $P\left(x_f, y_f, z_f\right)$. When the initial phase of the left and right undulating fins is zero and the waveform is completely symmetrical, the undulation equations are as follows:

$$\text{Left}: \begin{cases} x_f = s, \\ y_f = r \cdot \cos\left[\theta_{\mathrm{m}} \sin\left(\delta_1 \cdot 2\pi f_1 + \frac{2\pi}{\lambda} s\right)\right], \\ z_f = r \cdot \sin\left[\theta_{\mathrm{m}} \sin\left(\delta_1 \cdot 2\pi f_1 + \frac{2\pi}{\lambda} s\right)\right], \end{cases} \tag{4}$$

$$\text{Right}: \begin{cases} x_f = s, \\ y_f = r \cdot \cos\left[\theta_{\mathrm{m}} \sin\left(\delta_2 \cdot 2\pi f_2 + \frac{2\pi}{\lambda} s\right)\right], \\ z_f = r \cdot \sin\left[\theta_{\mathrm{m}} \sin\left(\delta_2 \cdot 2\pi f_2 + \frac{2\pi}{\lambda} s\right)\right], \end{cases} \tag{5}$$

where s is the natural coordinate in the baseline direction of the undulating fin, r is the natural coordinate in the fin direction, f_1 and f_2 are the left and right undulating frequencies, respectively, λ is the wavelength, θ_{m} is the wave amplitude, and δ_1 and δ_2 represent the wave transmission direction. When $\delta_i = -1$, the wave is propagated in the $-x$ direction; when $\delta_i = 1$, the wave is propagated in the $+x$ direction.

Let the origin of the left and right undulating fins in the local coordinate system be $O_{f1}(a, -b, 0)$ and $O_{f2}(a, b, 0)$, respectively. a and b are determined by the robot geometric shape.

The angles between the undulating fin and the body are φ_1 and φ_2. According to the rotation relationship, the expression of the fin surface in the body coordinate system is as follows:

$$\text{Left}: \begin{bmatrix} x_b \\ y_b \\ z_b \end{bmatrix} = \begin{bmatrix} 1 & 0 & 0 \\ 0 & \cos\varphi_1 & -\sin\varphi_1 \\ 0 & \sin\varphi_1 & \cos\varphi_1 \end{bmatrix} \begin{bmatrix} x_f \\ y_f \\ z_f \end{bmatrix} + \begin{bmatrix} a \\ -b \\ 0 \end{bmatrix}, \tag{6}$$

$$\text{Right}: \begin{bmatrix} x_b \\ y_b \\ z_b \end{bmatrix} = \begin{bmatrix} 1 & 0 & 0 \\ 0 & \cos\varphi_2 & -\sin\varphi_2 \\ 0 & \sin\varphi_2 & \cos\varphi_2 \end{bmatrix} \begin{bmatrix} x_f \\ y_f \\ z_f \end{bmatrix} + \begin{bmatrix} a \\ b \\ 0 \end{bmatrix}. \tag{7}$$

Furthermore, the coordinates of point $P(x_b, y_b, z_b)$ within the geographical system $P(x_e, y_e, z_e)$ are determined as follows:

$$\begin{bmatrix} x_e \\ y_e \\ z_e \end{bmatrix} = J_1(\eta) \begin{bmatrix} x_b \\ y_b \\ z_b \end{bmatrix} + \begin{bmatrix} X \\ Y \\ Z \end{bmatrix}. \tag{8}$$

In CFD simulation, the deformation motion of the undulating fins is defined by Equations (4) and (5). Through Equations (6)–(8), the coordinate of any points on the fin surface can be transformed between the global coordinate system, $O_eX_eY_eZ_e$, and the fin coordinate system, $O_fX_fY_fZ_f$. This kinematic relationship facilitates the definition and updating of dynamic mesh in fluid simulation analysis.

3.2.2. Underwater Dynamic Model

The mass of the robot is defined as m, and the inertia tensor is $\boldsymbol{J} \in \boldsymbol{R}^{3\times3}$. The moment of inertia of the spindle is J_x, J_y and J_z. The three-dimensional force and moment of the robot underwater in the local robot coordinate system are expressed as $\boldsymbol{F} = (F_x, F_y, F_z)^{\mathrm{T}}$ and $\boldsymbol{M} = (M_x, M_y, M_z)^{\mathrm{T}}$. The centroid dynamic equation of the robot is as follows:

$$\begin{cases} \boldsymbol{F} = \frac{m\mathrm{d}\boldsymbol{V}}{\mathrm{d}t} = m\boldsymbol{V}' + \boldsymbol{\omega} \times m\boldsymbol{V}, \\ \boldsymbol{M} = \frac{\mathrm{d}\boldsymbol{\omega}}{\mathrm{d}t} = \boldsymbol{J}\boldsymbol{\omega}' + \boldsymbol{\omega} \times \boldsymbol{J}\boldsymbol{\omega}. \end{cases} \tag{9}$$

Substitute the velocity and angular velocity components; we can obtain the acceleration of each direction.

$$\begin{cases} \dot{u} = vr - wq + \frac{F_x}{m} \\ \dot{v} = -ur + wp + \frac{F_y}{m}, \\ \dot{w} = uq - vp + \frac{F_z}{m} \end{cases} \tag{10}$$

$$\begin{cases} \dot{p} = \frac{qr(J_y - J_z)}{J_x} + \frac{1}{J_x}M_x \\ \dot{q} = \frac{pr(J_z - J_x)}{J_y} + \frac{1}{J_y}M_y. \\ \dot{r} = \frac{pq(J_x - J_y)}{J_z} + \frac{1}{J_z}M_z \end{cases} \tag{11}$$

When the force and the moment of the robot are solved, the acceleration of the robot can be obtained using Equations (10) and (11), and the linear and angular velocity can be obtained via the integral. Through rotation matrix transformation, the velocity, angular velocity, and displacement of the robot in the global coordinate system can be solved. Therefore, the hydrodynamic and motion performance of the robot in different swimming modes can be simulated.

3.3. Simulation Result

3.3.1. Surge Motion

When $f_1 = f_2$ and $\delta_1 = \delta_2$, the undulating fins on both sides propagate in the same direction with equal frequency. At this time, the two undulating fins produce forward or backward thrust at the same time, and the robot moves forward or backward linearly. Firstly, we simulated the in situ undulating motion of the robot under the condition of a fixed carrier. The change in the thrust force of the undulating fins under different frequencies is shown in Figure 8a, and the mean force is presented in Figure 8b. The propulsion force changed periodically with time, and the frequency of change was twice the frequency of the undulating fin. The higher the wave frequency, the greater the fluctuation amplitude and mean value of the propulsion force. The thrust generated by the undulating fin was proportional to the square of the wave frequency, and the direction of the thrust was opposite to the direction of the wave. Therefore, when the undulating fin transmitted the wave forward or backward, the generated force and motion were symmetrical.

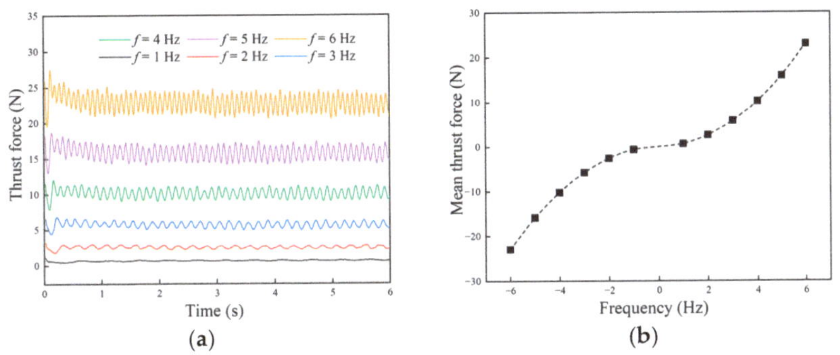

Figure 8. Simulated thrust force of undulating fins in fixed-body mode: (**a**) thrust force changing with time; (**b**) mean force at different frequencies.

Given $f_1 = f_2 = 2$ Hz, the linear motion of the robot in the fluid was simulated. The animation video can be found in the Supplementary Materials, and the pressure cloud image sequence of the robot is shown in Figure 9. The different colors represented the velocity in the fluid field. The result showed that the joint force of the robot was forward, and the robot moved linearly without lateral and steering motion. The speed and displacement curves of the robot are shown in Figure 10a. The robot passed through a stage of acceleration, then reached the maximum speed and maintained a constant speed, and the steady-state average speed was about 0.31 m/s.

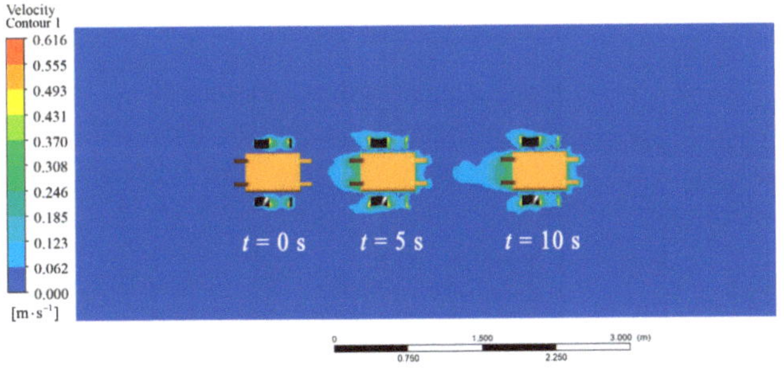

Figure 9. Velocity cloud image sequence of the amphibious robot in linear motion simulation.

The thrust and fluid resistance suffered by the robot during movement are shown in Figure 10b. The results show that the thrust of the undulating fin was the largest in the initial stage, and then the thrust decreased gradually with the increase in the robot's speed. At a stable stage, the thrust of the undulating fin also fluctuated around the average value.

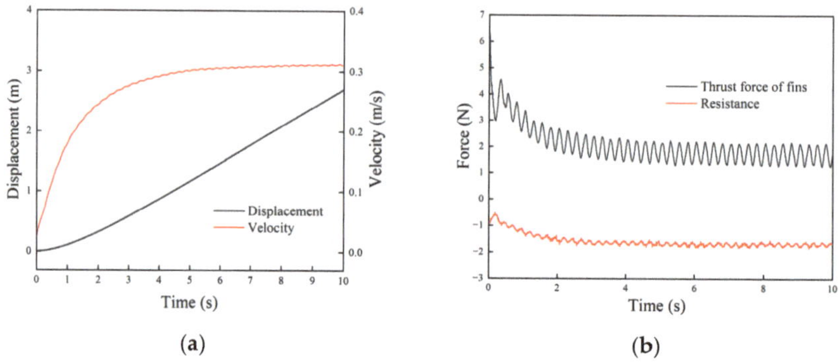

Figure 10. Simulation results of the amphibious robot in linear motion: (**a**) velocity and displacement; (**b**) thrust force and resistance.

According to the simulation result, the mean thrust force of the undulating fin can be simplified as follows:

$$F_T = \text{sign}(f)C_T(\lambda f - V_x)^2, \tag{12}$$

where C_T is the thrust coefficient, V_x is the linear velocity, and $\text{sign}(f)$ is the symbolic function.

The resistance of the robot can be expressed as follows:

$$F_D = -\text{sign}(V_x)\frac{1}{2}\rho S C_D V_x^2, \tag{13}$$

where ρ is the fluid density, S is the cross-sectional area of the carrier, and C_D is the resistance coefficient. In the steady stage, the resultant force is zero, which means $F_T + F_D = 0$. Therefore, the steady-stage velocity is solved as follows:

$$V_x = \frac{\lambda f}{1 + \sqrt{\frac{\rho S C_D}{2 C_T}}}, \tag{14}$$

The propulsion speed of the robot at different frequencies was simulated, and the results are shown in Figure 11. The steady-state average speed of the robot increased linearly with the wave frequency. The higher the frequency, the greater the starting acceleration, and the shorter the time required to reach the maximum speed. The simulation result was consistent with the law of Equation (14).

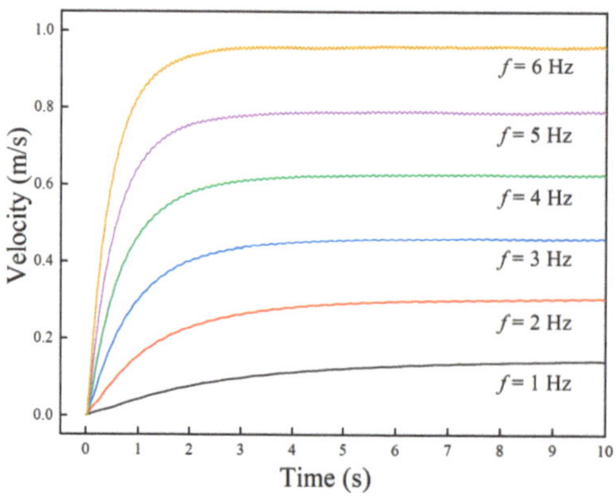

Figure 11. Propulsive velocity of the amphibious robot under different frequencies.

3.3.2. Steering Motion

The motion performance of the robot in steering mode was simulated. Given the left and right frequencies of f_1 = 2.5 Hz and f_2 = 2 Hz, the robot was supposed to deflect to the right side while moving forward. The trajectory of the robot obtained via simulation is shown in Figure 12a. The robot maintained steering during the movement, and passed through an arc trajectory in the *xz* plane. The tangential of any point on the trajectory corresponds to the combined speed directions of the robot.

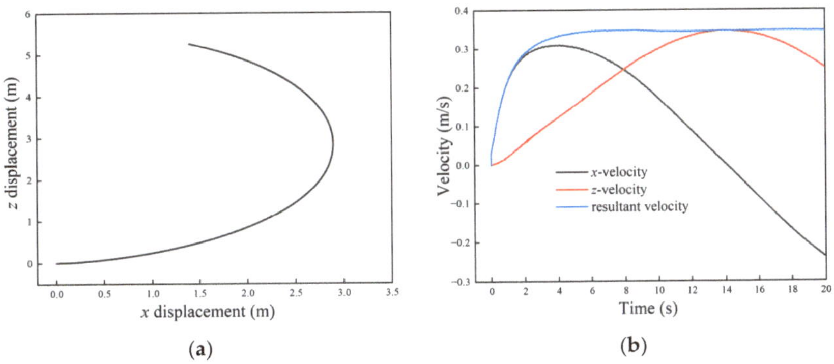

Figure 12. Velocity and displacement of the amphibious robot in steering motion simulation: (**a**) displacement in *x* and *z* direction; (**b**) velocity in *x* and *z* direction.

The sub-velocity in the *x* and *z* directions and the combined velocity of the robot are shown in Figure 12b. The *x*-velocity increased first and then decreased until it became negative. The velocity in the *z*-axis increased first and then decreased, corresponding to the two stages of the robot's positive and negative motion in the *x* direction. When t = 14 s, the *x*-velocity was zero, and the corresponding *z*-velocity was the largest. The total velocity of the robot reached the maximum value at about 6 s, and then remained stable at about 0.35 m/s.

The motion posture and spatial position of the robot are shown in Figure 13 (the animation video can be found in the Supplementary Materials). The result showed that the velocity of the flow field was small at the beginning. With the drive of the undulating fins, the velocity obtained by the robot from the fluid gradually increased and tended to be stable. Under the asymmetric waveform, the robot successfully realized the steering maneuver movement. The simulation results were consistent with the theoretical analysis.

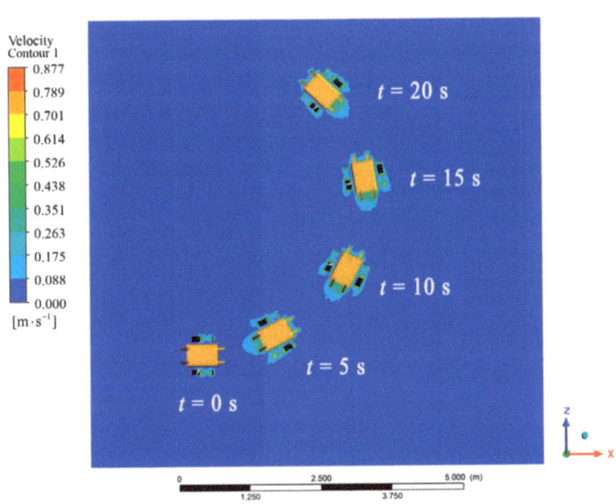

Figure 13. Velocity cloud image sequence of the robot in steering motion simulation.

3.3.3. Rotation Motion

The in situ rotation mode is a special case of steering motion, in which case the wave of the left and right fins transmit in the opposite direction. The thrust forces cancel each other out and form a rotating torque around the center of the robot. The frequencies are given as $f_1 = f_2 = 2$ Hz and $\delta_1 = -\delta_2 = 1$, and the motion of the robot's right turn in situ is simulated. The yaw angle and speed are shown in Figure 14a. The results show that the yaw angular velocity experienced two stages: starting acceleration and stable maintenance. The yaw angle gradually increased with time, and the oblique rate increased first and then remained unchanged, which corresponded to the regulation of angular velocity. The steady-state average yaw rate was about -0.58 rad/s.

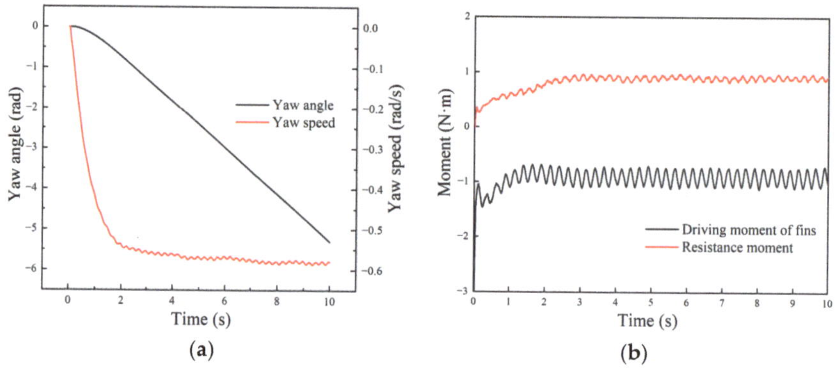

Figure 14. Motion and hydrodynamic moment of the robot in rotation motion simulation: (**a**) yaw angle and speed; (**b**) driven moment and resistance moment.

During the rotation motion, the relationship between the driving moment provided by the undulating fins and the fluid resistance moment on the carrier over time is shown in Figure 14b. The results showed that the driving moment was largest in the initial stage, then gradually decreased and eventually fluctuated around the stable value. The change regulation of the resistance moment was opposite. The resistance increased with the rotation speed. During the stable stage, the driving moment and the resistance were balanced, and the robot conducted uniform steering motion. We can find that the rotation motion presented similar regulations with the linear motion, which corresponded to the results of the dynamic analysis of the undulating fins and the theoretical analysis of the fluid resistance.

The velocity cloud image sequence of the rotation motion process is shown in Figure 15, which showed that the robot realized continuous rotation motion around the center of mass, and the position of the robot did not change, indicating that the robot only rotated in situ without translational motion. The animation video can be found in the Supplementary Materials.

Further, the angular velocity of the robot at different undulating frequencies was simulated, and the results are shown in Figure 16. The stable angular velocity of the robot increased linearly with the wave frequency. The higher the frequency, the greater the starting acceleration, and the shorter the time required to reach the maximum angular velocity. When the frequency was 6 Hz, the maximum rotation angular velocity reached -1.75 rad/s.

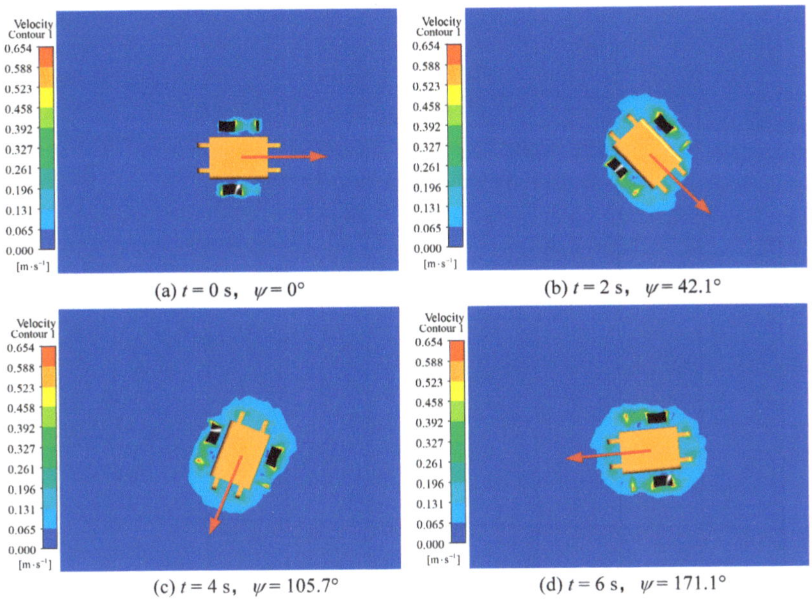

Figure 15. Velocity cloud image sequence of the robot in the rotation motion simulation: (**a**) t = 0 s; (**b**) t = 2 s; (**c**) t = 4 s; (**d**) t = 6 s.

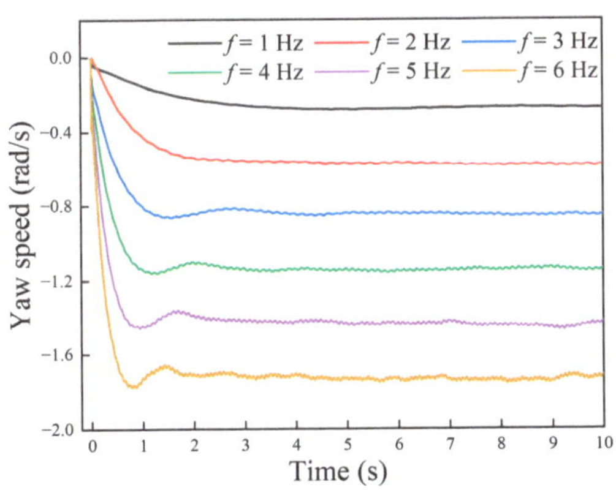

Figure 16. Simulated underwater rotation speed of the robot under different frequencies.

4. Experiment Validation

4.1. Robot Prototype

To verify the robot's hydrodynamic performance in the CFD simulation, a robot prototype was developed for physical experiments. We integrated hardware circuits and control software for the robot. The hardware scheme is shown in Figure 17a. The master control board was based on an STM32F407 microchip (ST, Geneva, Switzerland), featuring an abundant peripheral interface. The control software was programmed on the Keil MDK5.14 platform. The control tasks included motor control, sensor data acquisition, wireless communication, and human–computer interaction. The drive motors adopted M3508 (DJI, Shenzhen, China) and the motor MCU communicated with the control board through the CAN bus. The two steering engines were controlled by a PWM signal, which could change the deviation angle between the undulating fin and the robot body. The robot was equipped with various sensors, including water depth sensors, inertial measurement units (IMUs), power meters, and leakage detection sensors. The sensor data were read through IIC, UART and other interfaces. The robot adopted wireless communication. We assigned two serial ports to connect the receiver and the data telemetry, respectively. The robot could be controlled either by handheld remote control (RadioLink, Shenzhen, China) or by input commands from a portable computer. The former method was ideal for field operation, while the latter was suitable for data monitoring and program debugging. The robot was powered by a lithium battery (24 V, 6000 mAh).

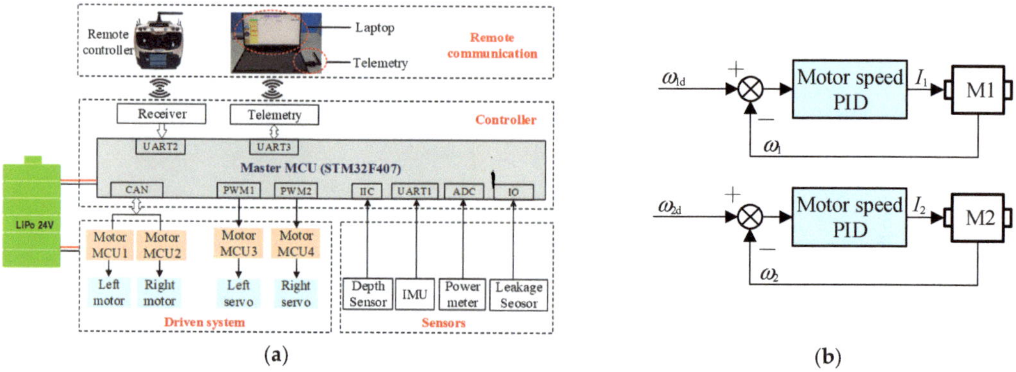

Figure 17. Hardware and control algorithms: (**a**) hardware scheme; (**b**) PID controller.

At the present stage, we primarily focused on the speed control of the robot, with the control principle illustrated in Figure 17b. The control board received the frequency command for the undulating fins and translated them into the desired angular velocity for each motor. Subsequently, a PID control algorithm was applied to regulate the motor speed. The actual speeds were transmitted by the motor driver via the CAN bus at a frequency of 1000 Hz. The master MCU then compared the target speed with the measured speed, executing separate PID controls for each motor and calculating the control current to be sent to the motor driver through the CAN bus. By managing the speeds of motors on either side, the robot could effectively conduct multimodal motion both on land and underwater.

4.2. Underwater Maneuverability

4.2.1. Underwater Linear Motion

To verify the robot's hydrodynamic and motion performance in the CFD simulation, experiments were conducted in outdoor aquatic environments based on the robot prototype. The underwater experiments video can be found in the Supplementary Materials. The linear motion experiment image sequence of the robot when the undulating frequency was 3 Hz is presented in Figure 18. The results prove that by controlling the wave direction of the undulating fins, the robot can generate vector thrust, thereby flexibly achieving forward and backward movements. In addition, the robot can maintain a relatively fixed heading angle.

The maximum velocity that the robot could achieve at different undulating frequencies was tested, and the result is shown in Figure 19. The squares were the measured velocity and the dotted line was the regression model. The curve illustrates that the maximum velocity was improved with the increase of in wave frequency, and is approximately linear to the frequency, which is consistent with the simulation results. When the frequency was 6 Hz, the speed of the robot was 0.54 m/s. As the length of the robot was 0.81 m, the maximum speed was 0.67 BL/s.

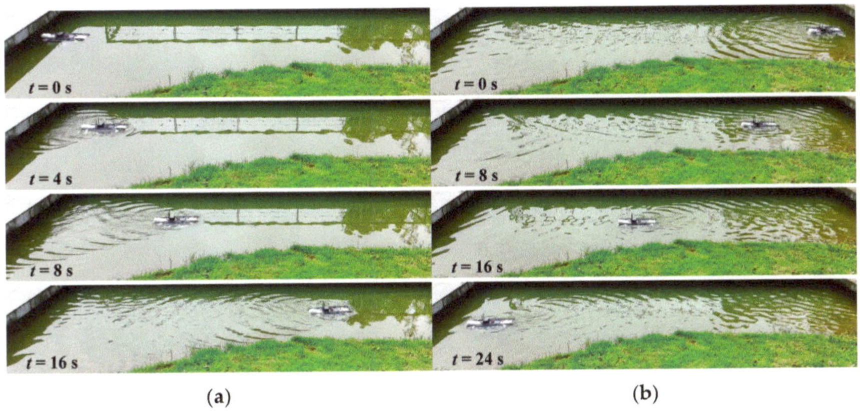

Figure 18. Experiment results of underwater linear forward and backward motion at 3 Hz: (**a**) forward motion; (**b**) backward motion.

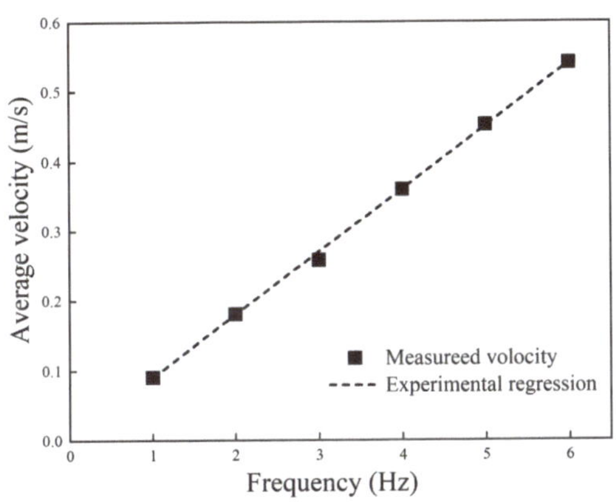

Figure 19. Experiment result of underwater propulsive velocity at different frequencies.

4.2.2. Underwater Steering and Rotation Motion

Given the undulating frequencies of f_1 = 3 Hz and f_2 = 2 Hz, the motion image of the robot is shown in Figure 20a, and the change in the yaw angle is shown in Figure 20b. The experimental phenomenon was consistent with the simulation analysis. The robot continuously deflected to the right side while moving forward. After about 4 s, the angular speed remained constant at about 13.5°/s. When the stop command was issued at 15 s, the robot continued to deflect at a certain angle due to inertia.

When the undulating frequencies were set as $f_1 = f_2$ = 3 Hz and the directions were opposite, the robot conducted in situ rotation locomotion. The motion image of the robot is shown in Figure 21a, and the yaw angle is shown in Figure 21b. The results indicated that the robot performed a continuous left turn movement and the position of the center of gravity remained unchanged. The robot initially accelerated until the driving moment was balanced with the resistance moment, at which time the robot attained the maximum speed. In the steady stage, the robot subsequently rotated at a constant speed of about 45°/s.

Further, the maximum rotation speed that the robot could achieve under different undulating frequencies was measured, and the result is shown in Figure 22. The average rotation speed of the robot was approximately linear to the increase in frequency, which is consistent with the theoretical analysis and fluid simulation result. When the undulating frequency was 6 Hz, the rotation speed of the robot could reach 84°/s.

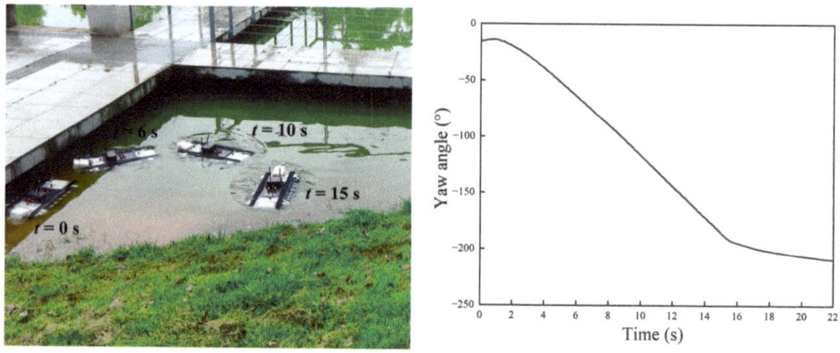

(a) (b)

Figure 20. Experiment results of underwater steering motion: (**a**) motion images; (**b**) yaw angle.

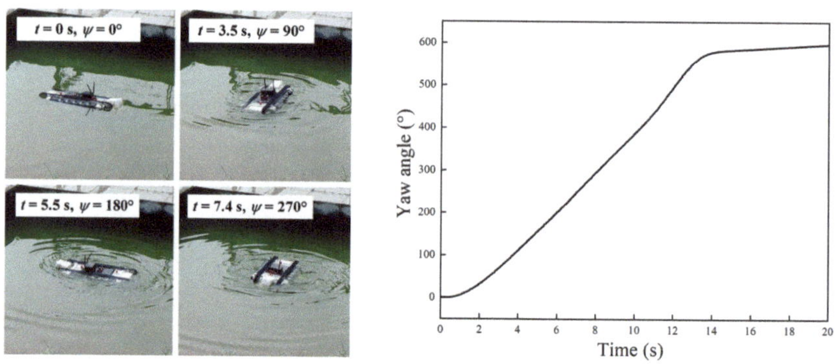

(a) (b)

Figure 21. Experiment results of underwater rotation motion. (**a**) motion images; (**b**) yaw angle.

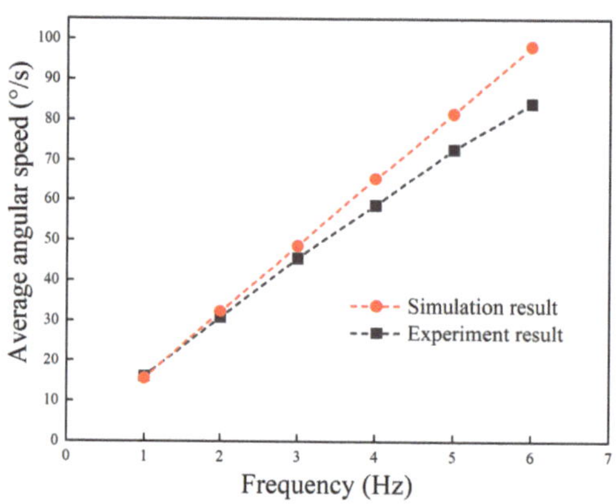

Figure 22. Experiment result of underwater rotation speed at different frequencies.

According to the experiment results, both the linear and rotation speed were relatively slower than those in the simulation results. This may be because the resistance force and moment of the real prototype were larger than those in the simulation model. The shape of the robot prototype was not continuous, while the simulation model was simplified for the convenience of grid division and calculation accuracy. Furthermore, the waveform of the mechanical undulating fin was fitted by the finite fin rays, and this fit was not perfect compared with that of the theoretical model in the simulation environment, thus leading to smaller driving force.

4.3. *Terrestrial Maneuverability*

4.3.1. Terrestrial Linear Motion

Through the previous analysis, the robot has the same control strategy on land and underwater, and we tested the robot's motion performance on land. The terrestrial locomotion video can be found in the Supplementary Materials. Setting the tracks' speed as $f_1 = f_2 = 3$ Hz ($\omega_1 = \omega_2 = 180$ r/min), and given that the directions were the same, the forward motion images of the robot are shown in Figure 23a.

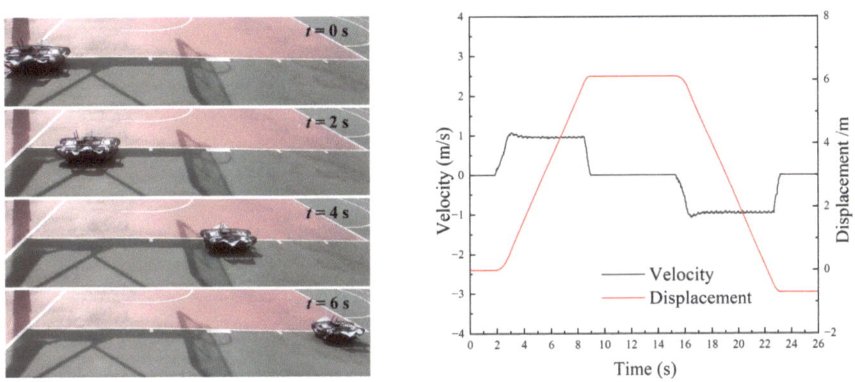

(a) (b)

Figure 23. Experiment results of land linear motion: (**a**) motion images; (**b**) velocity and displacement.

We utilize track speed and yaw angle as odometric parameters and calculate the robot's velocity and planar displacement. The centroid velocity of the robot is computed as follows:

$$V = (\omega_1 + \omega_2)R/2, \tag{15}$$

where R is the radius of the driven wheel.

By decomposing the displacement in each time step, the total displacement of the robot at time T in the x-axis and y-axis directions can be obtained:

$$\begin{cases} \mathrm{d}s = V\mathrm{d}t, \\ \mathrm{d}x = \mathrm{d}s\cos\psi \Rightarrow S_x = \int_0^T \mathrm{d}x, \\ \mathrm{d}y = \mathrm{d}s\sin\psi \Rightarrow S_y = \int_0^T \mathrm{d}y, \end{cases} \tag{16}$$

The velocity and displacement curves in linear motion are shown in Figure 23b. The initial phase of the robot experienced a period of acceleration until the motor reached the preset speed. Subsequently, the robot maintained a constant speed in a straight line. When commanded to stop, the robot braked promptly. After that, upon receiving the backward command, the robot moved backward to the starting position in a straight line at the same frequency.

Given different speeds of the track, the average speed of the robot walking on land was tested, and the results are shown in Figure 24. The black line is the speed measured by the external laser distance meter, and the red line is the theoretical speed calculated by the wheel odometer. The results showed that the speed of the robot was proportional to the motor speed, and the test speed was almost consistent with the theoretical speed, indicating that the robot had no slip with the ground and the motor control accuracy was excellent. Using the speed of motor feedback as a vehicle odometer would have had satisfactory accuracy. When the driving frequency was 7 Hz (the track speed was 420 r/min), the propulsion speed of the robot reached 2.26 m/s, corresponding to 2.8 BL/s.

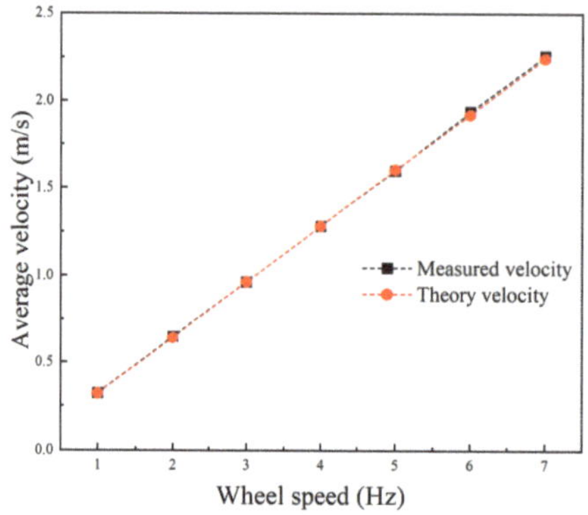

Figure 24. Experimental propulsion velocity on land at different wheel speeds.

4.3.2. Terrestrial Steering and Rotation Motion

The yaw motion on land was identical to the underwater case. When the left and right tracks rotated at different speeds, the robot would deviate to the lower-velocity side due to kinematic constraints. The steering locomotion experiment result is shown in Figure 25a, in which case we set $f_1 = 3$ Hz, $f_2 = 2$ Hz. The planar displacement of the robot is presented in Figure 25b, where the black curve represented the continuous position and the arrows indicated the velocity direction. The experimental result was consistent with the theoretical analysis. The robot conducted continuous right steering. The trajectory of the robot in the steady-state process was approximately a circle, which meant that there existed a fixed instantaneous center of rotation. Similarly, when the speed on the right side was larger, the robot carried out a left turn movement. By regulating the speeds of the left and right tracks, the robot could be controlled to rotate around any point.

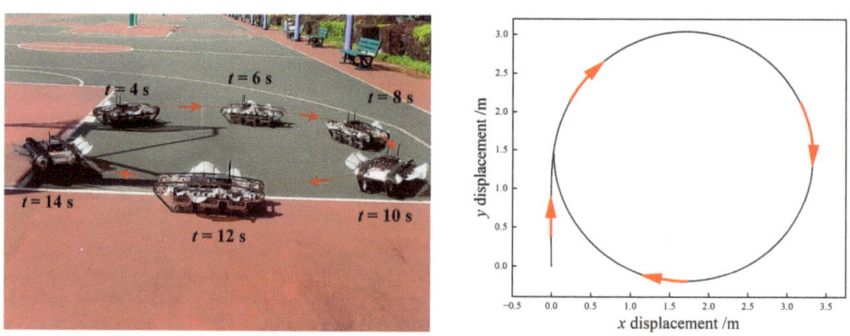

(a) (b)

Figure 25. Experiment results of land steering motion. (**a**) motion images; (**b**) displacement.

When the left and right tracks rotate in the opposite direction with the same frequency, the instantaneous rotation center falls on the geometric center of the robot, so the robot will rotate around itself. Given the left and right tracks frequencies of $f_1 = f_2 = 3$ Hz in opposite directions, the image sequence of the robot is shown in Figure 26a. The results showed that the position of the robot center was almost unchanged during the movement. As the motor started to accelerate, the yaw angular velocity gradually increased and finally stabilized at the maximum angular velocity. The rotation speed of the robot was 212°/s in the steady state.

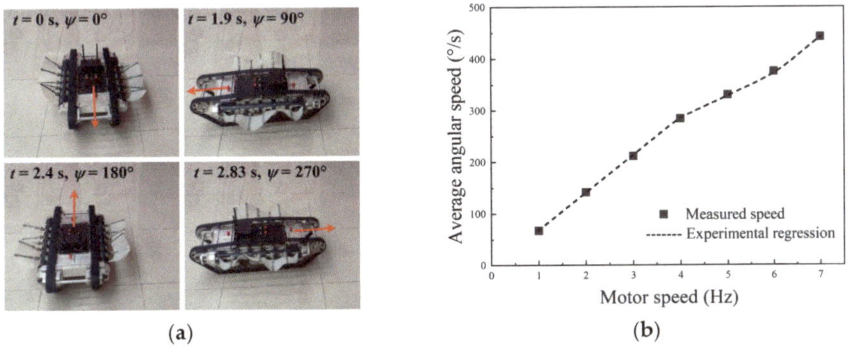

Figure 26. Experiment results of land rotation motion: (**a**) motion images at f = 2 Hz; (**b**) angular speed at different frequencies.

We tested the angular velocity of the robot at various differential frequencies, and the result is presented in Figure 26b. The square dots were the measured angular speed and the line was the regression model. As the track speed escalated, the robot's angular speed was significantly enhanced. When the track frequency was 7 Hz, the maximum speed reached 442°/s. The result also indicated that the angular speed was not linearly correlated to the track speed. However, within the range of 1–4 Hz, it could be approximately assumed that the steering speed was proportional to the frequency, facilitating robot motion analysis and trajectory control at medium and low speeds.

5. Conclusions

Conventional robots face challenges such as substantial size, excessive noise, limited adaptability to environments, and poor cross-media motion ability. To enhance their capability in executing tasks within complex and unstructured environments, it is essential to draw inspiration from the wide array of species that inhabit oceans, lakes, and river banks. Deriving inspiration from nature leads to a bioinspired system, which holds the potential to solve the technical challenges faced by traditional structures and actuators in existing robots. In this paper, we present a novel bionic amphibious robot that takes inspiration from stingrays. The robot was equipped with a pair of bionic undulating fins that propelled it in the water as efficiently as a fish. Furthermore, it also exhibited exceptional maneuverability and speed on land via the compound drive of tracks. We employed fluid simulation technology to accurately replicate the multi-modal motion of the robot underwater, thereby establishing a theoretical foundation and providing feasibility validation for subsequent prototype development. Through the prototype experiment, the performance of the robot's amphibious locomotion was demonstrated, which verified the robot's advancements in self-adaptive drive and self-organization control. The main conclusions are as follows.

(1) A biologically inspired amphibious robot was designed. Through a compound drive mechanism, a pair of bionic undulating fins and a pair of tracks were parallelly equipped on the robot, which were responsible for efficient locomotion underwater and on land, respectively. Sharing a same driving source enabled autonomous switching of the track and the fin, along with a unified motion control strategy both on land and underwater. Therefore, the robot did not need to judge in which kind of

environment it was, as the robot's locomotion and control principles such as forward, backward, and turning were the same in various environments.

(2) Based on the kinematics and dynamics model of the robot as well as the motion equation of the undulating fins, the hydrodynamic and motion performance of the robot under linear motion, steering motion, and in situ rotation motion were simulated using dynamic mesh method. The simulation results showed that the robot could generate vector thrust through the wave-like motion of the undulating fins. Both the linear swimming speed and the turning speed achieved by the robot were proportional to the wave frequency.

(3) The prototype experiment validated the amphibious motion performance of the robot. Using the same motion control strategy, the robot was capable of achieving unified forward and backward, steering, and in-place rotation locomotion both on land and underwater. The maximum linear speed and steering speed on land were 2.26 m/s (2.79 BL/s) and 442°/s, respectively. The maximum linear speed and steering speed underwater were 0.54 m/s (0.67 BL/s) and 84°/s, respectively.

Due to its remarkable mobility and high speed in both aquatic and terrestrial environments, the proposed amphibious robot will have broad application prospects, such as search and rescue, field reconnaissance, resource exploration, and environmental monitoring. In the future, we will focus on developing the robot's accurate position control and path planning functions.

Supplementary Materials: The following supporting information can be downloaded at https://www.mdpi.com/article/10.3390/biomimetics9100580/s1: Video S1: simulation_animation.mp4. Video S2: experiment_video.mp4.

Author Contributions: Conceptualization, M.X. and Z.L. (Zirong Luo); methodology, M.X.; software, M.X. and Q.Y.; validation, Q.Z., Z.L. (Zhongyeu Lu) and Y.Z.; formal analysis, M.X.; investigation, Q.Z.; resources, Z.L. (Zhongyeu Lu); data curation, Y.Z.; writing—original draft preparation, M.X.; writing—review and editing, Z.L. (Zirong Luo) and Y.Z.; visualization, M.X.; supervision, Z.L. (Zirong Luo); project administration, Z.L. (Zirong Luo); funding acquisition, Z.L. (Zirong Luo) and Q.Y. All authors have read and agreed to the published version of the manuscript.

Funding: This research was funded by National Natural Science Foundation of China, grant number 52105289 and 52475240, and Natural Science Foundation of Hunan Province of China, grant number 2023JJ40048.

Institutional Review Board Statement: Not applicable.

Data Availability Statement: The data presented in this study are available on request from the corresponding author upon reasonable request.

Conflicts of Interest: The authors declare no conflicts of interest.

References

1. Hammond, M.; Cichella, V.; Lamuta, C. Bioinspired Soft Robotics: State of the Art, Challenges, and Future Directions. *Curr. Robot. Rep.* **2023**, *4*, 65–80. [CrossRef]
2. Arena, P.; Bucolo, M.; Buscarino, A.; Fortuna, L.; Frasca, M. Reviewing Bioinspired Technologies for Future Trends: A Complex Systems Point of View. *Front. Phys.* **2021**, *9*, 750090. [CrossRef]
3. Wu, J.; Wu, M.; Chen, W.; Wang, C.; Xie, G. Multi-Modal Soft Amphibious Robots Using Simple Plastic Sheet-Reinforced Thin Pneumatic Actuators. *IEEE Trans. Robot.* **2024**, *40*, 1874–1889. [CrossRef]
4. Guetta, O.; Shachaf, D.; Katz, R.; Zarrouk, D. A Novel Wave-like Crawling Robot Has Excellent Swimming Capabilities. *Bioinspir. Biomim.* **2023**, *18*, 026006. [CrossRef]
5. Yin, Q.; Xia, M.; Luo, Z.; Shang, J. Adaptive Obstacle Climbing and Hydrodynamic Performance Analyses of the Amphibious Robot with Wheels and Flexible Undulating Fins. *Proc. Inst. Mech. Eng. Part C J. Mech. Eng. Sci.* **2022**, *236*, 5300–5317. [CrossRef]
6. Ren, K.; Yu, J. Research Status of Bionic Amphibious Robots: A Review. *Ocean Eng.* **2021**, *227*, 108862. [CrossRef]
7. Bai, X.J.; Shang, J.Z.; Luo, Z.R.; Jiang, T.; Yin, Q. Development of Amphibious Biomimetic Robots. *J. Zhejiang Univ.-Sci. A* **2022**, *23*, 157–187. [CrossRef]
8. Ge, Y.; Gao, F.; Chen, W. A Transformable Wheel-Spoke-Paddle Hybrid Amphibious Robot. *Robotica* **2024**, *42*, 701–727. [CrossRef]

9. Graf, N.M.; Behr, A.M.; Daltorio, K.A. Crab-like Hexapod Feet for Amphibious Walking in Sand and Waves. In *Biomimetic and Biohybrid Systems*; Springer: Cham, Switzerland, 2019; Volume 11556, pp. 158–170; ISBN 9783030247409.
10. Marquardt, J.G.; Alvarez, J.; von Ellenrieder, K.D. Characterization and System Identification of an Unmanned Amphibious Tracked Vehicle. *IEEE J. Ocean. Eng.* **2014**, *39*, 641–661. [CrossRef]
11. Yan, Z.; Li, M.; Du, Z.; Yang, X.; Luo, Y.; Chen, X.; Han, B. Study on a Tracked Amphibious Robot Bionic Fairing for Drag Reduction. *Ocean Eng.* **2023**, *267*, 113223. [CrossRef]
12. Shim, H.; Yoo, S.; Kang, H.; Jun, B. Development of Arm and Leg for Seabed Walking Robot CRABSTER200. *Ocean Eng.* **2016**, *116*, 55–67. [CrossRef]
13. Baines, R.; Freeman, S.; Fish, F.; Kramer-Bottiglio, R. Variable Stiffness Morphing Limb for Amphibious Legged Robots Inspired by Chelonian Environmental Adaptations. *Bioinspir. Biomim.* **2020**, *15*, 025002. [CrossRef] [PubMed]
14. Ayers, J. Underwater Walking. *Arthropod Locomot. Syst. Biol. Mater. Syst. Robot.* **2004**, *33*, 347–360. [CrossRef] [PubMed]
15. Chen, G.; Xu, Y.; Wang, Z.; Tu, J.; Hu, H.; Chen, C.; Xu, Y.; Chai, X.; Zhang, J.; Shi, J. Dynamic Tail Modeling and Motion Analysis of a Beaver-like Robot. *Nonlinear Dyn.* **2024**, *112*, 6859–6875. [CrossRef]
16. Yang, Y.; Geng, Z.; Zhang, J.; Cheng, S.; Fu, M. Design, Modeling and Control of a Novel Amphibious Robot with Dual-Swing-Legs Propulsion Mechanism. In Proceedings of the 2015 IEEE/RSJ International Conference on Intelligent Robots and Systems (IROS), Hamburg, Germany, 28 September–3 October 2015; pp. 559–566.
17. Hirose, S.; Yamada, H. Snake-like Robots [Tutorial]. *IEEE Robot. Autom. Mag.* **2009**, *16*, 88–98. [CrossRef]
18. Thandiackal, R.; Melo, K.; Paez, L.; Herault, J.; Kano, T.; Akiyama, K.; Boyer, F.; Ryczko, D.; Ishiguro, A.; Ijspeert, A.J. Emergence of Robust Self-Organized Undulatory Swimming Based on Local Hydrodynamic Force Sensing. *Sci. Robot.* **2021**, *6*, eabf6354. [CrossRef]
19. Crespi, A.; Ijspeert, A.J. AmphiBot II: An Amphibious Snake Robot That Crawls and Swims Using a Central Pattern Generator. In Proceedings of the 9th International Conference on Climbing and Walking Robots (CLAWAR 2006), Brussels, Belgium, 12–14 September 2006; pp. 19–27.
20. Yu, S.; Ma, S.; Li, B.; Wang, Y. An Amphibious Snake-like Robot with Terrestrial and Aquatic Gaits. In Proceedings of the 2011 IEEE International Conference on Robotics and Automation, Shanghai, China, 9–13 May 2011; pp. 2960–2961.
21. Zhang, S.; Zhou, Y.; Xu, M.; Liang, X.; Liu, J.; Yang, J. AmphiHex-I: Locomotory Performance in Amphibious Environments with Specially Designed Transformable Flipper Legs. *IEEE/ASME Trans. Mechatron.* **2016**, *21*, 1720–1731. [CrossRef]
22. Zhong, B.; Zhang, S.; Xu, M.; Zhou, Y.; Fang, T.; Li, W. On a CPG-Based Hexapod Robot: AmphiHex-II with Variable Stiffness Legs. *IEEE/ASME Trans. Mechatron.* **2018**, *23*, 542–551. [CrossRef]
23. Zou, X. Machine Learning-Based Robotic Design: A Case Study on Fin-Based Amphibious Robots. Ph.D. Thesis, The State University of New Jersey, New Brunswick, NJ, USA, 2021.
24. Baines, R.; Patiballa, S.K.; Booth, J.; Ramirez, L.; Sipple, T.; Garcia, A.; Fish, F.; Kramer-Bottiglio, R. Multi-Environment Robotic Transitions through Adaptive Morphogenesis. *Nature* **2022**, *610*, 283–289. [CrossRef]
25. Wu, M.; Xu, X.; Zhao, Q.; Afridi, W.H.; Hou, N.; Afridi, R.H.; Zheng, X.; Wang, C.; Xie, G. A Fully 3D-Printed Tortoise-Inspired Soft Robot with Terrains-Adaptive and Amphibious Landing Capabilities. *Adv. Mater. Technol.* **2022**, *7*, 2200536. [CrossRef]
26. Xia, M.; Wang, H.; Yin, Q.; Shang, J.; Luo, Z.; Zhu, Q. Design and Mechanics of a Composite Wave-Driven Soft Robotic Fin for Biomimetic Amphibious Robot. *J. Bionic Eng.* **2023**, *20*, 934–952. [CrossRef]
27. Yu, D.; Che, T.; Zhang, H.; Li, Y.; Sun, D.; Wang, Z. Terrestrial Locomotion Stability of Undulating Fin Amphibious Robots: Separated Elastic Fin Rays and Asynchronous Control. *Ocean Eng.* **2024**, *311*, 118853. [CrossRef]
28. Wang, G.; Liu, K.; Ma, X.; Chen, X.; Hu, S.; Tang, Q.; Liu, Z.; Ding, M.; Han, S. Optimal Design and Implementation of an Amphibious Bionic Legged Robot. *Ocean Eng.* **2023**, *272*, 113823. [CrossRef]
29. Dudek, G.; Giguere, P.; Prahacs, C.; Saunderson, S.; Sattar, J.; Torres-Mendez, L.A.; Jenkin, M.; German, A.; Hogue, A.; Ripsman, A.; et al. AQUA: An Amphibious Autonomous Robot. *Computer* **2007**, *40*, 46–53. [CrossRef]
30. Dey, B.B.; Manjanna, S.; Dudek, G. Ninja Legs: Amphibious One Degree of Freedom Robotic Legs. In Proceedings of the 2013 IEEE/RSJ International Conference on Intelligent Robots and Systems, Tokyo, Japan, 3–7 November 2013; pp. 5622–5628.
31. Ma, X.; Wang, G.; Liu, K. Design and Optimization of a Multimode Amphibious Robot with PropellerLeg. *IEEE Trans. Robot.* **2022**, *38*, 3807–3820. [CrossRef]
32. Yu, J.; Ding, R.; Yang, Q.; Tan, M.; Zhang, J. Amphibious Pattern Design of a Robotic Fish with Wheel-Propeller-Fin Mechanisms. *J. Field Robot.* **2013**, *30*, 702–716. [CrossRef]
33. Uddin, M.; Garcia, G.; Curet, O. Underwater Collision Avoidance Using Undulating Elongated Fin Propulsion. *Ocean Eng.* **2023**, *285*, 115335. [CrossRef]
34. Shi, X.; Chen, Z.; Zhang, T.; Li, S.; Zeng, Y.; Chen, L.; Hu, Q. Hydrodynamic Performance of a Biomimetic Undulating Fin Robot under Different Water Conditions. *Ocean Eng.* **2023**, *288*, 116068. [CrossRef]

35. Li, Y.; Chen, L.; Wang, Y.; Ren, C. Design and Experimental Evaluation of the Novel Undulatory Propulsors for Biomimetic Underwater Robots. *Bioinspir. Biomim.* **2021**, *16*, 056005. [CrossRef]
36. Wang, S.; Wang, Y.; Wei, Q.; Tan, M.; Yu, J. A BioInspired Robot with Undulatory Fins and Its Control Methods. *IEEE/ASME Trans. Mechatron.* **2022**, *22*, 206–216. [CrossRef]

biomimetics

MDPI

Article

Empirical Data-Driven Linear Model of a Swimming Robot Using the Complex Delay-Embedding DMD Technique

Mostafa Sayahkarajy * and Hartmut Witte

Group of Biomechatronics, Fachgebiet Biomechatronik, Technische Universität Ilmenau, D-98693 Ilmenau, Germany; hartmut.witte@tu-ilmenau.de

* Correspondence: mostafa.sayahkarajy@tu-ilmenau.de

Abstract: Anguilliform locomotion, an efficient aquatic locomotion mode where the whole body is engaged in fluid–body interaction, contains sophisticated physics. We hypothesized that data-driven modeling techniques may extract models or patterns of the swimmers' dynamics without implicitly measuring the hydrodynamic variables. This work proposes empirical kinematic control and data-driven modeling of a soft swimming robot. The robot comprises six serially connected segments that can individually bend with the segmental pneumatic artificial muscles. Kinematic equations and relations are proposed to measure the desired actuation to mimic anguilliform locomotion kinematics. The robot was tested experimentally and the position and velocities of spatially digitized points were collected using QualiSys® Tracking Manager (QTM) 1.6.0.1. The collected data were analyzed offline, proposing a new complex variable delay-embedding dynamic mode decomposition (CDE DMD) algorithm that combines complex state filtering and time embedding to extract a linear approximate model. While the experimental results exhibited exotic curves in phase plane and time series, the analysis results showed that the proposed algorithm extracts linear and chaotic modes contributing to the data. It is concluded that the robot dynamics can be described by the linearized model interrupted by chaotic modes. The technique successfully extracts coherent modes from limited measurements and linearizes the system dynamics.

Keywords: bio-inspired locomotion; soft robotics; bio-robotics; data-driven modeling; CDE DMD

1. Introduction

The agility and energy efficiency of animals' swimming have triggered roboticists and researchers to investigate bioinspired locomotion methods for the navigation of robotic systems within fluidic environments. Furthermore, miniature swimming robots [1,2] or soft robots [3] are designed to move based on biomimetic propulsion as the conventional propellers with electric motors are too bulky or heavy for such applications. Nevertheless, the aquatic locomotion of animals contains complex dynamics as a mirror of interacting neuromuscular activities and hydrodynamics. In anguilliform swimming, the entire body participates in the locomotion, which makes the body–fluid interaction sophisticated to simulate and analyze. Each part of the body contributes to propulsion, unlike in other forms of undulatory locomotion where only certain body parts might move. Anguilliform locomotion is generally slower than other types of swimming, such as in the other extreme of thunniform locomotion seen in tunas and sharks. However, it is highly energy efficient [4] and allows the animals to navigate complex environments like narrow crevices and dense vegetation (which is one of the main obstacles for technical drives with rotatory propulsion). The efficiency may be attributed to the phenomenon that a wake-induced drag vanishes while the vortex wake is absent [5]. The propeller drives for boats cannot be used in the overgrown edge zones of shallow waters, as the plants block the rotational movement very quickly.

The technical implementation of fish-like propulsion systems is an alternative to standard engineering solutions. Fish use muscle fibers to bend their bodies. These muscle fibers actively generate tension on the concave side while passively extending on the convex side. The most analogous artificial muscle is the McKibben actuator [6], a lightweight and flexible actuator well-suited for soft swimming robots. In [7], it is discussed that the pneumatic soft actuators' force generation characteristics are similar to the natural muscles. Furthermore, they are well-suited for underwater field applications because of the absence of metal components and operating with pneumatic actuation. Industrial pneumatic pressurization enables McKibben actuators to generate sufficient force to operate effectively, even in the high-density water environment.

In anguilliform swimming, three components including fluid dynamics, neural control, and the musculoskeletal system interact with each other. While studying the system as a whole is a formidable task, considerable research has been devoted to different aspects of anguilliform locomotion. This swimming mode is observed ubiquitously in single-cell organisms with tiny sizes up to lampreys and eels, covering Reynolds numbers on the order of 10^{-3} to 10^{3}. Anguilliform locomotion involves complex undulations as a result of superimposing traveling mechanical waves, as well as stationary waves [8,9]. A complex traveling wave can be conceptualized as an interplay of the configurations defined by the real and imaginary components of the wave [10]. The real part becomes apparent when the temporal modulation of the imaginary part is zero, while the imaginary part emerges when the temporal modulation of the real part vanishes.

Video recording, followed by video processing and particle image velocimetry (PIV), seems to be the standard technique to calculate the kinematics of anguilliform swimming of animals like eels and lampreys [11,12]. The kinematic model can be used for hydrodynamic simulations to investigate the underlying physics and interaction forces. The bending kinematic creates vortices in the environment producing negative pressure regions that pull the fish forward with the suction trust according to [11]. Transsected animals are additional subjects for studying altered kinematics within the locomotion mode. In healthy lampreys, the amplitude of the traveling wave increases as it travels along the body, but spinal-transacted lampreys exhibit a uniform wave with less amplitude and wavelength [12]. Nevertheless, studying transected animals has some complexities and is subject to further problems such as decreased muscle force and complex neural responses. On the other hand, bio-inspired robots can serve as manageable devices for mimicking animal kinematics. Some eel-like robots have been developed by different robotic laboratories worldwide. Examples include AmphiBot I [13] and II [14,15].

However, the robots consisting of serially jointed rigid links lack the compliance of natural swimmers. Commonly, the gesture of conventional robots cannot passively be affected by external forces. On the other hand, emerging rapidly in the last decades, soft robotics promote compliant-body systems without explicit joints and rigid links of conventional robots. In recent years, researchers have developed some eel-like soft robots with pneumatic [16–23] or cable-driven actuation [24,25]. A soft robot is structurally an infinite dimensional system that deforms due to both actuator force and the interaction force exerted by the environment. Due to technology limitations, manipulating only a limited number of active degrees of freedom (DOFs) is conceivable in soft robots. This makes soft robots underactuated systems that may not outperform real fish. Nevertheless, a soft-body robot is qualitatively more similar to its natural counterpart, making them better candidates for either investigating the swimming principles or replicating the natural performance within robotics.

Our previous study, ref. [7], showed that complex body–fluid interaction produces different modes that appear as exotic fluctuations in the swimmers' midline. The study employs numerically expensive fluid–solid interaction (FSI) simulations to extract the system modes. Extracting the rearward moving waves related to the anguilliform locomotion is generally difficult. Such decomposition and analysis are important for evaluating and modeling the locomotion of the robots, which are still far from the efficiency of their natural counterparts. Nevertheless, the FSI method has limitations, including modeling approximations and huge numerical costs. In this work, we investigate a method to analyze the experimentally measured data without any measurements or calculations of the fluid environment.

In the case of soft swimming robots, the generation of traveling waves is complex and comprises ostensibly noisy movements of the robot due to the softness of the robot's structure. Coherent structures cannot be visually observable when multiple chaotic modes or additional parasitic and noisy signals influence the system states. Machine learning (ML) methods, and especially linearization methods based on dynamic mode decomposition (DMD) and its variants, can help analyze anguilliform locomotion. Showcase studies include examining various data from mathematically expressed sinusoidal traveling waves to videos taken from natural swimmers like eels in experimental water tanks. DMD, originally derived from Koopman analysis, proposes a linear model for data sequences [26] and is investigated to analyze fluid dynamics [27,28]. The method acts as a modal decomposition technique with a separation of temporal and spatial variables, providing a linearized model that may be used for estimation or control purposes. In [29], the method is used to linearize a worm robot model. The ability of DMD-based data-driven modeling methods to extract spatiotemporal patterns from data makes them ideal for analyzing traveling wave phenomena. When partial state information is available, delay coordinates can be a better choice of observables to approximate the Koopman operator in establishing a DMD model.

In this work, the swimming kinematics of a soft eel robot is analyzed using pure experimental data. A pneumatic eel robot and its actuation kinematic equations are introduced. The robot contains six segments that are individually actuated using contraction-type pneumatic actuators. The robot is experimentally tested and a data-driven algorithm is proposed to inspect coherent and linear patterns within the exotic fluctuations in the robot. First, in Section 2.1, the robot structure and the relations that relate actuator contraction to the traveling mode are presented as the working principle. Section 2.2 explains the complex delay-embedding DMD algorithm (CDE DMD). Delay embedding is used to make data-driven models from partial measurements to understand the dynamics of partially observed systems. The simulation results are presented in Section 3. In Section 4, it is discussed that the algorithm revealed linear and chaotic modes within the data, enabling us to partition the system into linear and nonlinear dynamics. The conclusion is summarized in Section 5.

2. Materials and Methods

2.1. Structure and Actuation Kinematics

The soft robot structure contains a flexible beam and seven rigid parts for connecting the actuators and balancing the robot on the water surface. Twelve pairs of contraction-type pneumatic muscles provide lateral bending for each segment of the tethered robot. In contrast with similar rigid robots in which actuation torque can be exerted on the rotational joints, the soft robot is driven by linear (i.e., axial) actuators. This brings more similarity with natural swimmers but requires measurement of the muscle actuation needed for a desired kinematic. Inspired by primitive natural swimming, the kinematic relations are derived with a supposition that the neutral axis of the robot in the steady state follows a moving sine wave. The schematic kinematic model of the robot is shown in Figure 1. The horizontal line represents the inactivated initial state of the robot backbone midline. Seven equally distanced points, $A_1, A_2, \ldots A_7$, represent the position of rigid connectors (also used as floats). The R_1L_1, R_2L_2, …, R_7L_7 are rigidly connected to the flexible backbone, and the dashed lines represent actuation axes. The actuation state vector $\vec{q} = \{q_1, q_2, q_3, \ldots, q_{12}\}$ is defined with the Euclidean distances $q_i \quad i = 1..12$.

$$q_i = |R_iR_{i+1}| \quad , \quad q_{i+6} = |L_iL_{i+1}| \quad i = 1..6 \tag{1}$$

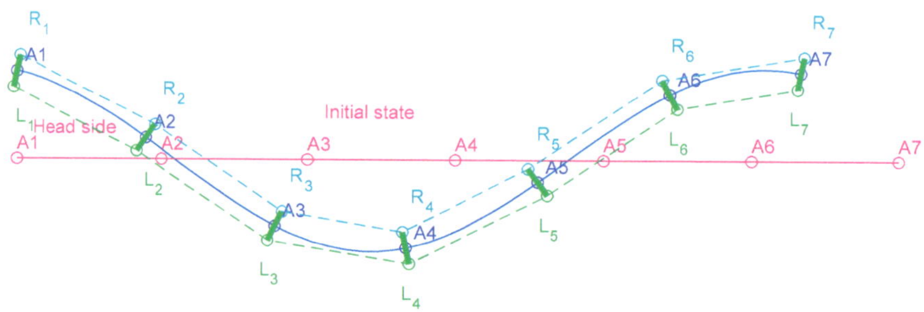

Figure 1. Geometrical model of the robotic fish. The solid lines represent the robot backbone midline, and the dashed lines show the actuators' nominal axes.

Geometric relations are used to obtain the actuator lengths for the given desired deformation to generate a locomotion pattern. Supposing the desired deformation of the neutral axis is shown as:

$$f(x,t) = \varphi(x)\,\mathrm{Sin}(2\pi m \frac{x - t\,v(t)}{l}) \tag{2}$$

where $\varphi(x)$ is the waveform, l is the backbone length, and $v(t)$ is the relative wave velocity. As the backbone length is constant during bending, we have

$$\int_{x_{Ai}}^{x_{A(i+1)}} \left(\begin{array}{c} 1 + (\frac{\partial \varphi}{\partial x}\mathrm{Sin}(2\pi m \frac{x - t\,v(t)}{l}) + \\ \varphi(x)\,(\frac{2\pi m}{l})\mathrm{Cos}(2\pi m \frac{x - t\,v(t)}{l}))^2 \end{array} \right) dx = l/6 \tag{3}$$

This equation is numerically solved to obtain the position of A_i points. Consequently, coordinates of R_i points are obtained readily as

$$\begin{array}{l} x_{Ri} = x_{Ai} + (\hat{n}.\hat{i})d_i = x_{Ai} - d_i \frac{\partial f/\partial x}{\sqrt{1+\partial^2 f/\partial x^2}} \Bigg|_{\,x = x_{Ai}} \\ y_{Ri} = y_{Ai} + d_i(\hat{n}.\hat{j}) = y_{Ai} + \frac{d_i}{\sqrt{1+\partial^2 f/\partial x^2}} \Bigg|_{\,x = x_{Ai}} \end{array} \tag{4}$$

Likewise, for the L_i points, we will have

$$\begin{aligned} x_{Li} &= x_{Ai} + d_i \frac{\partial f/\partial x}{\sqrt{1+\partial^2 f/\partial x^2}}\Bigg|\ x = x_{Ai} \\ y_{Li} &= y_{Ai} - \frac{d_i}{\sqrt{1+\partial^2 f/\partial x^2}}\Bigg|\ x = x_{Ai} \end{aligned} \tag{5}$$

Thus, the system states are obtained for i = 1..6 using

$$\begin{aligned} q_i &= \left(\left(x_{R(i+1)} - x_{Ri} \right)^2 + \left(y_{R(i+1)} - y_{Ri} \right)^2 \right)^{0.5} \\ q_{i+6} &= \left(\left(x_{L(i+1)} - x_{Li} \right)^2 + \left(y_{L(i+1)} - y_{Li} \right)^2 \right)^{0.5} \end{aligned} \tag{6}$$

Finally, the state vector $\vec{q}$ is numerically measured for one cycle in N steps and is saved as a matrix $Q = \left[\vec{q}_1 \middle| \vec{q}_2 \middle| \vec{q}_3 \middle| \ldots \middle| \vec{q}_N \right]$. Matrix Q is the desired actuators' length corresponding to the desired anguilliform gait. The matrix can be normalized to represent contraction rates for the McKibben actuators. The 'offline' calculated matrix is repeatedly given to the robot actuators with the desired frequency to repeat the gait. (Note that the relation between the actuator length (or the contraction rate) and pressure is supposed to be linear. The maximum contraction (10 percent) corresponds to maximum pressure (7 MPa) and the interval values can be expressed with pulse width modulation (PWM), which is given as the input signal to the solenoid valves.

The kinematic equations can be used for different reasons, especially explanations of the robot's working principles, and geometric design. The equations are used to demonstrate the robot's performance before fabrication graphically. They can be used for optimization, e.g., maximizing the bending amplitude. Nevertheless, a desired kinematic may differ from the actual observation. In practice, the robot's kinematic properties and motion arise from the interactions of modes and cannot be fully explained by considering the individual robot. Emergent phenomena often exhibit behavior or characteristics that are not predictable from the properties of the parts in isolation. On the other hand, control of a pneumatic system, i.e., air pressure control, is hard. The actual kinematics can be different from the desired kinematics due to random factors (such as leakage, fault, etc.) and the fluid interplay with the compliant body. Even the existence or dominance of a traveling wave within the robot swimming can be vailed under such circumstances. We propose data analysis to obtain insight into such phenomena.

2.2. Proposed CDE DMD Algorithm

Suppose the lateral position and velocity of s digitized points of the robot midline are recorded as $y \in \mathbb{R}^{s\times 1}$ and $v \in \mathbb{R}^{s\times 1}$ in discrete time intervals, Δt. A filtered velocity vector $\overline{v}$ is defined for the discrete signal v using a canonical discrete filter model

$$\overline{v}_{,k} \simeq \overline{v}[k] = \sum_{i=0}^{d-1} v[k-i].h[i] \tag{7}$$

where the filter kernel is chosen $h[i] = \frac{1}{d}$, $d \in \mathbb{N}$ for a moving average filtration. The complex state vector at time $t = k\Delta t$ is shown by $z_{,k} \in \mathbb{C}^{s\times 1}$ and is defined as

$$z_{,k} = y_{,k} + j\overline{v}_{,k} \tag{8}$$

Note that in this context, the ith entry of a vector like $z_{,k}$ is denoted as $z_{i,k}$ (and similarly, for representing the ith entry over all the times, we use $z_{i,}$). The time-resolved matrix, $Z \in \mathbb{C}^{s \times n}$, is constructed with the complex state vectors as its columns, that is

$$Z = \begin{bmatrix} | & | & & | \\ z_{,1} & z_{,2} & \cdots & z_{,n} \\ | & | & & | \end{bmatrix} \tag{9}$$

The time-delay embedded matrix $H \in \mathbb{C}^{s(d+1) \times (n-d)}$ with the embedding dimension number d is constructed as follows

$$H = \begin{bmatrix} z_{,1+d} & z_{,2+d} & \cdots & z_{,n} \\ z_{,d} & z_{,1+d} & \cdots & z_{,n-1} \\ \cdot & \cdot & & \cdot \\ \cdot & \cdot & \cdots & \cdot \\ \cdot & \cdot & & \cdot \\ z_{,1} & z_{,2} & \cdots & z_{,n-d} \end{bmatrix} \tag{10}$$

Based on a linear approximation of the Koopman operator, the DMD approach assumes a linear mapping that approximates snapshots of the system dynamics one timestep forward. The DMD algorithm takes the matrices of snapshots

$$X = \begin{bmatrix} | & | & & | \\ h_{,1} & h_{,2} & \ldots & h_{,n-d} \\ | & | & & | \end{bmatrix}, \; X^{+} = \begin{bmatrix} | & | & & | \\ h_{,2} & h_{,3} & \ldots & h_{,n-d+1} \\ | & | & & | \end{bmatrix} \tag{11a}$$

where $h_{,k}$ represents column k of the H. It is supposed that a linear operator, as matrix $A \in \mathbb{C}^{s(d+1) \times s(d+1)}$, maps the column data to the next snapshot as in $h_{,k+1} = Ah_{,k}$. The best-fit solution relates the data matrices as

$$X^{+} = AX \tag{11b}$$

Next, the method involves the SVD decomposition of the X matrix and order reduction, with truncation order r and matrices $U \in \mathbb{C}^{s(d+1) \times r}$, $\Sigma \in \mathbb{C}^{r \times r}$, and $W \in \mathbb{C}^{(n-d-1) \times r}$

$$X = U\Sigma W^{*} = \begin{bmatrix} | & | & & | \\ u_{,1} & u_{,2} & \ldots & u_{,r} \\ | & | & & | \end{bmatrix} \begin{bmatrix} \sigma_1 & & 0 \\ & \ddots & \\ 0 & & \sigma_r \end{bmatrix} \begin{bmatrix} | & | & & | \\ w_{,1} & w_{,2} & \ldots & w_{,r} \\ | & | & & | \end{bmatrix}^{*} \tag{12}$$

where W^{*} represents the conjugate transpose of matrix W. Note that the reconstruction of the data is given by

$${}_{a}^{b}X = \sum\nolimits_{i=a}^{b} u_{,i}\sigma_i \overline{w}_{i,} \quad , \quad a = 1\,, b = r \tag{13}$$

where $\overline{w}_{i,}$ represents the complex conjugate of vector $w_{,i}$. It follows that the DMD operator can be expressed as

$$A = X^{+}D \quad , \quad D = W\Sigma^{-1}U^{*} \tag{14}$$

We can suppose a partitioning of D into ${}^{L}D$ and ${}^{N}D$ so that

$$D = {}^{L}D + {}^{N}D \tag{15}$$

where

$${}^{L}D = \sum\nolimits_{i=1}^{k} w_{,i}\sigma_i^{-1}\overline{u}_{i,} \quad , \quad {}^{N}D = \sum\nolimits_{i=k+1}^{r} w_{,i}\sigma_i^{-1}\overline{u}_{i,} \tag{16}$$

Supposing ${}^{L}D$ contains selected modes with some properties of interest, in the linearity in this context, we can extract a ${}^{L}A$ matrix that captures the properties, using a partitioning $A = {}^{L}A + {}^{N}A$, so that

$$ {}^{L}A = X^{+}\, {}^{L}D \tag{17} $$

Then, an eigen decomposition ${}^{L}A = \Psi\Lambda\Psi^{-1}$ is performed, where Ψ contains the eigenvectors as its columns and Λ is a diagonal matrix of the eigenvalues. Following an order reduction, Ψ_m and Λ_m are rewritten, keeping only m first eigenvectors and eigenvalues correspondingly. To obtain $f_{,k}$, the system state in the eigenbasis, $h_{,k}$, is transformed into a new coordinate system defined by the eigenvectors with the following relation $f_{,k} = \Psi_m^{-1} h_{,k}$. Then, in the modal space, we will have

$$ f_{,k+1} = \Psi_m \Lambda_m f_{,k} \tag{18} $$

Note that by transforming back to the original (delayed) basis, the solution can be represented as

$$ h_{,k} = \Psi_m \Lambda_m \Psi_m^{-1} h_{,0} \tag{19} $$

Finally, a matrix $\widetilde{A} \in \mathbb{C}^{s \times s(d+1)}$ containing s rows of ${}^{L}A$ returns the predicted position data one timestep forward by the following relation

$$ y_{,k+1} \approx real(\widetilde{A} h_{,k}) \tag{20} $$

As a numerical method of discovering the patterns and underlying modes, the success of finding coherent structures depends, undeniably, on the physics of the system, in addition to the algorithm's ability and the data's quality and quantity. Therefore, simulation is an inseparable part of the method. The simulations and visualizations given in Section 3.2 are used to accomplish this exploration. In finding the delay embedding order, simulation results are intuitively useful. By increasing the number of delay coordinates, the embedding captures a larger portion of the system's past dynamics, which in turn enhances the ability to reconstruct the underlying state-space representation. This is particularly useful for nonlinear systems with multi-scale behavior or chaotic dynamics. Nevertheless, excessive delay embedding might introduce redundant information if the system dynamics are inherently low-dimensional.

3. Results

3.1. The Eel Robot

Like flying propulsors, anguilliform swimmers such as eels and lampreys have flexible bodies bent by the contraction of lateral muscles. The muscles on the concave side are actively contracting in concentric mode, while those on the convex side exhibit reduced activity in the eccentric mode, forming antagonistic muscle systems. Inspired by this principle, we propose a soft structure, as shown in Figure 2. The robot contains a highly flexible beam with seven floats (the green components). The structure includes six segments that bend by two pairs of McKibben soft actuators. Seven markers (the gray balls in Figure 2) are used to track the backbone's equally distanced points. The actuation geometry for steady swimming is simulated and graphically presented in Figure 3. Note that acceleration from rest, in contrast to steady swimming, can be associated with different kinematics, as reported about lampreys swimming [30]. Nevertheless, in this study, only a monotone input for steady swimming was applied to the robot from zero initial conditions.

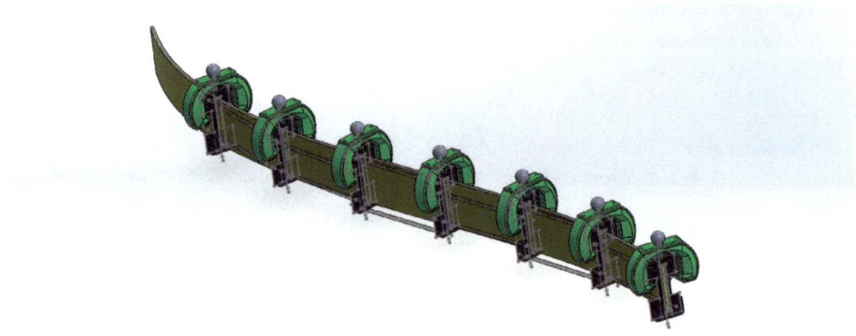

(**a**)

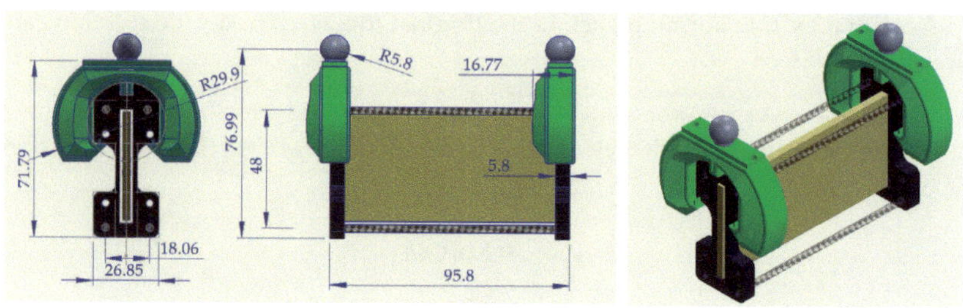

(**b**)

Figure 2. Mechanical drawings: (**a**) 3D model of the robot; (**b**) the segmental assembly parts and balancing bladders.

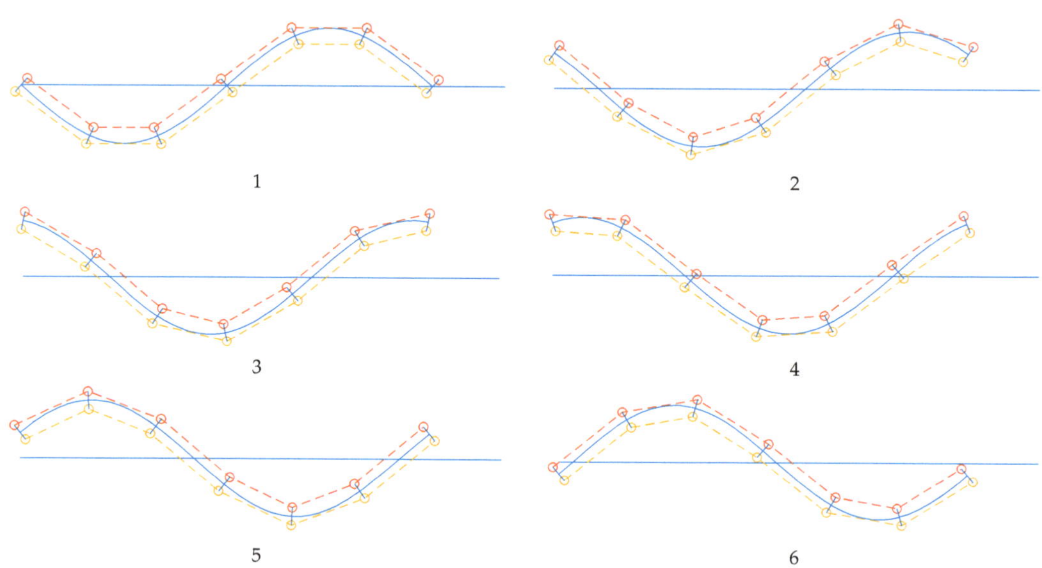

Figure 3. Geometrical representation of the desired actuator lengths calculations with a wave traveling left to right from 1 to 6. The blue curve represents the robot backbone axis versus the initial state, i.e., the straight blue line. The dashed lines represent the soft actuators' axes.

The graphs in Figure 3 present the MATLAB® R2022b (MathWorks®, Natick, MA, USA) simulations of the equations and relations in Section 2.1. They can be used to verify the functionality principle, particularly the structure design's compatibility. The dashed lines, representing the actuation axis, touch the 'backbone' at the maximum bending due to the opposite actuator. The minimum and maximum lengths of the dashed line are, respectively, equal to the length of an actuated muscle with the maximum pressure and the nominal length of an inactive muscle (as mentioned before, such a measurement can be performed within an optimization technique such as the genetic algorithm, which is not within the scope of this article). Nevertheless, the actual kinematics, i.e., the position and backbone curvature evolution, is highly influenced during a real swimming test. We are more interested in swimming kinematics and investigating the experimental data in this context. The experimental setup used for this purpose is explained in Appendix A. The data collected from a swimming test of the robot is illustrated in Figure 4, representing the tracking markers' lateral displacement and velocities in the time domain. The velocities were calculated using differentiation of the measured position data. The time series plots are given for the seven tracking markers, A_1 to A_7. The visualization suggests a periodic motion with overall repetition expected from steady swimming.

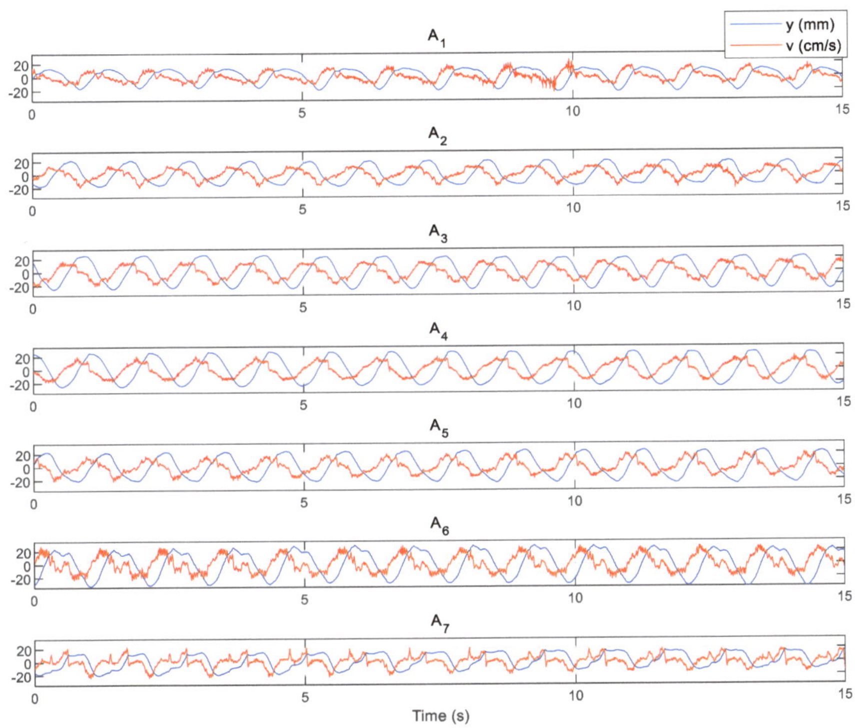

Figure 4. The experimental data. The robot's undulation period is 1.1 s.

A valuable tool for diagnosing mechanical behavior and analyzing vibratory systems is a phase plot. A phase plot is a graphical representation used in dynamical systems analysis to visualize the state of a system in phase space. It provides insights into the relationship between key dynamic variables, such as position and velocity when the system evolves in time. In the context of system dynamics, the phase plot is particularly useful for examining oscillatory or non-linear behavior, offering a concise way to understand the motion without explicitly referencing time. To construct a phase plot, one requires time-series data of the system's position and velocity. The real axis represents the position, while the imaginary axis represents the velocity. By treating position as the horizontal component and velocity as the vertical component, the trajectory in the phase space is plotted as the system evolves. Each point on the curve corresponds to a unique state of the system at a particular moment in time. This method is especially effective for oscillatory systems, where periodic motion appears as closed loops, indicating energy conservation, while spiraling trajectories may suggest dissipative effects or damping.

The phase plot of the experimentally measured values is illustrated in Figure 5. The graph represents the complex states $z_{i,}$ for the 7 tracking markers (i = 1..7). The advantages of using a phase plot include its ability to highlight the system's characteristics. It can visually reveal attractors, periodicity, or chaotic behavior. Unlike time-domain plots, which may obscure the interplay between variables, phase plots emphasize their relationship directly, enabling a more intuitive understanding of system dynamics. Nevertheless, the phase graphs of the experimental data do not illustrate coherent structures. At first look, it seems impossible to draw a deduction about the system dynamics or contribution of nonlinearity, stochastic, noise, or chaotic phenomena. In the CDE DMD, we look for a new coordinate system where coherent structures or patterns are revealed.

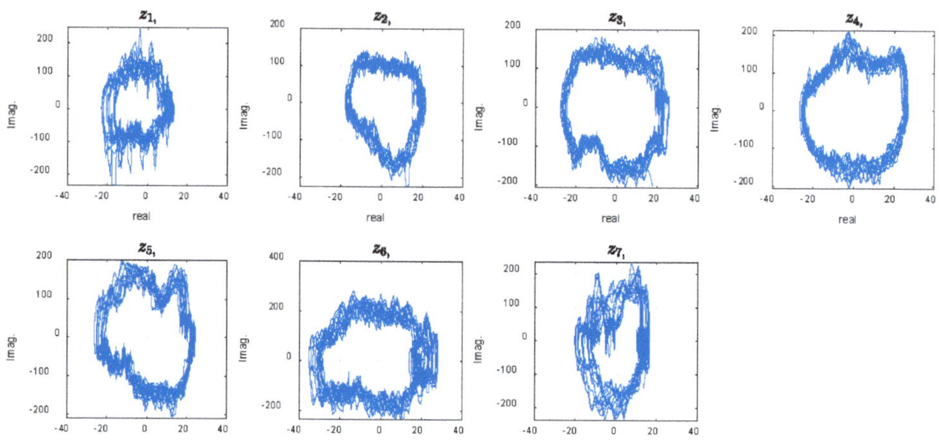

Figure 5. Phase plot of the experimental data.

3.2. CDE DMD Analysis Results

The phase plot reveals strange behavior of the system that, at first look, may be attributed to random, chaotic, or noisy phenomena within the robot or its interaction with the environment. We expect the phase plots to appear as perfect circular shapes for a perfectly swimming system exhibiting a smooth sinusoidal wave. On the other hand, with chaotic (but not random) systems, smooth trajectories where each point does not go back to its place in the previous period, resulting in repeating coherent curves, are anticipated. Previous studies attributed the exotic dynamic to the system modes are measured from the FSI model and simulations [7,31]. Such simulations are numerically expensive and need modeling assumptions and approximations. This study proposes the CDE DMD analysis described in the method to discover the underlying modes that may explain the system behavior based on the experimental data. For this aim, calculations with different embedding dimension, d, were conducted. The phase plot of the dynamic variables for $d = 30$ and 110 are shown in Figures 6 and 7. Increasing the delay steps yields more circular shapes, revealing coherent attractors in the leading modes in the CDE coordinates. Note that the sampling time was $\Delta t = 0.005$ s, and the robot actuation period was $T = 1.1$ s. Therefore, the delay number $d = 110$ corresponds to a half cycle of the undulation.

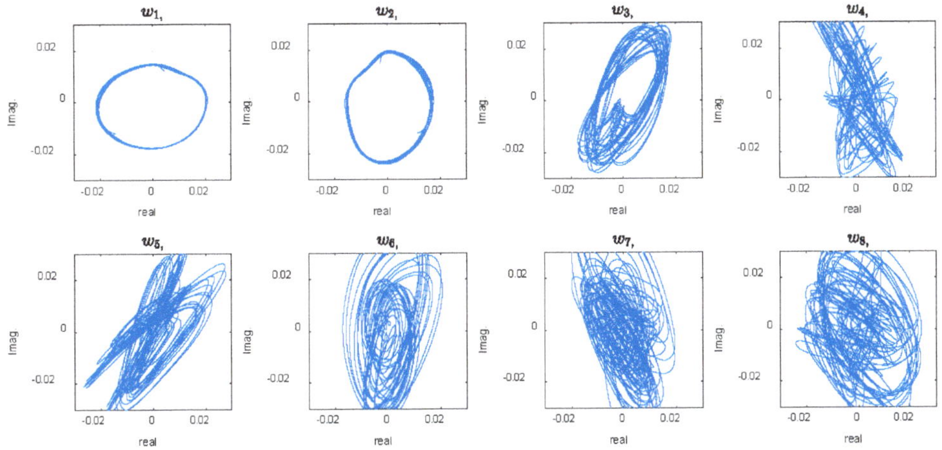

Figure 6. Phase plot of the dynamic variables with d = 30.

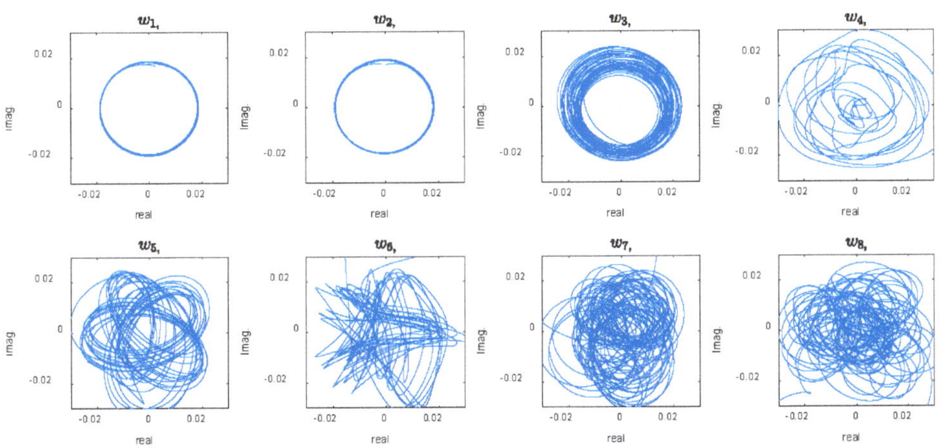

Figure 7. Phase plot of the dynamic variables with d = 110.

The circular graph of the first two variables, $w_{1,}$ and $w_{2,}$, in Figure 7 indicates the presence of two harmonic oscillations in the CDE coordinates. The phase plots represent the temporal modes that construct the overall dynamics. Coherent patterns and structures appear in the 'CDE coordinate system' (in contrast to the 'physical coordinate system' where the measurement is performed). Delay embedding contains some limitations, necessitating numerical assessment to verify the method for a specific application. High dimensionality, unknown delay number, and noise handling may be limitations, depending on the underlying physics of a particular application. Note that increasing the embedding dimensions by more delay steps increases the matrix size and unnecessary calculation complexity and noise. The right number of delay steps, which is often unknown, is estimated using simulations and observing the modes. The number is increased and the evolution of the first two modes is considered as an optimization criterion. As shown in Figure 8, by diverging from the optimal value (Figure 8e), the graphs start to show ovality.

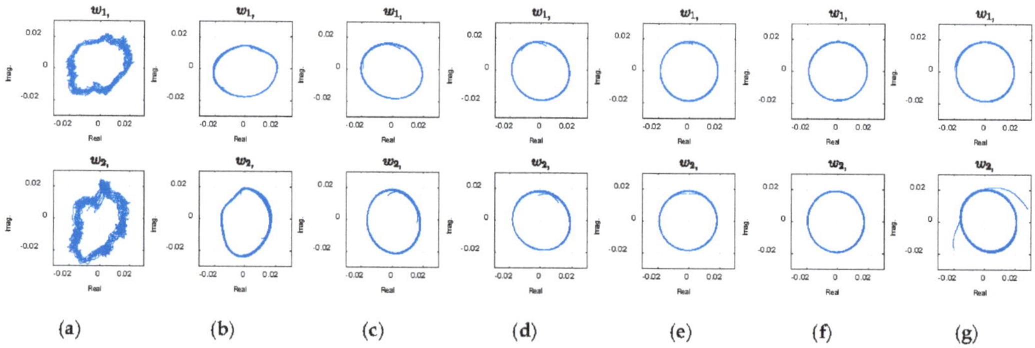

Figure 8. Evolution of the first two modes with embedding dimension: (**a**) Standard DMD; (**b**) CDE DMD with $d = 30$; (**c**) $d = 60$; (**d**) $d = 85$; (**e**) $d = 110$; (**f**) $d = 140$; (**g**) $d = 170$.

The working principle of the CDE DMD method is shown abstractly in Figure 9. The measured data, in the physical coordinate system, is first encoded by projecting the data to the CDE coordinates. Encoding involves increasing the dimensionality due to the large size of the Hankel matrix. Then, superimposing linear modes, in the decoding stage, provides the linear dynamics. Similarly, nonlinearities can be obtained by com-posing the nonlinear modes.

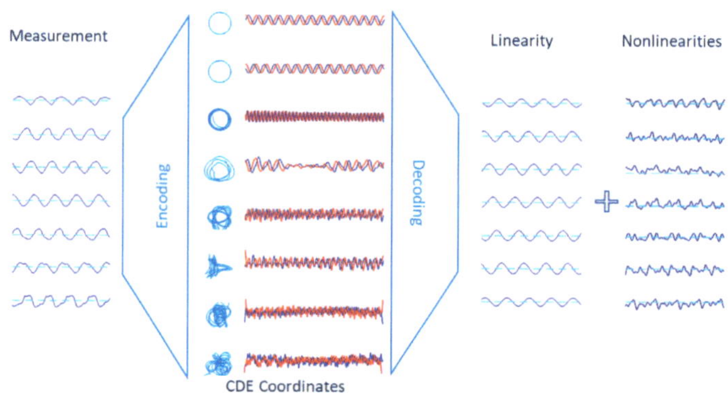

Figure 9. Conceptualization of the linearization technique. Measurement data are projected to the CDE coordinates, in the encoding stage. Consequently, the linear (and nonlinear) portion is obtained by superimposing the linear (and nonlinear chaotic) modes, in the decoding stage.

The corresponding time history of the variables visualized in Figure 10 indicates the linearity property within the dominant modes. The first two modes present smooth oscillations with a constant amplitude typical in mass-spring and linear vibration systems. The phase shift, comparing mode one and mode two, is attributed to traveling wave phenomena. It is inferred from the results that the temporal modes consist of nearly linear oscillators disturbed by chaotic modes. This fact suggests partitioning the system dynamics into an approximate linear subsystem and a highly nonlinear chaotic subsystem. Therefore, the reconstruction of the model with the linear modes results in a linear surrogated system model. Reconstruction of the data with the linear and nonlinear subsystems is shown in Figures 11 and 12. The linear dynamics, shown in Figure 11, is composed of the first two modes, corresponding to ${}_{1}^{2}X$ in Equation (13). In Figure 12, the full reconstruction, from ${}_{1}^{12}X$, as well as the nonlinearity portion, corresponding to ${}_{3}^{12}X$, is demonstrated. The graphs contain the experimental data for comparison.

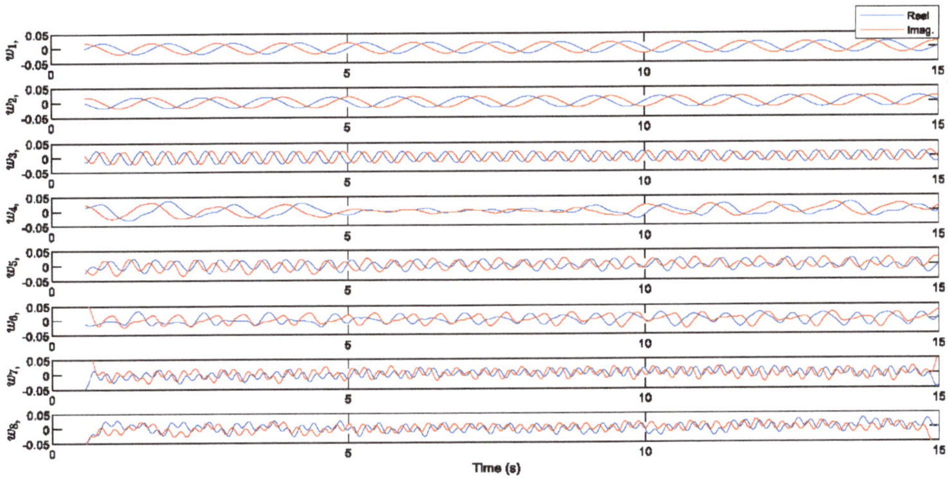

Figure 10. Time-series plot of the CDE dynamic variables with $d = 110$.

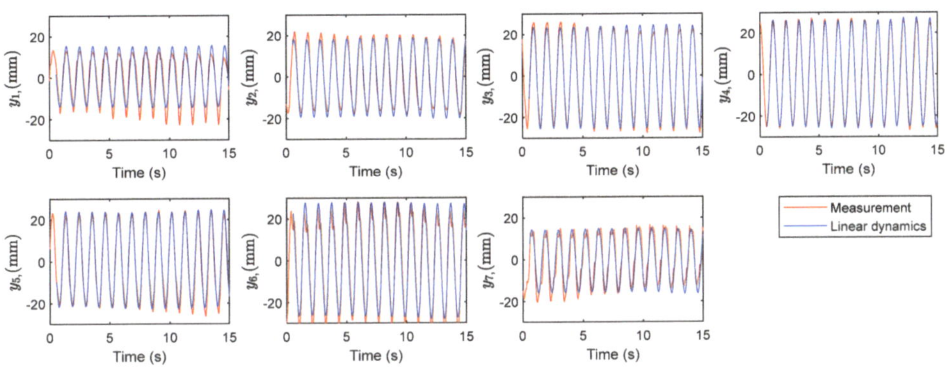

Figure 11. The linearized model versus the measured data.

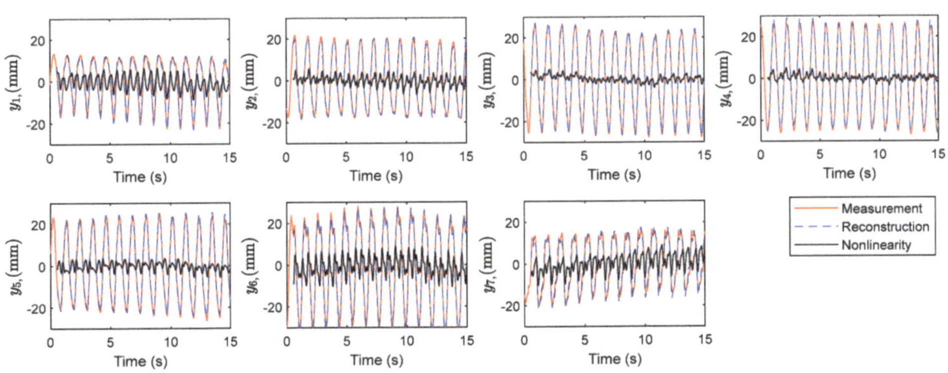

Figure 12. Comparison of the reconstructed data, the nonlinearity contribution, and the measurements.

Therefore, the CDE DMD provides a linearized model by including the first modes. The simulation results of the linearized model Equation (20) are shown in Figure 13. In the numerical calculations, H was a 777 by 2890 matrix with a rank of 777 calculated after the construction of the matrix. With 7 sensed points, $s = 7$, and 110 delays, $d = 110$, the number of rows of H became $s(d+1) = 777$. Likewise, with the measurement sampling time of 0.005 s and a total time of 15 s, $n = 15/0.005 = 3000$, the number of columns will be equal to $n - d = 2890$. The ${}^{L}A$ and A matrices are of size 777 by 777, while $\tilde{A}$ is a 7 by 777 matrix.

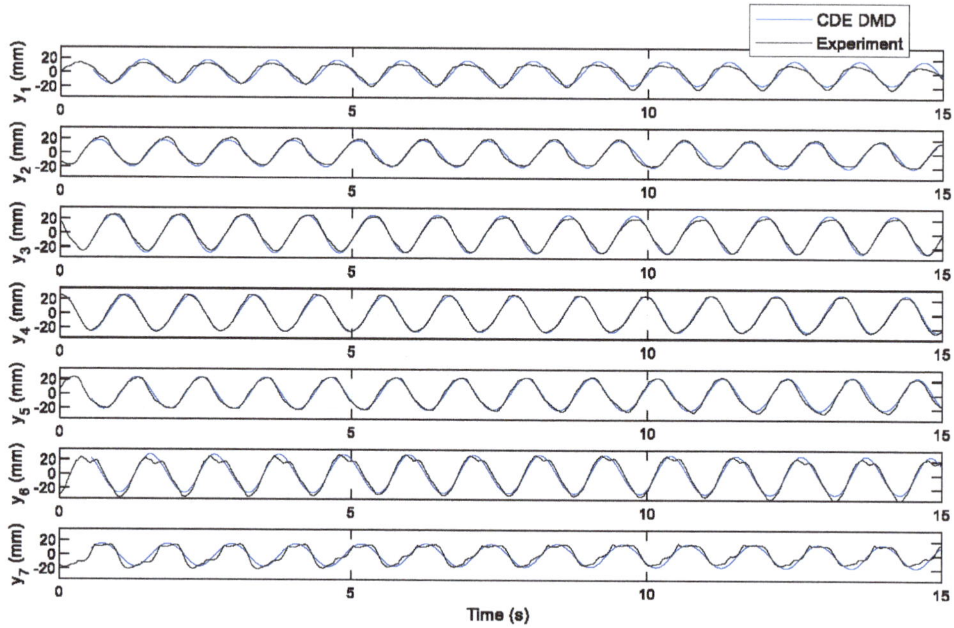

Figure 13. Comparison of the CDE DMD results with the original data.

3.3. CDE DMD's Noise Handling

The CDE DMD handles noise effectively by leveraging delay embedding to distribute noise across dimensions, applying low-rank approximations via SVD and isolating dominant modes. These properties make it a robust tool for analyzing noisy measurements, particularly when the signal-to-noise ratio is moderate to high. In this Section, the ability of the method to handle noisy data is examined by simulation. As the measurement was achieved with a precise laboratory setup, a random signal was added to simulate the corruption of the original signal by the artificial noise. The noisy signal and the probability density function (PDF) of the added noise are shown in Figure 14. In this case, the noise contains Gaussian signals with a mean of zero and standard deviations of 12.

The influence of added noise on the dynamic variables, i.e., the CDE modes, is shown by simulation. Figure 15 illustrates the results for the original signal (top phase plots) and the corresponding modes for the noisy signal (bottom plots). A comparison of the plots shows that the first two modes remain similar, before and after the noise injection. In fact, the noise is projected to higher modes due to its high-frequency nature.

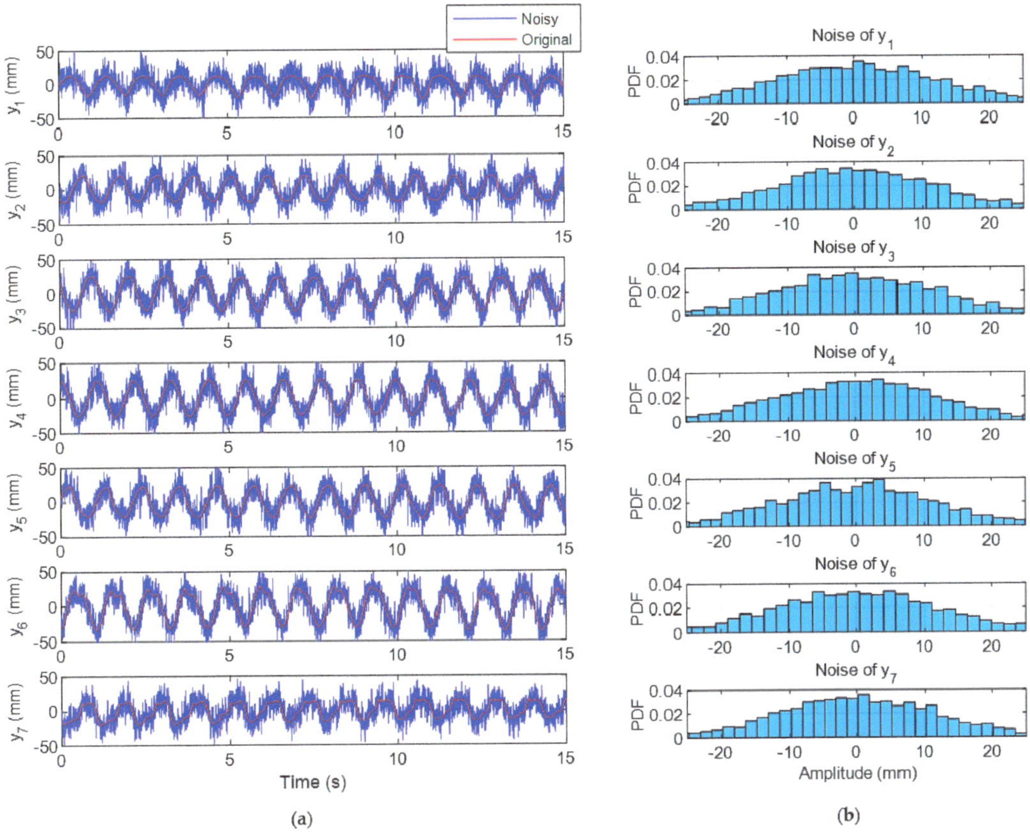

Figure 14. Data with injected noise: (**a**) The original and corrupted data; (**b**) Histogram of the added noise.

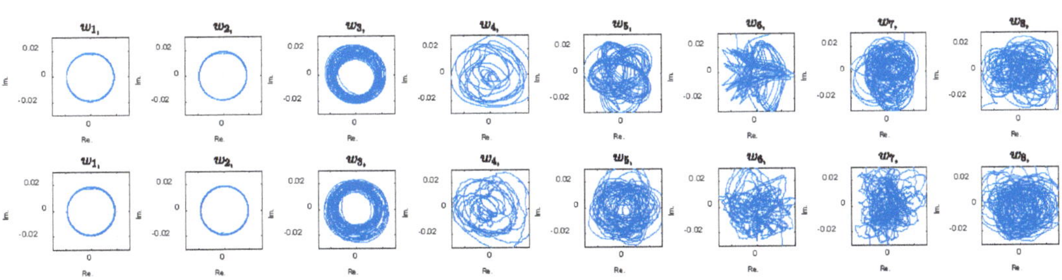

Figure 15. Comparison of the dynamic variables of the original (**top**) and the corrupted (**bottom**) data.

It is also worth mentioning that random and chaotic signals are distinguishable in such graphs. Chaos exhibits repeating and smooth modes while random signals do not show such deterministic and predictable patterns.

The performance of the CDE DMD linearized model with the noisy data is shown in Figure 16. The results represent the ability of linearization in the presence of highly noisy data. This advantage can be useful with inherently noisy measurement systems, such as inertial measurement units (IMUs), or in-field measurements where extra environmental noise and vibrations are inevitable.

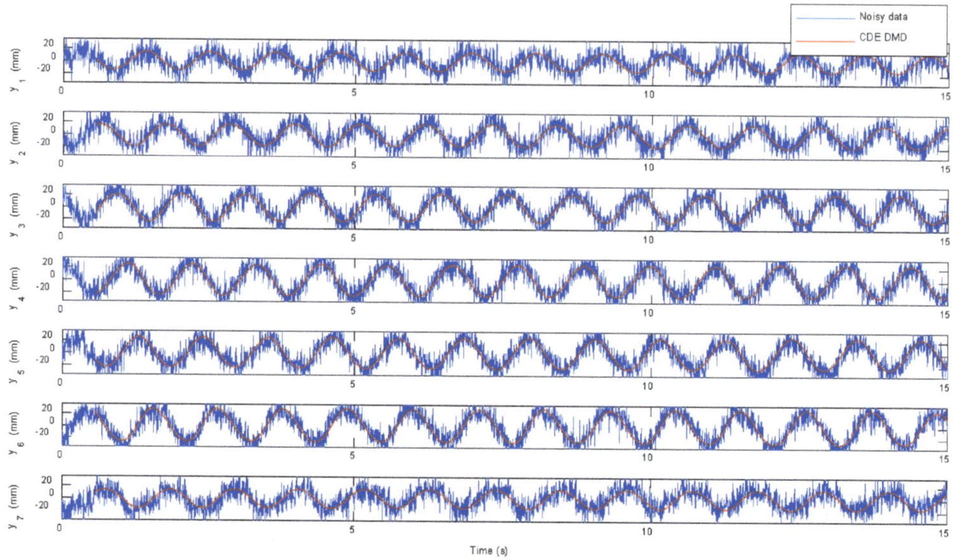

Figure 16. Comparison of the CDE DMD results with the noisy data.

4. Discussion

The results of this study demonstrated that the experimental (raw) data reflect the robot's compliance (i.e., influenceability by the external loads) but do not display smooth oscillations. As a side effect of compliance, the system is prone to deformations due to any external loads or faults within the system. Attributing such fluctuations to random or chaotic phenomena does not appear straightforward. The delay embedding results, on the other hand, reveal coherent structures and oscillators within the system dynamics. In particular, the phase plots reveal linear and chaotic influences that are typically observed in many fluid dynamic problems. Within the phase plot of the experimental data, coherent structures are not visible, and the interplay of noise or chaos is not discoverable. The delay embedding and recovering the model with the linear modes resulted in a linearized model of the system. The simulation results showed that the linearized model captures a large portion of the system dynamics, compared with the experimental results. Additionally, the method can separate the nonlinearity effect, which is useful in signal filtering.

From the phase lag within the time series plot of the first two modes, it is inferred that the modes construct the traveling wave along the robot body. This conclusion was drawn similarly in the previous research using intensive FSI simulations. The linearized model, introduced as the CDE DMD model, matches well with the experimental data. Such a model has different applications. The first goal of the linearization, in this study, was to investigate the existence of a traveling wave within the robot locomotion and explain the disorder. The linear model acts as a predictor that has many applications in robotics. Firstly, the predictor can be used for prediction purposes as in fault detection and outlier recognition. Faults in pneumatic actuators (leakage) and outliers (missed points during tracking) within measurements are common problems, which can be detected by comparison of the measured data with the predicted data. Secondly, the model is potentially useful for navigation (path following). Classic control deals with stabilization around zero, which is extended to control over a desired path. Similarly, the anguilliform swimmer oscillates along a curve (path), which is considered a straight line in the first step. The linear model works as a filter that can be used to eliminate nonlinearity effects. Another question that the results can explain, at least as a lateral conclusion, is related to the repeatability of the measurements. A classic question to be answered within any scientific measurement in general and for aquatic locomotion, in particular, is how the experimental observation is repeated in different runs. Chaotic and random signals are known to be notoriously unrepeatable. The results show that robot locomotion is mainly ruled by approximately linear dynamics disturbed by 'small' chaotic modes. This means that the gross motion of the robot is approximately predictable and repeatable, while the minor fluctuations are not. Simulations with added noise show the CDE DMD's capability of handling excessive noise. The method is proposed to extract the main modes of body undulation, especially for soft robots, which are highly influenced by additional modes and environmental vibrations. Soft robots, in contrast to rigid robots, are prone to deformation due to water interaction. Application for other types of aquatic locomotion is to be investigated as future work, but it is imaginable that carangiform and thunniform swimming contain a single linear mode (compared to two linear modes in anguilliform swimming).

5. Conclusions

This paper investigates the kinematics of a soft eel robot actuated by pneumatic muscles. The experimental results of the swimming test were used as the raw data for further analysis using a DMD-based algorithm. The algorithm employs the delay embedding technique to cope with the limitation of measurement points, compared to the infinite dimensionality of the system. The simulations showed that coherent structures are revealed with sufficient embedding order (here, 110 lags is equivalent to a half period of the robot undulation) in the complex delay embedded (CDE) coordinate systems. In particular, two linear oscillators were detected, which were attributed to the traveling wave (anguilliform undulation). The two modes compose the linearized model, which captures the gross motion of the robot. Additionally, the nonlinearity is composed of chaotic modes. It is discussed that the CDE DMD provides a prediction model that can be used in various applications such as data filtering, fault detection, or navigation applications as the future works. The CDE DMD's ability to analyze anguilliform locomotion characteristics in the presence of noise and body fluctuations makes it proper for studying soft-bodied swimmers, particularly soft robots.

Author Contributions: Conceptualization, M.S. and H.W.; methodology, M.S.; software, M.S.; validation, M.S. and H.W.; formal analysis, M.S.; investigation, M.S. and H.W.; resources, M.S. and H.W.; data curation, M.S. and H.W.; writing—original draft preparation, M.S.; writing—review and editing, H.W.; visualization, M.S.; supervision, H.W.; project administration, M.S. and H.W.; funding acquisition, H.W. All authors have read and agreed to the published version of the manuscript.

Funding: This research received no external funding.

Institutional Review Board Statement: Not applicable.

Data Availability Statement: The original contributions presented in the study are included in the article; further inquiries can be directed to the corresponding author.

Conflicts of Interest: The authors declare no conflicts of interest.

Appendix A

The experimental setup used in this research is shown in Figure A1. The measurement system contains a QualiSys® Tracking Manager (QTM) 1.6.0.1, which collects the position of tracking markers in $\Delta t = 0.005$ s intervals. The position data are saved as '.tsv' files and are post-processed offline using Python 3.8.19 within the Spyder 5.5.1 integrated development environment and imported in MATLAB® R2022b. The velocities were obtained by calculation (differentiation).

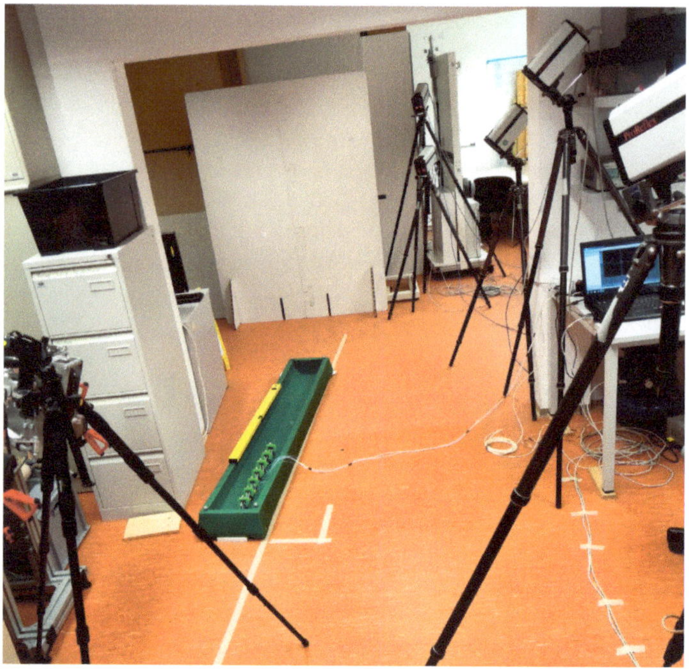

Figure A1. The measurement setup.

References

1. Wang, Y.; Chen, H.; Law, J.; Du, X.; Yu, J. Ultrafast miniature robotic swimmers with upstream motility. *Cyborg. Bionic. Syst.* **2023**, *4*, 0015. [CrossRef] [PubMed]
2. Hu, N.; Li, B.; Bai, R.; Xie, K.; Chen, G. A torsion-bending antagonistic bistable actuator enables untethered crawling and swimming of miniature robots. *Research* **2023**, *6*, 0116. [CrossRef]
3. Wu, M.; Afridi, W.H.; Wu, J.; Afridi, R.H.; Wang, K.; Zheng, X.; Wang, C.; Xie, G. Octopus-inspired underwater soft robotic gripper with crawling and swimming capabilities. *Research* **2024**, *7*, 0456. [CrossRef] [PubMed]
4. van Ginneken, V.; Antonissen, E.; Müller, U.K.; Booms, R.; Eding, E.; Verreth, J.; van den Thillart, G. Eel migration to the Sargasso: Remarkably high swimming efficiency and low energy costs. *J. Exp. Biol.* **2005**, *208*, 1329–1335. [CrossRef]
5. Vorus, W.S.; Taravella, B.M. Anguilliform fish propulsion of highest hydrodynamic efficiency. *J. Mar. Sci. Appl.* **2011**, *10*, 163–174. [CrossRef]
6. Sayahkarajy, M.; Witte, H.; Faudzi, A.A.M. Chorda Dorsalis System as a Paragon for Soft Medical Robots to Design Echocardiography Probes with a New SOM-Based Steering Control. *Biomimetics* **2024**, *9*, 199. [CrossRef]
7. Sayahkarajy, M.; Witte, H. Analysis of Robot–Environment Interaction Modes in Anguilliform Locomotion of a New Soft Eel Robot. *Actuators* **2024**, *13*, 406. [CrossRef]
8. Tytell, E.D.; Hsu, C.-Y.; Williams, T.L.; Cohen, A.H.; Fauci, L.J. Interactions between internal forces, body stiffness, and fluid environment in a neuromechanical model of lamprey swimming. *Proc. Natl. Acad. Sci. USA* **2010**, *107*, 19832–19837. [CrossRef]
9. McMillen, T.; Holmes, P. An elastic rod model for anguilliform swimming. *J. Math. Biol.* **2006**, *53*, 843–886. [CrossRef]
10. Feeny, B. A complex orthogonal decomposition for wave motion analysis. *J. Sound Vib.* **2008**, *310*, 77–90. [CrossRef]
11. Gemmell, B.J.; Fogerson, S.M.; Costello, J.H.; Morgan, J.R.; Dabiri, J.O.; Colin, S.P. How the bending kinematics of swimming lampreys build negative pressure fields for suction thrust. *J. Exp. Biol.* **2016**, *219*, 3884–3895. [CrossRef] [PubMed]
12. Fies, J.; Gemmell, B.J.; Fogerson, S.M.; Morgan, J.R.; Tytell, E.D.; Colin, S.P. Swimming kinematics and performance of spinal transected lampreys with different levels of axon regeneration. *J. Exp. Biol.* **2021**, *224*, jeb242639. [CrossRef]
13. Crespi, A.; Badertscher, A.; Guignard, A.; Ijspeert, A.J. AmphiBot I: An amphibious snake-like robot. *Robot. Auton. Syst.* **2005**, *50*, 163–175. [CrossRef]

14. Crespi, A.; Ijspeert, A.J. AmphiBot II: An amphibious snake robot that crawls and swims using a central pattern generator. In Proceedings of the 9th international conference on climbing and walking robots (CLAWAR 2006), Brussels, Belgium, 12–14 September 2006; pp. 19–27.
15. Liu, Y.; Liu, Z.; Fang, Y.; Liu, H.; Guo, X. A Novel Design Methodology of CPG Model for a Salamander-like Robot. *IEEE Robot. Autom. Lett.* **2024**, *9*, 6115–6122. [CrossRef]
16. Milana, E.; Van Raemdonck, B.; Cornelis, K.; Dehaerne, E.; De Clerck, J.; De Groof, Y.; De Vil, T.; Gorissen, B.; Reynaerts, D. EELWORM: A bioinspired multimodal amphibious soft robot. In Proceedings of the 2020 3rd IEEE International Conference on Soft Robotics (RoboSoft), New Haven, CT, USA, 15 May–15 July 2020; pp. 766–771.
17. Nguyen, D.Q. Kinematic evaluation of a series of soft actuators in designing an eel-inspired robot. In Proceedings of the 2020 IEEE/SICE International Symposium on System Integration (SII), Honolulu, HI, USA, 12–15 January 2020; pp. 1288–1293.
18. Chen, Y.; Wang, T.; Wu, C.; Wang, X. Design, control, and experiments of a fluidic soft robotic eel. *Smart Mater. Struct.* **2021**, *30*, 065001. [CrossRef]
19. Nguyen, D.Q.; Ho, V.A. Anguilliform swimming performance of an eel-inspired soft robot. *Soft Robot.* **2022**, *9*, 425–439. [CrossRef] [PubMed]
20. Nguyen, D.Q. Evaluation on swimming efficiency of an eel-inspired soft robot with partially damaged body. In Proceedings of the 2021 IEEE 4th International Conference on Soft Robotics (RoboSoft), New Haven, CT, USA, 12–16 April 2021; pp. 289–294.
21. Dang, R.; Gong, H.; Wang, Y.; Huang, T.; Shi, Z.; Zhang, X.; Wu, Y.; Sun, Y.; Qi, P. Bionic body wave control for an eel-like robot based on segmented soft actuator array. In Proceedings of the 2021 40th Chinese Control Conference (CCC), Shanghai, China, 26–28 July 2021; pp. 4261–4266.
22. Cervera-Torralba, J.; Kang, Y.; Khan, E.M.; Adibnazari, I.; Tolley, M.T. Lost-Core Injection Molding of Fluidic Elastomer Actuators for the Fabrication of a Modular Eel-Inspired Soft Robot. In Proceedings of the 2024 IEEE 7th International Conference on Soft Robotics (RoboSoft), San Diego, CA, USA, 14–17 April 2024; pp. 971–976.
23. Trinh, H.X.; Nguyen, B.N.; Nguyen, A.T.; Kien, H.T.; Nguyen, Q.D. Dynamics Modeling and Validation of a Bio-Inspired Soft Eel Robots for Underwater Motion. In Proceedings of the 2024 IEEE/SICE International Symposium on System Integration (SII), Ha Long, Vietnam, 8–11 January 2024; pp. 233–238.
24. Wang, Q.; Hong, Z.; Zhong, Y. Learn to swim: Online motion control of an underactuated robotic eel based on deep reinforcement learning. *Biomim. Intell. Robot.* **2022**, *2*, 100066. [CrossRef]
25. Hall, R.; Espinosa, G.; Chiang, S.-S.; Onal, C.D. Design and Testing of a Multi-Module, Tetherless, Soft Robotic Eel. In Proceedings of the 2024 IEEE International Conference on Robotics and Automation (ICRA), Yokohama, Japan, 13–17 May 2024; pp. 8821–8827.
26. Schmid, P.J. Dynamic mode decomposition and its variants. *Annu. Rev. Fluid Mech.* **2022**, *54*, 225–254. [CrossRef]
27. Meng, T.; Wang, Y.; Qin, S.; Liu, P.; Wang, Y.; Tao, C.; Liu, Z. Complex flow field analysis in Multi-Shaft stirred Reactors: Dynamics of Wave-Vortex coupling revealed by POD and DMD methods. *Chem. Eng. Sci.* **2025**, *301*, 120753. [CrossRef]
28. Long, Y.; Guo, X.a.; Xiao, T. Research, Application and Future Prospect of Mode Decomposition in Fluid Mechanics. *Symmetry* **2024**, *16*, 155. [CrossRef]
29. Rahmani, M.; Redkar, S. Optimal DMD Koopman Data-Driven Control of a Worm Robot. *Biomimetics* **2024**, *9*, 666. [CrossRef]
30. Du Clos, K.T.; Dabiri, J.O.; Costello, J.H.; Colin, S.P.; Morgan, J.R.; Fogerson, S.M.; Gemmell, B.J. Thrust generation during steady swimming and acceleration from rest in anguilliform swimmers. *J. Exp. Biol.* **2019**, *222*, jeb212464. [CrossRef]
31. Sayahkarajy, M.; Witte, H. Temporal Evolution of the Hydrodynamics of a Swimming Eel Robot Using Sparse Identification: SINDy-DMD. *J* **2025**, *8*, 2. [CrossRef]

MDPI AG
Grosspeteranlage 5
4052 Basel
Switzerland
Tel.: +41 61 683 77 34

Biomimetics Editorial Office
E-mail: biomimetics@mdpi.com
www.mdpi.com/journal/biomimetics

www.ingramcontent.com/pod-product-compliance
Lightning Source LLC
Chambersburg PA
CBHW041412210926
51850CB00013B/3250
* 9 7 8 3 7 2 5 8 4 5 4 6 0 *